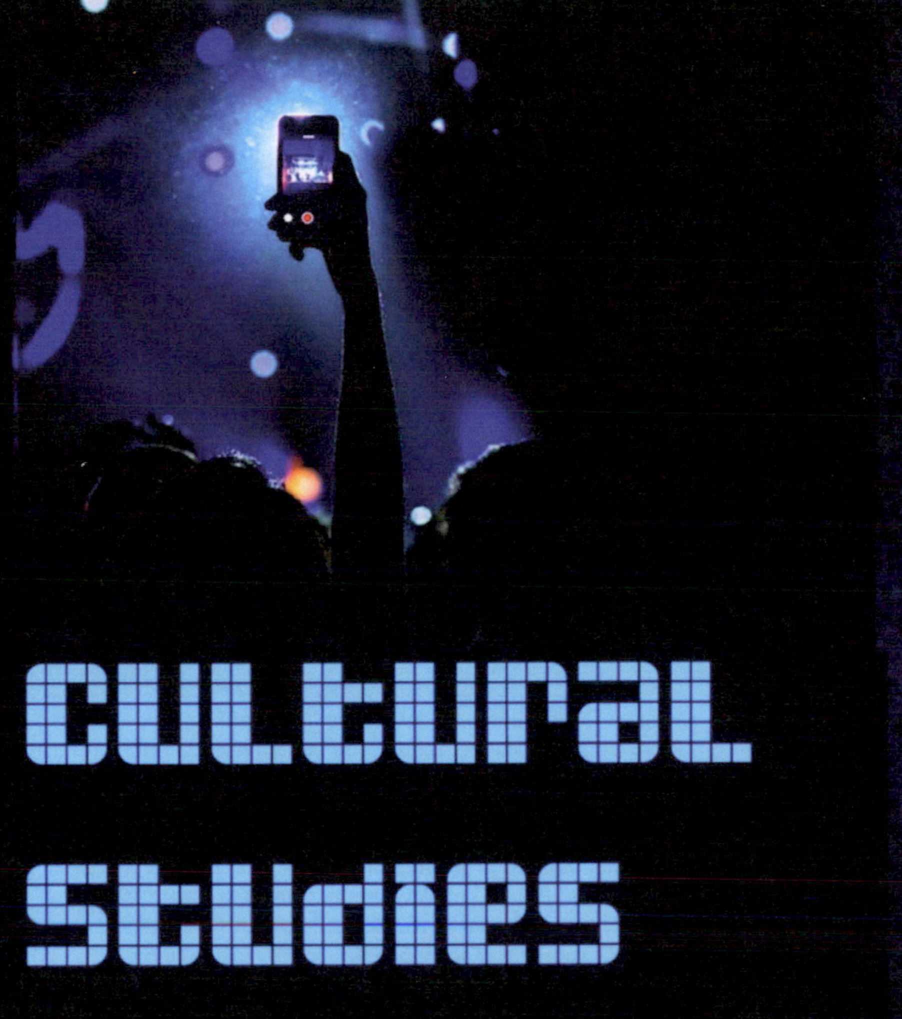

Cultural Studies in the Digital Age

An Anthology of 21st Century Interdisciplinary Inquiries, Postulations, and Findings

Edited by Antonio Rafele
Frederick Luis Aldama
& William Anthony Nericcio

Cultural Studies in the Digital Age: An Anthology of 21st Century Interdisciplinary Inquiries, Postulations, and Findings was curated and edited by Antonio Rafele, Frederick Luis Aldama, and William Anthony Nericcio, and is published by Hyperbole Books, an imprint of San Diego State University Press.

San Diego State University Press and Hyperbole Books publications may be purchased at discount for educational, business, or sales promotional use. For information write SDSU Press Next Generation Publishing Initiative (NGPI), San Diego, California 92182-6020. Substantive discounts are available should the book be adopted for use in university, college, and high school classes.

hype.sdsu.edu | sdsupress.sdsu.edu | facebook.com/sdsu.press

Cover and Book Design by Guillermo Nericcio García
Editing and Typesetting by Brian Frastaci & William Nericcio

The cover fuses digital creations by Daniel Gzz (unsplash.com/@danigzz90) and Markus Spiske (unsplash.com/@markusspiske) filtered through the imagination (and the iMac) of Guillermo Nericcio García using software by Apple, Preview, and Pixelmator by the Pixelmator team in Vilnius, Lithuania.

ISBN 978-1-879691-31-5

 First Edition. Published 1, January 2021, by Hyberbole Books, an SDSU Press imprint.

Printed in the U.S.A. by Books International.

FIRST EDITION PRINTED IN THE UNITED STATES OF AMERICA

Table of Contents

SOCIETY

MEDIA

DIGITAL

GRUNDIG

#DigitalProwess In the Time of COVID-19 World Optics

A FOREWORD

FREDERICK LUIS ALDAMA

From Mexico City to New York, megalopolises whisper an eerie silence. They've flatlined smells, sights, sounds, touch and tastes of life. Gone: most social interactions, many *cariños*, and all and in-person spaces of learning and creativity: libraries, museums, and schools.

Gone, but not disappeared. To a very large extent life's heartbeats have been squeezed into and pushed across pixelated screens, digitized algorithms, and 5G data streams.

Today's quotidian: Zoom. Facebook. Twitter. Snapchat. TikTok. Instacart.

This is not a sci-fi flick. Neither is this an academic exercise nor a philosophical musing. Debord's 1967 formulation of our move from tactile experiences and authentic existence to life lived in its consumption of things and things of things. Baudrillard's brain-pretzel formulation of a negative ontotheological world whereby the real—truth—is a simulacrum; whereby "truth" conceals that there is none and that only the simulacrum is true. Think the Wachowski sisters's *The Matrix* (1999).

No. This massive mainlining of the digital is our actual lived reality today. It's our digital-optic world in the time of COVID-19 when lockdown is the order of the day and digitization our daily nourishment.

This is our *nueva normalidad.*

Not so long ago, interrogating our increasingly digital world seemed easy. Media pundits, academics, and lay people either considered digital tech as a means for otherwise pushed out and locked out people to access new and ever wider spaces for artistic creation and truth-finding or as an obstacle further deepening socioeconomic divides and proving destructive generally to our humanness: lack of tech access, erosion of empathy, dopamine Insta-rewards, cyberbullying and suicide.

Already, on both sides of the table, opinion pieces, coupled with more informed and researched scholarly texts fill whole libraries—digital libraries with their infinite spaces. Scholars who come readily to mind include Mark Bauerlein, Steven Johnson, Nicholas Carr, Don Tapscott, Douglas Rushkoff, Maggie Jackson, Clay Shirky, Jean M. Twenge, and Todd Gitlin. Articles that I've taught include: "Is Google Making Us Stupid"; "Go F ck Your Selfie"; "The Rise of the Self and the Decline of Intellectual and Civic Interest." And, books I've assigned to students that come readily to mind include, *Post-Digital Rhetoric and the New Aesthetic* and *The Digital Divide.*

Today's waking hours fill with all things mediated and created in and through the digital optic: Twitter, Facebook, Snapchat, e-published books and their virtual book-launch parties, asynchronous and synchronous lectures filled with hundreds of Zoomed-in students (complete with Zoom-bombing ideologues), and Google-Meet family reunions.

More than any other time, this digital blitzkrieg has intensified the polar oppositions—no, extreme paradoxes—of our zeitgeist. Denial vs. Proof. Ideology vs. Science. Fascism vs. Democracy. White vs. Other. Haves vs. Have-nots. Big Brother vs. Privacy. Narcissists vs. Altruists. Corporate vs. Worker. Murder vs. life.

More than ever, at play is a digital optics that seeks to create a funhouse mirror to warp and distract us from the planetary destruction of our social tissue.

More than ever, we need the common sense, tools, and knowledge to develop a critically engaged digital optics that sees with decisive clarity—and that allows us to progressively act.

Digital dissemination of misinformation is shifting our planet's *de regula iuris* from one based on human rights to the work-or-die policies that protect the capitalist 1% and its oligarch cronies. It's helping to crush the working poor. Within a matter of a few months, Mexico City has seen the obliteration of seven hundred thousand permanent jobs and millions of informal jobs: merchants and others *all* forcibly removed. Not one is left. In the US, there's a record 36 million people who filed unemployment. Those who still have jobs will be forced to return, protection or no protection. From factory to mining industry, teachers to store clerks—all will have to risk their lives to COVID-19 or lose their jobs forever. From municipal workers to K-12 children and university students, all will be forced to return, creating ever enlarging hotspots of death.

Instead of a digital optics that makes clear all of the above, we see the deep drive to make omniscient a digital myopics that divides our populations, especially the working poor. In the US, we see armed militia stand outside beauty salons, declaring their right to freedom of hair stylistic action and expression. Asian Americans are spit on, pummeled, and kicked—at the least. Neighborhood vigilantes hunt down and kill African American joggers.

Murder. Erasure of livelihoods. Trillions of dollars thrown at national and corporate debt. For what? To gild-coat the toilets of the 1%.

All that is solid need *not* melt into air. Clearsighted, science-based knowledge and solidarity networks built across the nations can be our critically engaged way forward. This means dashing to the ground the noxious and nefarious anti-intellectualism and anti-science—and anti-creativity and anti-aesthetics. Our critically engaged digital optics must steadfastly insist that all forms of aesthetic activities and creative and scientific domains should be supported—and totally independent of the powers that be. It means using all at hand including our digital critical optics to share accurate information and to communicate with one another to create a unified front against current attacks against the workers *in toto*.

Closer to home at the universities across the nation, we see many admin pushing the do-or-die policy. To open campuses to students, staff, and faculty without safety precautions in place in order to hold at bay college's mostly preordained impending debt doom and to preserve the respective university's prestige.

We know different.

Opening universities this fall will create coronavirus hotspots; it will turn these spaces of learning and progressive transformative action into places of death and destruction, penetrating all levels of our community. Universities will lose status and lose money in ways that they may never recover from.

Our clearsighted, critical optics must seek to see through this and state firmly that we need our campuses to be safe and healthy. We need universities where life flourishes—and not death. Among our priorities we should count specific measures and strategies to save middle and higher education in the US. We must use all appropriate digital and other means to stand together—all of us—to fight for the funds and resources to ensure that the American university continues to belong to all of the people, in the US and beyond.

Sure, I hate it when I click to a news piece and it tells me the average time it takes to read, telling me that knowledge has a 3-minute tops time-limit. Sure, I hate that my usual spaces of learning—classrooms, archives, libraries, and museums—are vault-tight shut. Sure, I hate that I don't have choices today that I did yesterday; that I'm forced more and more to touch, smell, and feel the world through its digital algorithms and pixelated-screen surfaces. I hate the feeling that I'm living more and more in an episode of *Black Mirror*.

But as the knowledge made in this volume powerfully attests, we are not passive victims living end of days in the thick of a digital zeitgeist. We can and will use our #DigitalProwess to study carefully—even interrogate—the self, the mind, the brain as shaped in this blitzkrieg of digital media in lockdown that forces us to live in social isolation and to self-reflect and self-interrogate to come out of the experience with better, sharper tools. We can and will use our #DigitalProwess to understand better our brain in society and its role as shaper of society. We can use our #DigitalProwess and activist ground-stomps to curtail the ripping apart of our social tissue. We can and will use discipline to turn habits of self-preservation into acts of solidarity, friendship, love, and respect for the Other. No law, no decree, no official intervention, no militarization of society has the power to counteract this new development.

I'm optimistic. I think that we can and will make decisions and actions that can and will wrest control of the narrative in this digital-optic world in the time of COVID-19.

ABOUT THE BOOK
CULTURAL STUDIES IN THE DIGITAL AGE

BRIAN FRASTACI

Cultural Studies in the Digital Age is divided into four sections: *Society*, *Image*, *Media*, and *Digital*. The names of the sections are short for good reason—it would be foolish to sharply delineate the themes of this book, as the reader will see that the topics of any one chapter may use and blend the theories, approaches, and subjects of the other sections. But there is a utility to offering a general framework in dealing with the overall question of digital cultural studies.

The first section is *Image*. Here, our scholars investigate a variety of case studies in the power of images, both pre- and postdigital, to impact and influence the viewer. **Federico Tarquini** engages with previous experiences of image and links those experiences to the effects of modern social media. **Antonio Rafele** exposes on the aesthetics and impact of instant replay in sports, using the example of soccer in particular to note the disjunction that replay imposes on the viewer's perception of time. **Tito Vagni** examines the phenomenon of "food porn," a social media–fueled fetishization of edibles, arguing that it is a part of the broader rise of "food media" in modern culture. **Gwendolyn Spring Kurtz** tackles the problem of gender and the image, using examples as varied as Marie Antoinette's breasts, Tatyana Fazlalizadeh's New York street art, and Facebook images to note the different loci where the image of the woman becomes empowerment or commodification. **Kristal Bivona** expands on the role of street art for those in oppressive situations, noting the similarities and differences in the contexts of street art in South Central Los Angeles and Bogotá, Colombia. **Luca Acquarelli,** by analyzing Alejandro Gonzalez González Iñárritu's art exhibit *Carne y Arena*, showcases how the real images of tangible objects and the simulated images of virtual reality combine to force the viewer to confront the experiences of the migrant at the US-Mexican border. **William Nericcio** treats readers to a sextet of Surrealist masters—from the 20th and 21st centuries. Lastly, **Katie Waltman**, *Lord of the Rings* connoisseur, closes the section with a forensic exploration of history and architecture as it relates to Tolkien and more.

The second section, *Society*, asks what we can say about the ever-evolving role of images in our increasingly digitalized, compartmentalized world. **Massimo Cerulo** provides the appropriate groundwork for this book's multifaceted approach, linking the sociology of emotion with "emotional capitalism" and digital technology. **Lorenzo Bruni** further remarks on the sociology of emotions by engaging with the work of sociologist George Herbert Mead. And **Alberto Abruzzese** questions the impact of the internet's advent on society, grappling with the difficulty of a sea change in societal perception in real time.

The next section is *Media*. Here, our writers present the interactions of various media with social and cultural issues. **Jennifer Carter** offers a brief history of slam poetry, cousin to hip-hop, and explores several of its facets, such as its function for performer and audience and its utility for articulating the experience of the oppressed. **Ralph Clare** analyzes the presentation of finance and credit in post-2008 films and advertisements, noting how none of the more salient examples properly address the social and political issues behind the US economy before and after the crash, and indeed how the commercials in particular push a dangerous and exploitative vision of credit and debit on viewers. **Nello Barile** explores his concept of "ontobranding" in fashion, the uniting of fashion with the blurring of reality and virtuality in the experience of the consumer. **Katlin Sweeney** compares the sensationalistic and dehumanizing reactions of the news media to the 2016 Pulse nightclub attack with DC's *Love Is Love* comic collection, demonstrating how the latter offers a model for a more humanistic exploration of the reactions of QTPOC to the tragedy. And **Bonnie Opliger**, using the case studies of Disney's *Cinderella* and *The Princess and the Frog*, notes how though that company has improved in portraying diverse assortments of main characters, the figure of the comedic sidekick remains a place for stereotypical representations of the racialized Other.

The final section, *Digital*, deals with media in mutation—the post-analog. **Matteo Treleani** grapples with the relation of history and digital archives, critiquing the digital "domestication of the past" and its effect of "presentifying" prior events. **Vanni Codeluppi** treats the phenomenon of "web stardom," whereby internet fame seems attainable to regular people yet perpetuates inequal relationships. **Guerino Bovalino** expounds on the concept of "imagocracy," analyzing the modern political with relation to the power of digital images in shaping reality. **Agnese Pastorino** presents a study of adolescents from Salerno, Campania to outline youth perception and use of online pornography. Finally, **Carlos Kelly** analyzes the video game *Hellblade: Senua's Sacrifice* through the lens of Gloria Anzaldúa's concept of "liminal third space," using his own experiences playing through the game to identify numerous facets of colonization, liminality, and identity. Editor **Antonio Rafele** then checks in with a lucid afterword.

To conclude this short summary, note that nearly everything about the production of this book, too, was digital. Society and culture are still in the beginnings of the digital life—perhaps this volume can help navigate some of its aspects.

IMAGES

TOWARDS A NEW VISUAL CODE
MEDIA AND EVERYDAY LIFE

FEDERICO TARQUINI

MEDIA TEACHING US A NEW VISUAL CODE

> Humankind lingers unregenerately in Plato's cave, still reveling, its age-old habit, in mere images of the truth. But being educated by photographs is not like being educated by older, more artisanal images. For one thing, there are a great many more images around, claiming our attention. The inventory started in 1839 and since then just about everything has been photographed, or so it seems. This very insatiability of the photographing eye changes the terms of confinement in the cave, our world. In teaching us a new visual code, photographs alter and enlarge our notions of what is worth looking at and what we have a right to observe. They are a grammar and, even more importantly, an ethics of seeing. Finally, the most grandiose result of the photographic enterprise is to give us the sense that we can hold the whole world in our heads—as an anthology of images.[1]

This quotation marks the beginning of one of the most important reflections on photography of the twentieth century. In her famous collection of essays *On Photography*, Susan Sontag sums up the issues that this medium has historically raised in the field of image studies. First, the author describes a relevant *effect* of photography: since the invention of this medium, "everything has been photographed, or so it seems." Susan Sontag follows the well-known theory of Walter Benjamin according to which after the Industrial Revolution *quantity is overtaken in quality*,[2] and she consequently observes how the quantitative increase of images made possible by photographic devices has crucially altered the forms of perception widespread among the public. Therefore, the author immediately warns the reader: "Being educated by photographs is not like being educated by older, more artisanal images." The introduction of a new device like photography, asserts Sontag, implies an immediate modification of the human condition. Being *educated* by photography gives us a different visual code from the past. The gaze of persons tends to adapt itself to the *eye* of the photographic device. Persons see what could be photographed with a simple eye movement, in the same way of

[1] S. Sontag, *On Photography*, Ferrar Straus and Giroux, New York, 1977, p. 1.

[2] W. Benjamin, L'Œuvre d'art à l'époque de sa reproductibilité technique, Allia, Paris, 2003.

ON PHOTOGRAPHY

SUSAN SONTAG

ethics of the gaze, elevating photography to one of the fundamental elements in understanding the condition of modernity.

The issues raised by Susan Sontag, about forty years ago, are still of great relevance nowadays. The affirmation of digital media proceeds, though with minor deviations, according to the trend described by Sontag. The incredible production and distribution of digital images fosters a significant transformation of the contemporary *grammar of seeing*, now fundamental to knowing the world. This article therefore aims to reflect on the impact of the digital image in the contemporary experience of people.

The importance of seeing in Western culture began before the massive spread of digital devices as well as the productive and distributive processes of images helped by them. The story began in the nineteenth century when metropolitization processes bloomed. The success of the metropolis as an urban model has historically and radically modified the dimensions of space and time through the consequent break in the existing continuity in the premetropolitan urban models between the social environment and the individual. The results of this procedure value individual intellectual ability, and above all seeing, essential in order to recreate the meaning of the individual's routine. For the most intellectually sensitive authors such as Georg Simmel,[3] it seemed clear that the metropolis's inhabitant reacts first through his intellect in order to defend himself from the constant flow of solicitations and images met "across the street." Otherwise the perception of the surrounding world in the premetropolitan city seemed to refer mainly to sentimentality. In order not to be overwhelmed by the profusion of incentives and images, the individual is forced to reduce his sensitivity to the phenomena he sees as well as their capability to touch the human emotion by differentiating between what is important for his own life and what flows indifferently at his horizon.

The experience of metropolis, and the meaning of the gaze, is masterfully described in the following words by Michel de Certeau:

> Seeing Manhattan from the 110th floor of the World Trade Center. Beneath the haze stirred up by the winds, the urban island, a sea in the middle of the sea, lifts up the skyscrapers over Wall Street, sinks down at Greenwich, then rises again to the crests of Midtown, quietly passes over Central Park and finally undulates off into the distance beyond Harlem. A wave of verticals. Its agitation is momentarily arrested by vision. The gigantic mass is immobilized before the eyes. It is transformed into a texturology in which extremes coincide—extremes of ambition and degradation, brutal oppositions of races and styles, contrasts between yesterday's

[3] G. Simmel, *Les grandes villes et la vie de l'esprit. Suivi de Sociologie des sens*, Payot, coll. « Petite Bibliothèque Payot », Paris, 2013.

> buildings, already transformed into trash cans, and today's urban irruptions that block out its space. Unlike Rome, New York has never learned the art of growing old by playing on all its pasts. Its present invents itself, from hour to hour, in the act of throwing away its previous accomplishments and challenging the future.[4]

With these few lines, de Certeau synthesizes most of the issues, conflicts, and images arising from the development of the metropolis in Western modernity. The great French scholar shows in particular the paradoxical condition that this urban dimension impresses on all its elements, material or immaterial. It is the capacity of inventing, from time to time, its present on the ashes of the past, within a context of life characterized by the brutality of contradictions. The paradox, as it is conceivable, is all in the peculiar stability that this system finds in its constant change and in its diametrical internal oppositions.

A first question arises as follows: how can persons live in such a changing environment? De Certeau suggests a precious element to understand the metropolis and its paradoxes. The experience of this context, as well as any kind of knowledge exercised within it, is possible only by *seeing*. In front of so many stirrings and suggestions, of races and styles, of skyscrapers, streets, masses, signs, the citizen of New York produces a selection and distinction of values not dissimilar from that described by Simmel about eighty years earlier. This seems to show that the experience of metropolis perceived by the Berlin philosopher has not completely settled, at least in comparison to the New York image of the Seventies. Seeing is also a fundamental way of understanding the diffusion of media images within metropolitan flows, from those produced by the press and photography, to those *in motion*, generated by the cinema and later on television.[5] During the period in question, the sense of sight became the most important route to comprehension, and therefore knowledge, of the world that involved the urban individual at each period. We can easily understand the deep link between sight, the resulting knowledge, and the typical characteristics of the modern European metropolis. From the monuments to the toponymy to the large boulevard—in other words, from the denser aspects of symbolic meanings to the simple orientation tools—the visual aspect qualifies the metropolis's organisation as well as its experiential form in a deep and crucial way—we could say, its aesthetic form. Through sight, the urban man selects, discerns, notes the details, draws routes. Through his point of view, he recreates as well as makes sense of his life and the surrounding world. The anthropologic conditions for mechanized image development (photography) as well as kinetic images (cinema, television) existed before the

---

[4] M. de Certeau, *The Practice of Everyday Life*, University of California Press, Berkeley, 1984, p. 143.

[5] M. McLuhan, Understanding Media. The Extensions of Man, Gingko Press, New York, 1964.

cultural factory of the twentieth century. Today, objectivity of the image is crucial for the completion of individual subjectivity.

Therefore, being a spectator is a typical condition, maybe the typical condition, of the urban experience. Hence it is one of the foundations of the twentieth-century cultural factory's evolution so strongly based on the image, as well as the collective desire to live through the metropolis elsewhere.[6]

Seeing as an activity—that is, the image experience—during the twentieth-century cultural factory is a connotative cause of a given human connotation. At the same time, the citizen is a viewer, as though the customer is the audience. In the next section, we will see how the effect of the media on twentieth-century culture was a crucial cause of this change.

McLUHAN AND MASS MEDIA

> Media, by altering the environment, evoke in us unique ratios of sense perceptions. The extension of any one sense alters the way we think and act – the way we perceive the world. When these ratios change, men change.[7]

The development of the modern metropolis highlights the close link between the environment, the media, and the individual. With deeper investigation, it is possible to affirm that the modern media are at the same time crucial causes for the social, cultural, and economic changes in the West over the last two centuries. The quote with which this section opened derives from an essay by Marshall McLuhan. It clarifies that it is crucial to set the study of social and cultural change in the context of the development of media devices.

Contrary to most of his peers, McLuhan highlights that social and cultural change is created by a deeper change that comes from the individual's perceptive and cognitive abilities through the media. Therefore: "Effective study of the media deals not only with the content of the media but with the media themselves and the total cultural environment within which the media function."[8]

The Canadian critic produced a bottom-up study. The research starts with the moment of reading a book or watching a movie and reaches the general dimension where the experience—that is, the awareness and knowledge systems—affects the individual's stimulated senses. In the

---

[6] A. Abruzzese, Lo splendore della TV. Origini e destino del linguaggio audiovisivo, Costa & Nolan, Genova, 1995; G. Fiorentino, L'Ottocento fatto immagine. Dalla fotografia al cinema, origini della comunicazione di massa, Sellerio, Palermo, 2007.

[7] Q. Fiore, M. McLuhan, The Medium Is the Massage. An Inventory of Effects, Bantam Books, New York, 1967, p. 41.

[8] M. McLuhan, "The Playboy Interview: Marshall McLuhan," *Playboy Magazine*, March 1969.

McLuhan essay, medium and cultural environment—the two elements we are analyzing—look so much alike that the reader feels that the communication device is the space where life really plays out.[9] Therefore, the history of media should be compared to the "history of nations" and imposed as the basis and crucial cause of a specific method for the historical disciplines.

In his well-known Playboy interview, McLuhan makes clear that a proper media study must take into consideration the link between technology, cultural environment, and the individual. In that case, the interpretation of the media could also be the interpretation of historical time. If it is true that a given medium could be perceived as the environment where people act, it is also plausible that our knowledge systems adapt to the space's qualities and, as a consequence, define the measure of our historical understanding. History dresses the part of a given medium (and its characteristics) and could be conceived through that medium.

PHOTOGRAPHY AND THE NOW OF WHAT WE COULD KNOW

Thus far we have tried to recall the permanent features that have ferried the analogic image and its social meanings to the present. The development of the metropolis and its experiential forms, the success and the spread of photographic devices together with their social and economic practices, and the arrival of advertisements, cinema, and television, have all quickly given prominence to the image. Similarly, we have seen with McLuhan that the medium's qualities cause, once they touch the human senses, processes of social, cultural, and economic change as well as transformation of the individual's perceptive and knowledge structures. On the image surface, it is therefore possible to track the processes observed until now. For our reflection, it is urgent to introduce another element linked to the image and its growing importance. We can do it through the words of Walter Benjamin:

> What distinguishes images from the "essence" of phenomenology is their historical index. (Heidegger seeks in vain to rescue history for phenomenology abstractly through "historicity). These images are to be thought of entirely apart from categories of the "human science", from so-called habitus, from style, and the like. For the historical index of the images not only says that they belong to a particular time; it says, above all, that they attain to legibility only at a particular time. And, indeed, this acceding "to legibility" constitutes a specific critical point in the movement at their interior. Every present day is determined by the images that are synchronic with it: each "now" is the now of a particular recognizability. In it, truth is charged to the bursting point with time. (This point of explosion, and nothing else is the death of the *intention*, which thus coincides with the birth of authentic historical time, the time of truth). It is not that what is past casts its light on what is present, or what is present its light on what is past; rather, image is that wherein what has been comes together in a flash

[9] A. Abruzzese, *Analfabeti di tutto il mondo uniamoci*, Costa & Nolan, Genova, 1996.

> with the now to form a constellation. In other words: image is dialectics at a standstill. For while the relation of the present to the past is purely temporal, the relation of what-has-been to the now is dialectical: not temporal in nature but figural. Only dialectical images are genuinely historical—that is not archaic—images. The image that is read—which is to say, the image in the now of its recognizability—bears to the highest degree the imprint of that critical, dangerous moment that lies at the ground of all reading.[10]

As is typical of him, the ideas in Benjamin's few lines represent a crucial landmark for this paper. The image, as we see, is a fundamental element for understanding history's effect thanks to its unique and irregular nature. As Benjamin wrote this amazing passage, maybe he had photographic devices in mind. Thanks to photography, the synchrony between the present and the image settles for the first time as a specific and concrete form: an object. Therefore, the photographic image must be read during the period of the modern West, that is, the society with the economic and cultural conditions that we have described above. Thanks to photographic technology, the urban epoch (and the experience) is readable through the image surface.

Following the Benjamin school of thought, we can understand that the achievement of knowledge is the result of a process based on our "experience of media," such as, in this case, the photographic image. However, as we have seen, this process is not linear, not even despite strict cause-and-effect relationships. Past experiences build the basis for the birth and the development of a given technological medium. At the same time, the interaction with products of the media, in this case the photographic image, builds the characteristics of knowledge experiences. In other words:

> Modern experience is the immediate presence of a perception in the consciousness of the subject, which is realized in a unique and unrepeatable form, but dependent on the immersion of the subject itself in a world of historical meanings ... it is the perceived "thing" and together the "meaning" that it assumes in relation to the one who perceives it.[11]

Through interaction with photography, experience and knowledge change. "Object" and "meaning" meld with "the one who perceives," thereby defining a temporary form of experience. Photography marks an inevitable discontinuity in the history of the experience of images. Our knowledge will surely change. Everything has been fully demonstrated by numerous studies on the social meaning of analogic photography.[12]

---

[10] W. Benjamin, *The Arcade Project (Das Passagen-Werk)*, Harvard University Press, Cambridge, 1999, p. 463.

[11] P. Jedlowski, *Il sapere dell'esperienza. Fra l'abitudine e il dubbio*, Carocci, Roma, 1996, p. 64.

[12] W. Benjamin, *The Arcade Project*, op. cit.; R. Barthes, *La chambre claire. Note sur la photographie*, Gallimard, Paris, 1980; G. Fiorentino, *L'Ottocento fatto immagine*, op. cit; A. Rafele,

Now it is necessary to put the theories explained above to the test through the use of the digital image as a case study.

TOWARD A NEW SEEING GRAMMAR: INSTAGRAM, THE SELFIE, AND EVERYDAY LIFE

The quantity of images that people each second of every day take and spread digitally is an unmeasurable phenomenon. Due to its extraordinary contribution, it is like counting the grains of sand in a desert. Together with the certainly interesting numerical-quantitative data, reflection on the growing meaning of these social activities of taking and sharing images (for the individual as well as the community) is needed. This introduction first of all provides a snapshot of today. A huge number of reports from the most important research institutes tell us about the growing number of uploaded as well as shared images on social networks (Nielsen, Ericson Consumer Lab, Pew Research Center). The reports suggest, more or less intentionally, "how" the analyzed data appears in the social contexts by telling us the "quantum" of a given percentage. The "how"—that is the communicative nature of the contemporary digital image—raises many doubts and provides the focus for much research today. The productive growth of images helped by the devices' dematerialization—a topic worthy of a separate publication—doesn't just appear as mere quantitative data. It also indicates the cultural, communicative, and social growth of activities and processes more and more relevant today—product together with action.

Therefore, we will continue with a dual-track approach by proposing again, maybe in a paradoxical way, Walter Benjamin's old adage: *quality has been transmuted into quantity*.

In order not to run into a specious ideology of modernity, it's best to specify that some of the social meanings we attribute to the image as well as its technological-mechanical nature or even its copious presence in the public/private dimension have remained stable in Western culture since one and a half centuries ago. Since the spread of photographic devices, the image rules our routine in a such an amazing way so as to push us to raise herein the main theoretical dilemma: after writing and speech, is the image the crux of the "grammar" of our knowledge today?

Daily, millions of people take pictures using digital devices and then instantly share the photos on Facebook, Twitter, or Instagram timelines. It is most common to "immortalize" an event or situation through digital photography. The selfie phenomenon is indicative. As Lev Manovich argues:

La métropole. Benjamin et Simmel, CNRS Editions, Paris, 2010; S. Sontag, *On Photography*, op. cit.

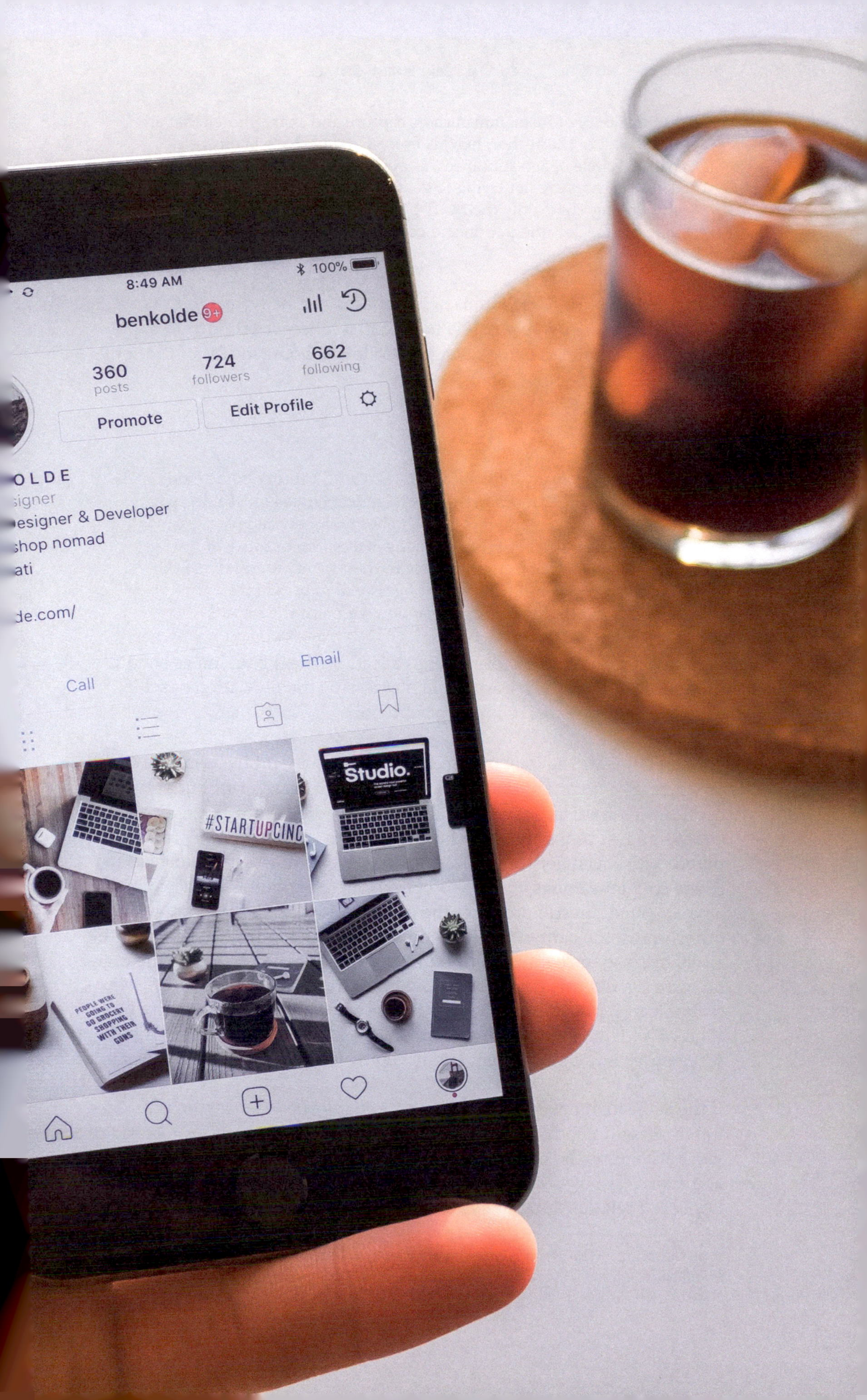
8:49 AM
100%
benkolde
9+
360
posts
724
followers
662
following
Promote
Edit Profile
& Developer
nomad
Call
Email
Studio.
PEOPLE WERE GOING TO GO GROCERY SHOPPING WITH THEIR GUNS

> The majority of Instagram authors capture and share photos that are of interest to the author, her/his friends and perhaps family or expanded circle of acquaintances, as opposed to complete strangers. These authors are not trying to get tens of thousands of followers, not do they share only their very best photos. Instead, they use Instagram for documentation and communication with people they know.[13]

In this sense, as Manovich himself does based on the results of the *Cultural Analytics Lab*, it is possible to note that "a larger proportion of people in many countries using Instagram follow a 'home mode' of the 20th century photography."[14]

At the same time though,

> there are also many differences between 20th century home mode and Instagram. "Traditional subjects" now include food, selfies, parties, etc. The demographics of both photographers and people we see in photos also changed—in many places, the majority of Instagram users and subjects are people in their teens, twenties and thirties as opposed to older authors of personal photos in the 20th century. But the essence of home mode remains the same.[15]

The meaning of the transformations caused by the digital image should therefore not be simply searched in narcissistic manifestation of selfie, but in the new "mode" of image production.

However, it is wrong to approach the subject as an "innovation censor," or worse, as a "guard of traditions." A piece of research without preconceptions—if it exists—should understand the digital image factors of change and, at the same time, the ones in continuity with the past. As noted, with social network images we don't see a drastic transformation in the social meanings of the photograph. On the photo's surface, this amazing medium still produces the present of what we could know. In other words, the content of the photograph wasn't radically changed precisely because the content is the surrounding world. Conversely, the digital form is a radical change of the medium that, thanks to new technological possibilities, doesn't change the contents' subjectivity but their objectivity. Having the picture of a loved one as your smartphone's screensaver is not so different from keeping his/her photo in your wallet.

If the subjectivity of the content doesn't change between analogic or digital images—in both cases we are talking about a loved one's photo—what changes is the objectivity. The digital image frees itself from paper and film so we can "shoot" more as well as spread pictures as content of our social relationships. Why? What does that have to do with

---

[13] M. Manovich, *Instagram and Contemporary Image*, www.manovich.net, p. 6.
[14] Ibid., p. 5.
[15] Ibid.

knowledge processes? Thanks to this quantitative increase, we can find the meaning of the image—and its production—within the special dimension of an event (parties, sport events, travels, etc.) as well as the ordinary daily dimension (selfie, profile picture, etc.). Within this last dimension, we can study the digital image. Daily life is the union of environments, practices, and relationships that we use in order to recreate our sense of life. The image becomes the content of routine, daily experience and puts flesh on its different moments. That is possible thanks to the technological characteristics of digital media such as connectivity, devices, and dematerialization of contents.

Connectivity, interaction, and production of "dematerialized" contents are not simple technological qualities. Instead, they are peculiarities of the kind of experience we have every day with the media as well as cultural, social, and communicative processes that the media create. There is therefore a process of combined development between medium and experience that values the image's function differently from the past. So I take and send a picture to a friend because I have chosen a medium that easily allows me to share the contents and the meanings that I want. That practice, as in the past with other media, presents some benefits within the relationships built during the digital era. The image, rather than the written word, seems to be well adapted to the present. Therefore, as McLuhan observes:

> The effects of technology do not occur at the level of opinions or concepts but alter sense ratios or patterns of perception steadily and without any resistance. ... For the man in a literate and homogenized society ceases to be sensitive to the diverse and discontinuous life of forms. He acquires the illusion of the third dimension and the "private point of view" as part of his Narcissus fixation.[16]

There is, therefore, a crucial link between the digital image, today's experiential forms and perceptions. This quantitative increase in photographic practices, as Jedlowsi highlights, shows a kind of historicized perception—that is, knowledge. Maybe the ways to perceive and know are adapting to the image's peculiarities instead of adapting to the written word. For this reason, the image becomes the crucial factor of a new knowledge "grammar."

References

Abruzzese, A. (1995). Lo splendore della TV. Origini e destino del linguaggio audivisivo. Genova: Costa & Nolan.

Abruzzese, A. (1996). *Analfabeti di tutto il mondo uniamoci*. Genova: Costa & Nolan.

Barthes, R. (1957 [2010]). *Mythologies*. Paris: Éditions du Seuil.

[16] M. McLuhan, *Understanding Media. The Extensions of Man*, Gingko Press, New York, 1964, pp. 19–20.

——— (1980). *La chambre claire. Note sur la photographie*. Paris: Gallimard.
Benjamin, W. (1982 [1999]). *The Arcade Project (Das Passagen-Werk)*. Cambridge: Harvard University Press.
Benjamin, W. (1936 [2003]). *L'Œuvre d'art à l'époque de sa reproductibilité technique*. Paris: Allia.
Borrelli, D. (2010). *Pensare i media. I classici delle scienze sociali e la comunicazione*. Roma: Carocci.
Chéroux, C. (2009). *La photographie qui fait mouche*. Paris: Carnets de Rhinocéros.
Cometa, M. (2012). *La scrittura delle immagini. Letteratura e cultura visuale*. Milano: Raffaello Cortina.
Debray R. (1992). *Vie et mort de l'image : une histoire du regard en Occident*. Paris: Gallimard.
de Certeau, M. (1980 [1984]). *The Practice of Everyday Life*. Berkeley: University of California Press.
De Rosnay, J. (1995). *L'homme symbiotique*. Paris: Seuil.
Fiorentino, G. (2007). *L'Ottocento fatto immagine. Dalla fotografia al cinema, origini della comunicazione di massa*, Palermo: Sellerio.
Flichy, P. (2001). *L'imaginaire d'Internet*. Paris: Éditions La Découverte.
Jedloswki, P. (1996). *Il sapere dell'esperienza. Fra l'abitudine e il dubbio*. Roma: Carocci.
Jenkins, H. (2007). *Convergence Culture: Where Old and New Media Collide*, New York: NYU Press.
Maffesoli, M. (1985). *La Connaissance ordinaire. Précis de sociologie compréhensive*. Paris: Librairie des Méridiens.
——— (2008). *Iconologies. Nos idol@tries postmodernes*. Paris: Albin Michel.
Manovich, L. (2017). *Instagram and Contemporary Image*. Manovich. Manovich.net.
McLuhan, M. (1964). *Understanding Media: The Extensions of Man*. New York: Gingko Press.
——— (1969). "The Playboy Interview: Marshall McLuhan." *Playboy Magazine*, March 1969.
McLuhan, M., and Q. Fiore (1967). *The Medium Is the Massage. An Inventory of Effects*. New York: Bantam Books.
Mirzoeff, N. (2015). *How to See the World*. London: Pelican.
Nielsen (2016). "The Social Media Report 2016." Nielsen. www.nielsen.com.
Pinotti, A., and A. Somaini (2016). *Cultura visuale. Immagini, sguardi, media, dispositivi*. Torino: Einaudi.
Pew Research Center (2016). "State of the New Media 2016." Pew Research Center. www.pewresearch.org.
Rafele, A. (2010). *La métropole. Benjamin et Simmel*. Paris: CNRS Edition.
Simmel, G. (2013). *Les grandes villes et la vie de l'esprit. Suivi de "Sociologie des sens", Payot, coll.* Paris: Petite Bibliothèque Payot.
Sontag, S. (1977). *On Photography*. New York: Ferrar Straus and Giroux.
Virilio, P. (2009). *Le futurisme de l'instant. Stop-Eject*. Paris: Galilée.

INSTANT REPLAY AND THE VIEWER
SPORTS IMAGES AND THEIR EFFECTS

ANTONIO RAFELE

Here I present a study on the pleasure and memory of the spectator, for whom sports images are a vivid, burning topic. The minute, singular discontinuities of time, in a sudden and immediate convergence between eye and image, push to the peak of the aesthetic experience. They are, even in their variety and nuance, the recurring motif, "the whole that returns at once," of reflection. Thus, the instant replay is configured as the spectator's time, pleasure, and memory as the dilation of an instant, attention as a temporary interruption of the temporal continuum, and the dream as the progressive thinning of consciousness to which corresponds the imposition of a clear, lucid image of the moment just passed. Johan Cruijff's maneuvers and images, and more generally the Dutch football of 1974, are elevated to the ideal birthplace of an aesthetic experience that grows and expands to include video games. These images are the place where memory and thought are born and then radiate, without it being possible to distinguish the historical images from the pace of reflection.

1. THE OPTICAL UNCONSCIOUS

I begin from a famous passage by Walter Benjamin contained in the *Little History of Photography* (1931):

> Whereas it is a commonplace that, for example, we have some idea what is involved in the act of walking (if only in general terms), we have no idea at all what happens during the fraction of a second when a person actually takes a step. Photography, with its devices of slow motion and enlargement, reveals the secret. It is through photography that we first discover the existence of this optical unconscious [...] photography reveals in this material physiognomic aspects, image worlds, which dwell in the smallest things—meaningful yet covert enough to find a hiding place in waking dreams.[17]

Photography sharpens the perception of the moment; it reveals minute variations and temporal intervals that the naked eye, "closed" in its perspective vision, can only perceive in a minimal way, as interferences, but not as *highlights*. With photography, the vision tends to precipitate in moments, in essential fragments.

At the same time, the viewer is pushed into the center of the scene; only the eye of the viewer completes the image, discovering how a dormant or hidden detail can be overlooked. The photographic image transcends the

[17] W. Benjamin, *Selected Writings. Volume 2 1927–1934*, Harvard University Press, 2005, pp. 511–512.

limits of aesthetic fruition to open up the time and the rhythm of everyday life: photography is a simulation of the ever-renewed links between the flows and suspensions of time, continuous and discontinuous.

The experience described by Benjamin takes place outside the domain of fashion; the discovery of a detail that relives in the mind of the viewer, even if for short fractions of a second, is the experience of shock, which immediately brings to light, as a living presence, a past, a just-passed. The object itself therefore seems to escape from the background, because it takes place as a living memory of the viewer; at the center are the ways in which it reaches the point of impressing the perceptive faculties of the viewer.

Photography is an experience and at once a representation of the ephemeral,[18] the irremediable transience of things. But the immobility, this mortuary fixity in which a fleeting and transient moment is broken, together with the uninterrupted memory of the viewer, has the effect of dilating time and the duration of an instant. With the addition of consciousness, that moment becomes a piece and a fragment of individual psychic history.

The extraordinary nature of Benjamin's passage also resides in an interpretation of photography that seems to come from the future, and television instant replay seems to be eerily present in these lines. McLuhan, thirty years later, would privilege the anti-aesthetic nature of television, updating the first romantic intuitions (Poe, Leopardi, Kierkegaard, Novalis, Wordsworth, and, the first historical completion of the romantic experiments, Baudelaire) on the literary work as a medium open to the reader[19]: a work whose material boundaries seem to dissolve suddenly in the mind of the reader and his memory.[20]

2. THE INSTANT REPLAY

---

[18] The reference is to Benjamin's monumental study of the *arcades* (cf. W. Benjamin, *The Arcades Project*, Harvard University Press, 2002); I refer in particular to the joint reading of the Konvolut B (Fashion), Y (Exhibitions) and N (Theory of History). The ephemeral frees itself from any implication of negativity towards the eternal, to become a principle of rereading and recreation *ex nihilo* of the ancient in the light of an absolute relativism, based precisely on the discovery and affirmation of the value of change, therefore of the work of time, of fashion.

[19] It is the point of observation, *reductio ad unum* of the forest of romantic thought, which organizes Benjamin's study of early romanticism: W. Benjamin, *Selected Writings, 1: 1913–1926*, Harvard University Press, 2004. Highlighting the disseminated proximities of aesthetic and gnoseological order, which combine in a theoretical configuration the romantic aesthetics and the theory of communication, is an essential step in the refined investigations of Claudio Colaiacomo: C. Colaiacomo, *Camera obscura*, Liguori, Napoli, 1992 and Ibid., *Il poeta della vita moderna. Leopardi e il Romanticismo*, Luca Sossella, Roma, 2013.

[20] The literary work is from the beginning, in the instant in which the idea impacts the mind, an image, "open" to the reader. McLuhan will extend (cf. M. McLuhan, *Understanding Media*, McGraw-Hill, 1964) these first romantic intuitions into a broader reflection on the concept of medium (prostheses, configurations, narcosis, amputations): the redefinition of the boundaries of the work—it operates as a "filter" that reaches its actuality and its completion in the reading—is already a new image of the communicative processes.

I cannot find a more fitting example of the effects that sports images have on us: out of an endless archive of images, I choose a maneuver by Johan Cruijff in 1974, at the semifinal of the World Cup between Holland and Brazil, which can be thought of as the ideal birth of the instant replay as an art form.

> Playback, the instant replay that focuses attention on the immediate process of the game is certainly one of the greatest art forms of our time. I asked various football players what that actually meant for their game and they tell me they had to change the game, open it a bit, so that people could watch it as if it were a replay.[21]

Sixty-five minutes later, Resenbrink and Cruijff hit on the fly, leaving the defender in place and displacing the goalkeeper. Cruijff's move breaks the balance of the game and the quiet of the spectator in an instant. It is an extraordinarily effective play, very complicated (the impact with the ball and the advance on the goalkeeper), but masks the effort that it truly takes—a rapid and unexpected movement, which appears and immediately dissolves, leaving the viewer astonished.

Let us consider a line from Gumbrecht: "There seem to be players whose performance capitalizes on this reactivity to sudden movements, which seem to arise from nowhere."[22] *To rise from nowhere*: here we find, in the sense of estrangement that assails the spectator, both the success of the play and the pleasure of its viewing. But it is a state completely complementary to the cut made by the camera. The apparent discontinuity that

---

[21] M. McLuhan, *TV as a New Mythic Form*, TV Ontario, 1970.

[22] H.U. Gumbrecht, *In Praise of Athletic Beauty*, Harvard University Press, 2006, p. 33.

separates the two moments—the cross and the blow—arrests the gaze on the body for an instant: the viewer's eye follows the action without ever being able to distance itself from it, modeling itself on the elastic and sudden movements of the body. At the same time, the gesture appears by superimposing, concealing the passages that precede the final event. Thus, only with the instant replay does the viewer reconstruct and complete the image.

Gumbrecht's lines here are again apropos:

> The unexpected appearance in space of a body that suddenly takes on a beautiful form, which just as quickly and irreversibly dissolves, can be thought of as a kind of epiphany. These epiphanies are the source of the joy we feel when attending an athletic event, and they mark the culmination of our aesthetic response.[23]

As fast as the frames that portray them, Cruijff's gestures deliver to the viewer a succession of extraordinary, unexpected moments. At the center is a particular experience of time: movements that take the viewer away from the predictable course of the game. Equally, one's vision tends to break up into a few significant moments, those moments in which the game falls into a series of unexpected events.

The link that joins the first television broadcasts of football to the changes in the game introduced by Holland in the seventies is the origin of a transformation that affects both the viewer and the performer. Pressing, insertions, first touches, sudden and continuous changes of scene, discontinuity of time, and finally movement in all directions: with Holland, the game becomes a *medium* aimed at reproducing, in the very rapid sequence of touches and movements, the same speed with which the television frames follow one another—a game catering to the eye of the viewer, which expands a little more the pleasure of vision. With replay, football and more generally sports become moments of television life, of the always-renewed links between flows and suspensions of time, ordinary and extraordinary.

3. THEORY OF SPORTS IMAGES

From Cruijff's image, I get four reflexive nodes on the relationship that connects, at the limits of an unlikely coincidence, the self and the images.24 In my considerations it acts, structuring its movement, as literature acts constituted by the romantic aesthetic (Leopardi, Poe, and

---

[23] Gumbrecht, *In Praise of Athletic Beauty*, cit., p. 34.

[24] On the "arcane" background of the images: that is how the images join in a real "graft" to the muscles and senses and to the flesh, of which the first years of cinema (and its background) constitute the mythical origin and are essential: Benjamin, *Arcades Project*, and McLuhan, *Understanding Media*.

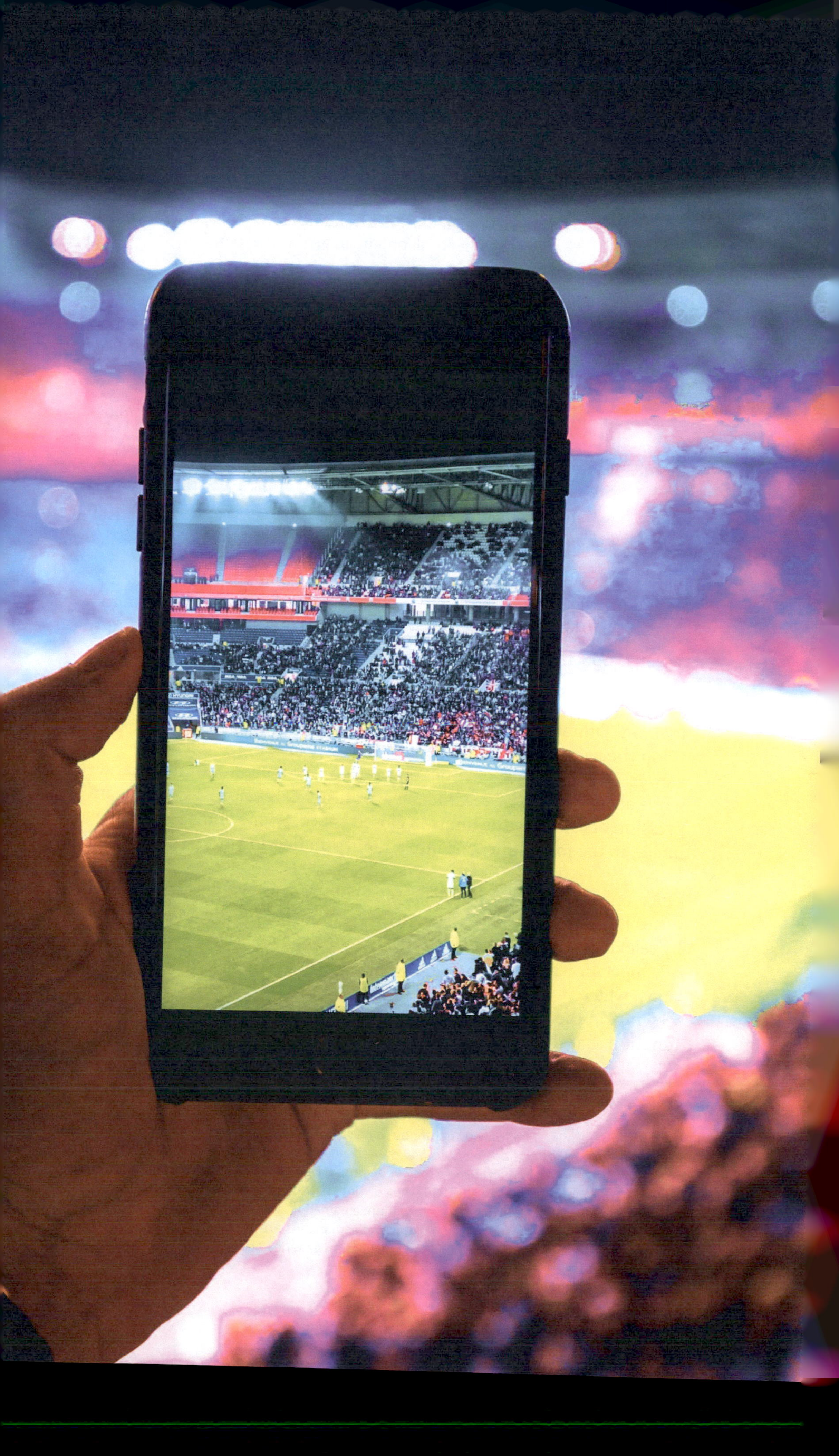

Kierkegaard) and by the theory of communication (Simmel, Benjamin, and McLuhan).

3.1. SCREENS

> Through mirrors extending along walls, these establishments all obtain an artificial expansion.
>
> Walter Benjamin, *The Arcades Project*, 1922–1940
>
> Puppets require the ground only to touch on, and by that momentary obstruction to reanimate the spring of their limbs.
>
> Heinrich von Kleist, *On the Marionette Theatre*, 1810

There is no proper opposition between the stadium and television, as if it were possible to distinguish two perceptions in contrast with each other, on the one hand an image modeled on technology, on the other a vision that is, so to speak, "natural."

From the very beginning, the gaze is fixed on the medium in which it takes place;[25] if on stage, a centralized image prevails: on television, to the contrary, the perception of the moment becomes more acute. But the replay is now the way in which even in the stadium the spectator approaches the sporting event.

London, Emirates Stadium: Right from the entrance, large screens envelop the passer-by, who is instantly in the position of a television viewer. The use of glass in buildings evokes the splendor of merchandise, while at the same time creating, in a fantastic extension of view, the impression of escaping the limits of closed space. Shops and shop windows invite, in the meantime, one into a rapid and fleeting *flânerie.*[26]

Once one is in the stands, the presence of the screen emerges and imposes itself in the absence. The memory, modeled on the archives built by television channels or videos posted on YouTube, arouses a complex of expectations about the protagonists, gestures, movements, and actions, which invoke in the mind of the user the residues of television viewing. In them, the retrospective moment of the replay is implicit, a function that defines the time-experience of the viewer: the rereading,

---

[25] "The mode of human sense perception changes with humanity's entire mode of existence. The manner in which human sense perception is organised, the medium in which it is accomplished, is determined not only by nature but by historical circumstances as well" [W. Benjamin, *The Work of Art in the Age of Mechanical Reproduction*, Penguin, 2008, p. 24].

[26] Cf. A. Abruzzese, *Forme estetiche e società di massa*, Marsilio, 1973 and A. Abruzzese, *Lo splendore della tv*, Costa & Nolan, 1994. The current shape of the stadiums, the image they provide for passers-by, is similar to the experience gained in the passages or between the pavilions of the Universal Expositions. Exhibitions and esplanades are the most important historical antecedent of television, even though today television galleries are a small-scale reproduction, a residue.

the endless rereading, completes the image and at the same time expands, always with new details and nuances in an accumulation of pieces, the sports memory.

The always-latent sense of the unlimited that the viewer feels in the very rapid overlapping of the scenes (and the infinite abundance, with the replay, of possibilities to amplify the pleasure of the vision), closes, at the stadium, in the sight of a circumscribed scene (as the possibilities to make the lived scene his own are reduced). In the same way, the ready and light pace of the protagonists, which on the screen vaguely recalls the movements or the lightning-fast appearances of video games, is broken in the now-dominant image of footballers planted on the ground and with a limited range of action. The centralized vision, stripped of details and interference, raises an unbridgeable distance with the game. In the replay, those details, which at the stadium remain hidden or in the shadows, offer a gateway to the most hidden mechanisms of the sporting event. It is a close focus, which shatters the sporting event. In the fragments there is a coincidence, which defies the limits of verisimilitude, between the viewer and the player, as in the image of Cruijff.

The proximity to the playing field—which the stadium pursues in every perspective point, almost as if to create the illusion of being close to the players—is only the mere attempt to reproduce the intensity of the individual moments of play, another essential prerogative of television enjoyment. Here a strong sense of estrangement is insinuated for the user accustomed to the snapshots of the game: the sensation of having now a limited spectrum of emotions and, moreover, of *reduced degree*. I am referring to the centrality that all those points that—far or near—appear, in the very rapid succession of frames, like the moments from which the plot originates. In the fragments, in the temporal discontinuities that they manifest, lies the origin of pleasure: a pleasure that is therefore everywhere, without a center.[27] That simple and raw movement in the direction of the goal is lacking, because what is decisive is the way in which the goal is reached. To this aim, Ajax and the recent variants of Sacchi, Zeman, Klopp, Spalletti, and Guardiola are devoted—not to the conclusion, but to the reproduction of a time of play. The spectator witnesses the unfolding of a dense menagerie of choices, triangles, and insertions, which denies a clear and manifest hierarchical intention: that the net is only a moment, and not the center of a sophisticated temporal configuration, is evident from the fact that the goals, as well as the countless errors committed under the goal (which are perhaps the unquestionable confirmation), almost never have the features of an *isolated* play. The Dutch or the Bohemian teams repeat the same pace of play in attack and in defense; thus, in an apparent reproduction of daily life, they outline an "infinite" work without contours. It is the pace of television, the image of an uninterrupted flow, which simulates and intensifies, like a snapshot, ordinary time. The spectator tunes in to that rhythm, presenting the specular form of an expanded and constantly solicited body.

3.2. TIME AND EMOTIONS

> Not only performing such acts, but also seeing them, being a spectator of active, energetic, rapid things,

[27] The succession of images, the elements of sports design, are the result of maniacal research into the possible reactions of an elongated eye on the screen and contribute to dilating the pleasure of vision: the touches, dribbling, and running give the impression that the footballers rise from the ground in a reactivity without mediation or obstacles. The protagonists, figurines of time, rather than gears of tactics, play as if suspended one meter from the ground, barely touching each other. Boots, t-shirts, and balls give the game the vague appearance of a dance show—a feeling that "envelops" the entire game, waiting for a play that can break or contract time, reopening the scene a little more.

> lively movements, and strong, difficult, motions, puts the soul in a certain active disposition, and communicates a certain inner activity.
>
> Giacomo Leopardi, *Zibaldone*, 1821

Cruijff's gestures break in an instant the time of play and the movements in space, making equally pleasant and essential, from any corner of the field, a pass, a dribbling, a fake, a cross, or a shot. These interrupt the linearity of the game, the mere movement in the direction of the goal; in the center, there is a sophisticated interpretation of time, a dense plot of "advances" that excludes or hides the conclusion to align on the eye of those who, between discards and sudden extensions, attend the game.

It is a peculiar tendency of modern football, which distinguishes Cruijff from the heroic, statuesque image of champions of the past, such as Di Stefano: an extraordinary strength of mind and superb physical qualities, as well as an endless technical repertoire, such as the containment, in a single point, of a living reproduction of the history of football, and jointly reflecting a static image, in which ideas and their executions seem to manifest at the same time. It is as if Cruijff had broken the wall that divides the images of the mind from their material realization, making appear as simple and natural that which on the contrary hides an unavoidable detachment.

With Cruijff, individual time naturally seems to become, without breaking or finding resistance, collective time. A way of constantly "opening" the game—sudden touches, verticalizations, fakes, filtering passages, dribbling, insertions—that attracts to itself, like a gravitational center, the players (the moves and movements) and the spectators, making the game interesting and unpredictable all of a sudden. This is the case with a perfectly disguised difference: individual and collective time come together to form an inseparable whole. The leap, which the spectator perceives and grasps, without being able to give a gradual explanation, is configured as the original (and irreducible) trait of the display.

It is therefore not by chance that Cruijff lacks a codified and recognizable style of play. His gestures remain a field of possibility, without ever closing in a desolating—and finally moribund—repetition. They are equivalent to the cuts made by the camera, because they multiply the scenes by eliminating the contours. Thus, the viewer has the impression that he is faced with a powerful textuality, still extensible, entirely involved in a succession of cuts and emotions that do not seem to have a limit or a predictable trajectory.

With Holland, the identification between viewer and performer is almost total: advances and discards elude the defenses of the opponent and at the same time constitute the main source of pleasure for the viewer. In the jumps scattered along the succession of images, the viewer enters the heart of the game: defenseless, he lets himself be overwhelmed by the rhythm of the scene until he identifies with it; immediately after, due to the pleasure of the unpredictable gesture that displaces expectations of the vivid impression that that moment has caused in his mind, the viewer completes the replay image, extending the sporting memory. But at that precise moment, the different planes of the text—the protagonists, the game, the camera, and the viewer—align to the point of concealing the differences. Language and emotions form a single and unique rhythm, a perfect and light machine, in which each action on the field seems to correspond to a reaction of the viewer. It is an experience that pushes out of the ordinary, at the height of the aesthetic experience.

3.3. DREAMS, ALLEGORIES, AND SHOCKS

> As soon as consciousness put in its appearance, it turns out that these mirages were not the idea. Now that the consciousness has awakened, if the imagination once again is nostalgic for those dreams, the mythical steps forth in a new form.
>
> S. Kierkegaard, *The Concept of Irony*, 1841

Cruijff's move "stole" the life of the spectator, but at the same time it opened a gap in the scene: a void and a kind of amnesia, compared to what has just been experienced, "touch" the viewer as dreamlike elements with the abrupt interruption of a dream—a temporary suspension of time, which expands the sensory capacities providing a vivid pleasure. The traces of that dream persist on awakening, bereft of the previous enchantments: an image that, due to the features that still remain "dormant" or "half-closed," invites to a new reading. The residue, that alienating sensation that the passage from one scene to another has imprinted in the mind of the viewer, is the tangible sign of an "open" game on the viewer, because only the viewer, taking in the maneuver that occurred but was hidden by Cruijff's technique, completes the image.

The images that make up the spectator's plot are captured in the moment in which they stratify as traces of the mind. The retrospective gaze places, in a dizzying twist backwards, the origin of the sporting event in the moment in which the image releases its effects: the sudden, almost instantaneous passage from the football play to the memory of the viewer. It is not a preexisting position to the analysis, but of the already-happened, in concomitance with the replay, and the centrality of the spectator. Because of the jumps, it disseminates during the game, the instant replay configures the images as allegories of the image, on which it is necessary to come back and linger to reconstruct the lived experience. Pleasure and memory are the two inseparable moments of the image, the constitution of a way of being entirely projected on the present. The images live in the mind of the viewer, in daily consumption, beyond the boundaries imposed by the sporting canon.

While one watches a game with Ajax or Holland, the fast and changing flow of images stratifies a repertoire of minute interference. Details that appeared and followed in a flash solicit a memory of the moments just passed. Those memories are not occasional digressions from the screen, but rather the elements that push towards a final resolution: the appearance of a consciousness that clears the field and makes clear the causes of the shock, a strong and vivid impression that the image has come to procure in the viewer. In the retrospective look that the viewer imposes on his own past, the time of fruition increases, expanding the expressive potential of the image. The observer, who lingers and returns to his footsteps, finds and isolates it until he dismembers, under the scrutiny of a magnifying glass, a detail that shortly before had managed to impress the mind and body fleetingly—but no less profoundly, it will be discovered later. When consciousness takes over, those mirages, dreamlike elements that persist upon awakening, recompose themselves into a puzzle. But this happened in a jump from the plane of the indistinct, the perception of a daydream, to the blinding surface of consciousness.[28]

---

[28] This is an essential moment of the gnoseological doctrine developed by Walter Benjamin: cf. W. Benjamin, *Epistemo-Critical Prologue*, in *The Origin of German Tragic Drama*, Verso, 2009.

3.4. FASHION AND DEATH

> Fashion: I put these orders and customs into the world, so that life itself, out of respect for body and soul, should be more dead than alive.
>
> G. Leopardi, *Dialogo della moda e della morte*, 1824

The overlapping of image and spectator, the unlikely coexistence of action and reaction of the lasting success of the shock technique, is justified, mirroring the imposing sensory transformations that the metropolis introduces into everyday life. The accelerated pace of fashion has protected individuals from the evils and "calluses" of repetition—a form of life that appears to the inhabitant of the metropolis as narrow and oppressive: the image of a rigor mortis in which the mind precipitates, stiffening itself into codified habits that deprive the pleasure of its expansive dynamics—but at the same time has exposed them to an intensification of nervous life, where life is "more dead than alive."

There is no unbridgeable distance between the plan of the work and the more general plan of daily life, but rather, as in a "circular room of mirrors and reflections," a profound correspondence. The two moments, the ordinary and the extraordinary, establish a productive overlap between them: a very close interdependence. Whether there is an evident opposition, rejection, or presumed coincidence or appearance, the varied world of emotions descends from their relationship, as does the rhythm and form of experience in the changed sensory conditions of modernity.[29] The communion between work and reader, of which we read here indirectly, transcends the circumscribed boundaries of taste to open itself to the pace and rhythm of daily life. That relationship, which concerns the cumbersome world of things, takes place in the domain of shock. The shock is the moment in which things come to impress, with ever new surprises, the mind of the observer: a body made weak and sickly, close to "death," by the rapid and changing succession of stimuli.

In such conditions, the distraction—the individual leftovers, suspended and dreamy among a thousand lights, images, glances, and shop windows—is not a degraded form of experience but an active and effective tool: it allows the individual to follow, without excessive internal upheavals, a fast and dispersive rhythm. But, while managing to quickly "slip off" all these stimuli, the individual suffers over time, due to the same overload, a progressive weakening of his sensory capacities.[30] The attention that an image can awaken in the spectator, therefore, takes place in an interruption of the continuum: a momentary and sudden suspension

---

[29] Cf. G. Simmel, *The Philosophy of Money*, Routledge, 2011.
[30] Cf. Simmel, *Philosophy of Money*.

of time through which the individual is immersed, even if for short moments, in a new illusion.[31]

At one point, the event dilates time until it "burns," disappearing into the frenetic rhythm of history; but the leap, which the viewer has made without opposing defense or resistance, is the result of an irresistible compulsion on the part of the work, of the circumstances.

In order to break down the wall of indifference and of the "already-lived," the images must intervene with ever new surprises—that is to say possess the semblance of a sensorial and emotional shock—which is at the same time a technical innovation and an unprecedented "reshuffle" of the imaginary.

The circular relationship established between distraction and attention constitutes the psychic space into which images, their production, as well as their success in the context of the metropolis and the media creep.

The technique of shock creeps into this dialectic of long historical duration in search of an effect, a gap that takes the viewer out of the ordinary.

[31]Here I summarize a more articulated theory of the media. In McLuhan's interpretation, the media, like the shocks of everyday life, subtract from the empty and homogeneous course of time to inaugurate new and further addictions: cf. McLuhan, *Understanding Media*.

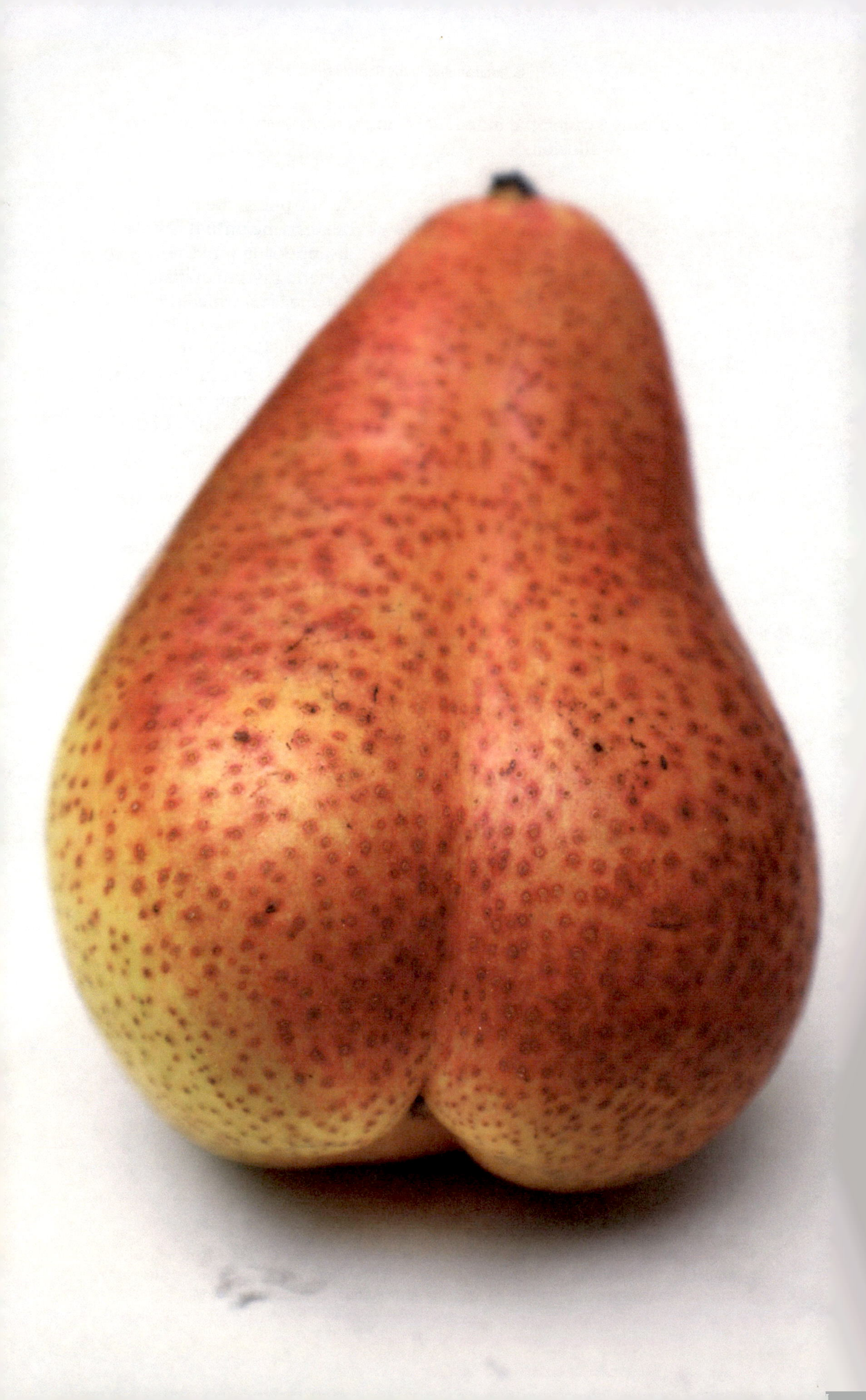

FOOD (IS NOT) PORN
VISUAL MEDIA AND THE CONCEALMENT OF THE WORLD

TITO VAGNI

FOOD MODERNITY AND THE CENTRALITY OF VISUAL

As this essay goes to press [18, September 2020], there are 239,940,360 photos published on the social network Instagram accompanied by the hashtag #foodporn. This suggests that the expression has entered the contemporary lexicon, but, at the same time, the variegated series of images accompanied by this tag constitutes a protean corpus with extremely blurred boundaries, which captures a cultural trend,[32] perceived and shared by users, without identifying a specific aesthetic and stylistic trait. #Foodporn is therefore not intended by its users as a genre: every food can be placed in this vast container as long as it is photographed, filmed, or narrated. The interpretative key we intend to propose is that the cultural trend in question is the growing importance of the visual and narrative aspect of food, which is surmounting its nutritional status. Food porn is therefore the indicator of a tension between food to be eaten and food to be looked at, read, and listened to.

The reconstructions of the birth and diffusion of the term food porn (Finkelstein 2014; Brown 2014; McBride 2010; Rousseau 2012) agree in placing the origin of the cultural trend in the works of Roland Barthes, in particular in some famous pages of his *Mythologies* (1957) dedicated to "ornamental cooking." In the essay, the scholar compares the photos of foods published by two French magazines, *Elle* and *L'Express*, identifying different models of exposure of the dishes. *Elle* is a generalist magazine aimed at a mass audience, so the photos published there have as their distinctive feature "the covered"—each dish seems to be wrapped in a patina that prevents you from distinguishing the elements of which it is composed, but at the same time the cover projects the dish into a dream world, precisely because of the beauty built art and the mystery of its composition. The public of *L'Express*, composed mainly of the upper middle class, has a higher purchasing power, so they do not need to dream of captivating dishes, because they can reproduce and consume them without worries. For this reason, the magazine offers them recipes and "real" photos, which can serve as a model for a domestic repetition. Barthes' reflection on the ornamental kitchen is the founding moment of

32 In the gastronomic field, the association between food and culture refers to anthropological issues. In this essay, instead, food is referred to as a cultural object and/or as a cultural environment.

a new way of studying and understanding food, because it addresses it as part of consumer culture, placing it beyond the edible dimension.[33] Despite the paradigmatic importance of this theoretical acquisition, Barthes's dichotomy risks being misleading because it focuses semiologically on the distinctive signs of culinary images, eventually overshadowing the sociologically relevant aspects, namely the construction of a media discourse about food *tout court*, the interest of the cultural industry in food, and its mutation into mass culture. Barthes, dwelling on the differences between the two culinary images, keeps silent the important point of agreement of the two expressive forms, that is to say the expository nature of food and therefore the entirely new relationship, of a visual order, that the consumer has with it. A few years after the *Mythologies*, in his essay *Pour une psycho-sociologie de l'alimentation contemporaine* (1961), the French semiotician introduces new elements into his reflection, referring to "food" (*nuriture*) as a "system of communication and a body of images" (Barthes 1961, p. 979). In this subsequent study of food, references to cultural nuances between social classes disappear, and food is attributed, without any distinction, to a communicative power and expressive capacity that is not at all self-referential but refers to the particular cultural context and to the consumers of which it provides much information. The "covered" and the "real" are no longer a revealing dichotomy, because they are understood as aesthetic traits, communicative choices addressed to different audiences. According to Barthes, when a need such as that of food is institutionalized, it is no longer possible to distinguish between its function as a sign and its specific function; it is no longer merely nourishment or pure communication, but it assumes both connotations at the same time. The Barthesian perspective on the value-sign of goods has been collected and ideally continued by Jean Baudrillard's studies in *La société de consommation* (1970), which describes the changes of goods—and therefore of food—as a "culturalization": a process in which each object becomes entertainment and an instrument of distinction, an accessory that adds itself to the panoply of other goods and manages to produce meanings that go beyond its specific function. In particular, the culturalization of food is a distinctive trait of "food modernity" (Teti 2015), that historical-social moment that came after long periods of scarcity in much of the West until the 1950s.

Thanks to Baudrillard's perspective, it is possible to understand that the importance of food today does not lie entirely in its physiological function but is given by the dialogue that food establishes with other goods and with consumers. As Roberta Sassatelli (2010) writes, reacting to Baudrillard's critique, consumption does not involve the satisfaction of

[33]Let me refer to my analysis on the birth of a sensitivity of the kitchen as an aesthetic experience—"Degrowth from Utopia to Lifestyle"—in which I try to explain how the onset of a kitchen devoted to pleasure and not to nutrition has its matrix in the fourth century BC in the writings and practices of Archestrato da Gela.

2 PM Wed Aug 12 58%

#foodporn (236,082,212)

sofiakosm_
8 minutes ago

thebrandondm
8 minutes ago

roxypastry88
8 minutes ago

cici.foodaholic
8 minutes ago

comidaboapet...
15 days ago

antony_marslp
8 minutes ago

mparrillero

bernardo_calc...

bhavsbites

needs but rather takes on the meaning of an act of access or inclusion in a system of signs that has replaced that of material contingency—a substitution that does not give rise to a disappearance, but to a multiplication that generates an augmented reality in which each material referent incorporates his memory and his myth, which represent the true seductive qualities of the objects. The myth of reality and its consequent amplification are produced by modern communication systems that, through their own language, perform a gigantic work of mediatization of reality.

The ornamental cuisine described by Barthes in his analysis of postwar magazines, as well as the most recent practices of sharing digital images through the hashtag #foodporn, indicate that the process of culturalization of food is closely linked to its mediatization, or rather to its encounter with the cultural industry, which, by extending its interest in food, has produced a phantasmagorical form, which has reached visibility with the explosion of photographic practices in digital environments. The use of the term "food porn" therefore indicates a sharing practice typical of contemporary social networks and at the same time a sociocultural phenomenon linked to the development of the consumer society. But the question of research that drives the interest of this essay is why we use this term to describe the current obsession with food and the practices that derive from it.

Research related to the spread of the hashtag #foodporn on Instagram (Mejova, Abbar, and Haddadi 2016) has shown that the most common photos on the social network are of foods with a high calorie and sugar content. Chocolate and cakes are the subjects most shared by users around the world. This leaves researchers to say that the association of the word "pornographic" to food is due to a perverse effect of food exposure that promotes incorrect lifestyles. But research has also shown that images of voluptuous foods are frequently associated with other hashtags which evoke positive feelings and healthy lifestyles. The conclusion is that "the feeling associated with #foodporn indicates that hashtaging is also used to promote healthy lifestyles. This redefinition of the gastroporn in social media communities is a further demonstration of the social effect that the media have on our vision and conceptualization of ourselves and what concerns us" (p. 9).

In the light of what has been written, it is necessary to understand why the cultural trend described is synthesized by the users with the expression "food porn," which is not at all neutral; on the contrary, it has with it a considerable number of sociocultural implications.

THE OBSCENE AS THE THEORETICAL BASIS OF FOOD PORN

If you look at the images most frequently associated with the hashtag #foodporn on the photographic social network Instagram, you can see how the tag converges with the Baudrillard concept of obscenity, declined in three different ways: a) images that show the overabundance of food; b) images that through manipulation reach an effect of reality higher than reality itself; c) images so close as to highlight details that an in-person vision would not be able to see.[34] Observing the heaps of goods, the shop windows, and the advertisements, Baudrillard sees a form of consumption that is no longer directed towards need, no longer a specific action that focuses on an object, but a self-sufficient mode: shopping.

> You take away the dangerous pyramid of canned oysters, meats, pears or asparagus, buying a small portion. Buy the part for the whole. And this metonymic, repetitive discourse, of consumable matter, of goods, comes back, through a great collective metaphor, to the image of the gift, of the inexhaustible and spectacular prodigality typical of the party (Baudrillard, 1970, p. 5).

The characteristic of the new form of consumption is the flanking of the goods, an aimless loss in search of continuous sensory stimuli that necessarily come from the visual dimension, therefore a sign. In this fragment, we can understand how Baudrillard understood the culturalization of food as a pornographic form starting from the display of the goods in drugstores and shopping malls, places where there is an "amalgamation of signs", a confusion of them justified by the "total opulence" typical of industrial capitalism. Consumption is no longer a transitional form; it has become a spectacle of goods, in the same way that food no longer satisfies a need—it has been transformed into the experience of food porn.

But Baudrillard's reflection on obscenity and pornographic aesthetics is more focused in his later works, particularly in *De la séduction* (1979). According to the French scholar, the technologies of representation produce an excess of closeness between the eye and the object and, consequently, an abundance of truth, which abolishes all forms of seduction, leaving room for an integral transparency. Baudrillard and other scholars who move in the same direction attribute this characteristic to pornographic aesthetics.

> Obscenity itself burns and consumes its own object. You see it too closely, you see what you have never seen—you have never seen it work, your sex, neither so closely nor generally in any other way, fortunately for you. Everything is too true, too close to be true. It is this that is fascinated, the excess of reality, the hyperreality of the

[34] There exist some other macrotypes, though less relevant, of photos considered by Instagram users as food porn: the images of foods edited so as to appear artificial, and images of voluptuous foods.

> thing. The only ghost of photography, if there ever was one, is therefore not that of sex, but that of reality and its absorption into something different from reality, the hyperreal (1979, p. 37).

Photography, cinematic image, and television flow-produce the effect of bringing the observer closer to everyday objects, making reality disappear and replacing it with a profusion of images that begin to enjoy a superior effect of reality, thanks to the approach of the eye to the object observed and the many details that this new condition allows to shine through. Technology reveals unknown elements, ending up producing an imaginary world made of copies that have submerged their prototypes, becoming a model of reality itself—a simulation of reality that acquires a more intense and involving character than its prototype, a simulacra of absolute perfection that reverberate their logic of functioning in everyday life with a viral mode from which it is impossible to escape.

The hyperreality to which Baudrillard refers is an effect of visual technologies that tend to transparency, or rather to extend the scope of knowledge to every detail of everyday life. In this sense, writes Baudrillard, instead of being seductive, this exhibition technique becomes pornographic, because unfolding every ripple of everyday life through the technological eye, it deprives the world of its eroticism, sweeping away the areas of shadow and all the imperfections from which the human eye has always been attracted. "Hyperrealism is not surrealism, it is a vision that leads a ruthless hunt against seduction by force of visibility" (Baudrillard 1979, p. 38).

On this consideration of Baudrillard, it is possible to draw a parallel once again with Barthes' reflections on photography (1980) and in particular on the unary image. According to Barthes, each image generates different intensities of pleasure and participation. When the gaze dwells on an image, it probably does so because it meets the viewer's taste or interest. However, even higher levels of involvement can occur, in which case the image has a "punctum." Some photographs, by virtue of the particular way in which they are displayed, the simultaneity,[35] offer the eye a conspicuous sequence of details that capture the viewer's attention. Punctum is the element that "starting from the scene pierces me" (Barthes 1980, p. 28) and is able to sting, generating an emotional wound that pushes to a strong attention. The punctum is the sign of a strong relationship of the observer with the image, a relationship that can only be very personal, as it is linked to the interests and emotions of each. It is that element that ensnares the user, stealing his or her gaze, transforming his or her interest and making him or her pass from to like to love, that is to say, to a more involving participation.

---

[35] On this particular expressive mode of photography, see McLuhan's fundamental essay in his *Understanding Media*.

Barthes's reconstruction also foresees the existence of images without interest, of any hold on which the observer can focus his attention, images that the French semiotician defines as unary. The example par excellence of this type of photography is precisely the pornographic image, described as "a photo always naive, without ulterior motives and without calculation, just like a showcase that displays illuminated a single jewel, the pornographic photo is entirely constituted by the display of one thing only: sex" (Barthes 1980, p. 42). As for Baudrillard, for Barthes too, photography without secrets is not a seductive photograph; the extreme transparency to which it aspires is the first suspect of the observer's disinterest. What makes it indifferent is the absence of a filter, of an element of disturbance that stands between the look and the evidence of a nakedness deprived of seduction. The terms "transparency" and "pornography" are used by both French scholars as synonyms or in a continuous way, as if they were adjacent modalities and concepts.

The unary image lacks the possibility of entry of the user, what McLuhan (1964) called the "completion of the image"; the images are so exhaustive at first glance "to the point that you have nothing more to add, that is, to give in return" (Baudrillard 1979, p. 39). The lacerating relationship with images capable of pricking and hurting is lost, weaving a high intensity relationship with the observer. The visibility that has become a paradigm bans shadow zones, making only well-lit spaces acceptable, both in the public and private spheres.

It is in this Baudrillardian analysis of transparency and obscenity that a particularly relevant theoretical basis emerges—one that combines food with pornographic aesthetics. The dishes photographed and then shared on social networks or blogging platforms constitute a gastronomic imaginary that tends to become a rule. For this reason, food is addressed to the machine that mediates its relationship with the public and, like a probe, drops the dish into the bowels of contemporary imagination (Vagni 2016). The dish is composed to attract the eye of the camera; before the photo is taken, the food is taken care of in every detail: raw materials, colors, arrangement, and quantity of food are altered to better adhere to the telephoto lens and compete to attract the gaze of the public. Moreover, photography is often so close as to bring out new details, normally hidden by resorting to the dogma of visibility and knowledge of every area of everyday life. Following Baudrillard's analysis of the images, the virtual entry of the eye into the plate produces a reality effect, typical of pornographic aesthetics, because it shows something that daily practice has historically preserved in the background.

VISUAL MEDIA AND FOCUSED GAZE

The common use of the term "food porn" by contemporary users reproduces the typical atmosphere of electronic media, that of complete transparency, but if, in the meaning of Baudrillard, this equates to a deprivation of elements of interest, charm, and seduction, then, for contemporary users, the exhibition value of the objects has completely lost its negative stigma, and food porn has become a hashtag that underpins the identity of each, evoking a hedonistic lifestyle, but also a refined and modern one. Over time, other analyses of the power of images and visibility have been added to Baudrillard's, proposing a different interpretation of the phenomenon from that of the French scholar, which could be useful to explain the enthusiasm with which we refer today to the food porn.

In particular, Susan Sontag's study of photography (1977) is particularly useful in questioning the relationship between transparency and pornographic aesthetics proposed by Baudrillard. In a long and accurate analysis of photography, Sontag argues that when this technology stops being used as an artistic form and spreads among mass consumers, it also begins to become part of a ritual, individual and collective, cataloguing the world and experiences. The photo is the parallel and successive form of an object; it is its testimony and its memory, the proof of an experience and a skill, the possibility, in each of these cases, of its public sharing. As Sontag writes, "[t]he photographer, behind his camera, creates a new tiny element of another world: the world of images, which promises to survive all of us" (Sontag 1977, p. 11).

Here, Sontag and Baudrillard's arguments appear, at first reading, in harmony because the images seem to have a higher value than the referents. But this is not the intention of the American scholar, the nerve center of her reflection is the vision and the way in which photography alters its potential. In particular, Sontag focuses on the fact that an experience, even a private one such as the making of a dish or its consumption, becomes true the moment it takes on a collective dimension. In this sense, photography disseminates "spectral traces" around the world (p. 8); i.e., it keeps alive moments that would have quickly vanished from intimate and collective memory. Photographing a moment means prolonging the possibility of viewing it beyond the instant of its happening, observing an event even if you are not eyewitnesses because of the spatial or temporal distance. In the specific case of gastronomy, therefore, through photography the consumer creates a personal anthology of fugitive memories, which paradoxically acquire truth when subjected to the scrutiny of the camera, which in turn removes them from their ephemeral nature.

The possibility of avoiding the oblivion of experience is only one aspect of photographic power; according to Sontag, it also determines the moments that can enter among the number of memorables based on their grammar and their photogenicity. Sontag therefore places the introduction of photographic practice at the origin of a new way of seeing: "by teaching us a new visual code, photographs alter and expand our notions of what is worth looking at and what we have the right to observe" (Sontag 1977, p. 3). Moreover, the dissemination of photographs produces a "world view that denies continuity, but that gives every moment the character of a mystery" (p. 21). Reality becomes "anatomic"—it fragments into a mosaic of frames that steal particularly significant episodes with the passing of time. The principle of photography's functioning breaks the linearity of time, while the materiality of the photographic image returns an idea of solidity, manipulable, and monumental. But in Sontag's text, it is immediately clear that the photographic image is never a form of transparency; it appears rather as an intervention of technology on the world:

> Immediately after 1840, the versatile and ingenious Fox Talbot not only composed photographs in the genres of painting ... but also pointed the camera at a sea shell, the wings of a butterfly (enlarged with the help of a solar microscope), a portion of two rows of books in his studio. But these subjects are still recognizable as a shell, the wings of a butterfly, books. When the usual visibility was further violated—and the object was isolated from its context, making it abstract—new conventions on what is considered beautiful came into force. Beautiful becomes what the eye cannot see: that shattered and dislocated vision that only the machine can give (p. 79).

From the first photographic representations of the nineteenth century, this emerging new medium reveals hidden corners, multiplies the points of observation, and favors a violation of the usual visibility. Photography also provokes a detachment from nature, because one begins to think of life as a concatenation of potential frames "so"—writes Sontag—"one of the perennial successes of photography has been its ability to transform living beings into things and things into living beings" (p. 86). One example of this is Edward Weston's photographs, particularly those of peppers. The pepper of the modernist photographer Weston (1929–1930) can only be identified by a caption "Pepper number 30" (Fig. 1, *opposite*).

The close position of the photographer, the play of lights, and the angle make the image indistinguishable, which rather evokes two bodies that join in a sinuous way. According to Sontag, the inability to identify the portion of the world on which the photographer's eye has rested generates an erotic attraction, because the viewer is disturbed by an unrecognizable image.

Photography, with the intention of approaching its object and stripping it of its secrets, generates the opposite effect and cloaks it in a mysterious aura into which the curiosity of the observer creeps. The approach is not revealing; on the contrary, this way of photographing distances the possibility of perception by creating the right distance in which the user can

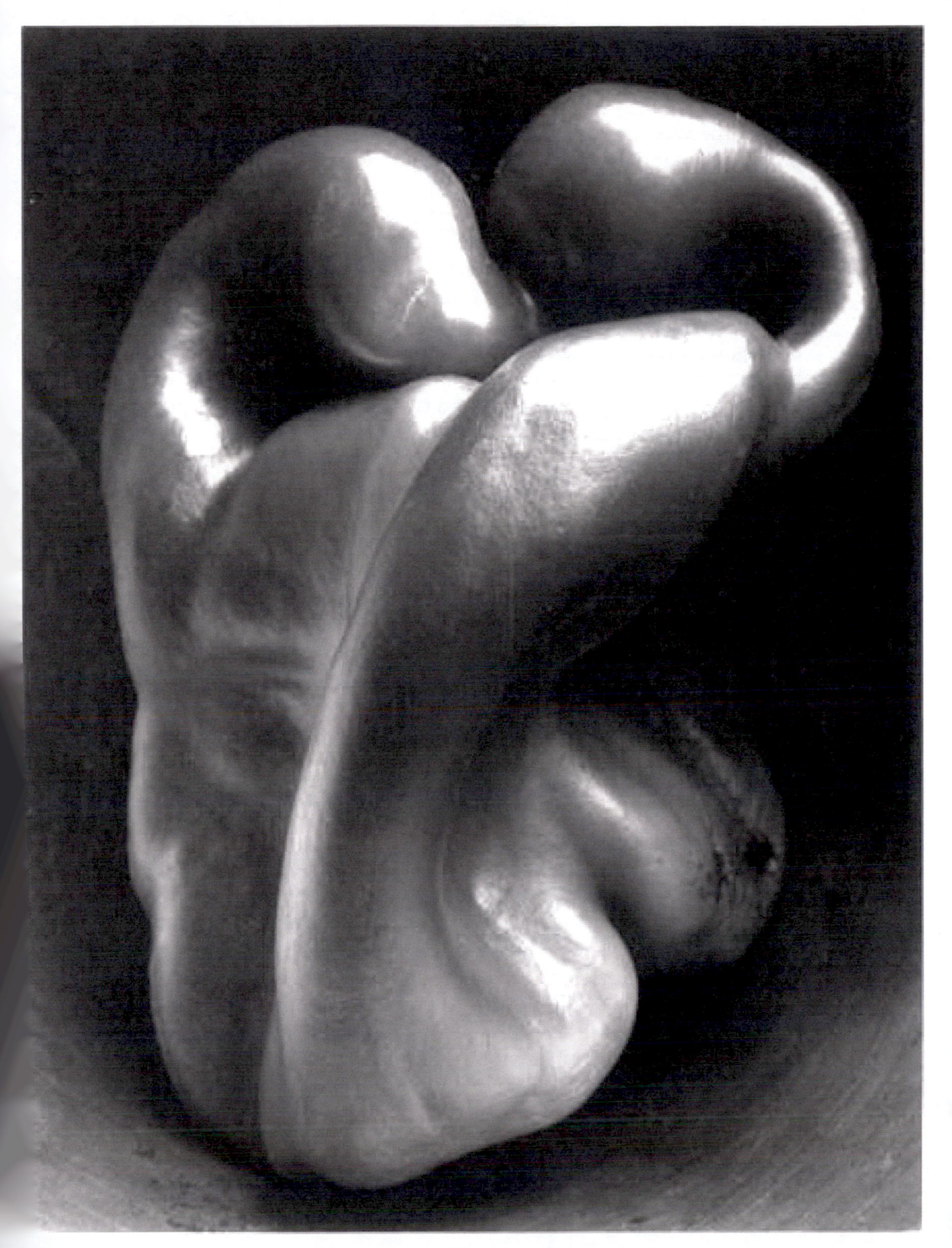

Figure 1. "Pepper number 30."

place himself to appropriate the image. This creates a tactile relationship with the image, and it is in this complicated relationship that the fascination of photography is expressed.

Today, the Weston pepper could be accompanied by the hashtag #foodporn because it presents many of the characteristic features of this expression: it portrays a voluptuous food; the lens approaches the object almost to the point of touching it; it expresses a communicative will, suggested by the work on the object. All these elements, as we have seen, can have a double reading. Following Baudrillard's analysis, they indicate an action of visibility on the object that deprives it of all seductive qualities, reducing it to an image with a pornographic aesthetic. Following Sontag, instead, it is possible to find in the image an erotic trait, given by the perturbing effect it generates on the observer because of its indistinguishability. Looking at the image and taking into account its caption, one could also say that the photograph surrounds the pepper in a perimeter of visibility, removing it from its context—a typical feature of this medium that sheds light on some portions of the world to the exclusion of all others from knowledge. The horizon of disappearance evoked by Baudrillard (1995) regarding technological images is overturned here: the accumulation of images does not stifle the world; it does not submerge it behind a systematic attempt at simulation—what determines its disappearance is instead the specific way in which the audiovisual media circumscribe the field of their observation and thus determine an internal (knowable) and an external (inscrutable), through the simple selection of the lens. The focused gaze of the media becomes the gaze through which the observer knows the world.

The caption of Weston's photo, "Pepper number 30," reveals such a characteristic: the observer might be led to wonder what happened to the previous twenty-nine shots of the pepper—are they fragments of the world lost because of the photographic selection that prevented a vision of them. Photography and visual media in general do not always produce transparency; in many cases they simply simulate it while constantly operating exclusions, generating huge areas of shadow.[36] This is well understood by analyzing gastronomic images: often the shots focus on food, playing continuously with the background that disappears with the technique of shading. The food gains the foreground, becoming the object of the vision, while everything that is next to it is faded until it becomes incomprehensible and insignificant. This particular exposition technique highlights a nature of the room, that of restricting the world. It seems paradoxical, because one of the characteristics of visual media is to show the great complexity of modern life (Vattimo 1989), leading the public beyond the narrow spaces of the physical place and direct knowledge. The media are windows on diversity, on novelty, on the other that we do

[36] In the field of communication research, one thinks, by way of example only, of the theory of the agenda setting, which reveals this particular functioning of modern media, and in particular of television.

not know. Yet they produce, at the same time, a reduction of the world, made arbitrarily through the focus of particular and momentary interests. Visual media, starting with photography, were born with the inverse intention of simulating transparency (Elsaesser, Hagener 2007), showing the centrifugal force of its representations—hence the interest in Instagram's images and their reductionist vocation, which tends to multiply the sources, points of view, and narratives of the world, a shared explosion of a multitude of individuality, aesthetically isolated, but included in the collective ethic of participation and sharing.

FROM FOOD PORN TO FOOD MEDIA

The zoom-in on the details typical of visual media and their focused gaze let emerge a fetishistic attitude of the users, understood here as the particular way of seeing and feeling of those who lose sight of the overall horizon because they are unconditionally attracted only by a detail, which makes sense in itself and not in the relationship with the scene. As Sontag's analyses show, the detail becomes so intrusive that it fills the screen, until it overflows from the frame, and breaks the natural harmony of the complete image, that is, the image of an order of things. Visual fetishism is deprived of order; it sees in a de-formatted way: its view is restricted to a detail, a part with which a very close bond is created. The characteristic form of its fruition is similar to that of the contemporary user, who is put in the condition of disinterest of the whole (a plate, a film, a television series) to devote himself to the fragments that involve him intensely. This way of seeing and feeling inscribed in the functioning mechanisms of modern media is also reproduced by the attitude of an emblematic figure of the dimension typed in: the fan, who through his tireless work, tries to fill the gap between himself and the work—think of the pioneering reflections of Henry Jenkins (2006) on the spoilers of *Survivor* or the intensity of participation of fans of *Star Trek* described by Robert Kozinets (2016)—working for a process of psychological and emotional rapprochement with the object of worship. The fan has a look focused on his passion and, thanks to his actions and the incessant accumulation of information, becomes one with it. It is in this climate that we can rethink the spread of the expression "food porn" in terms of a media coverage of the gastronomic experience. To do so, however, it is necessary to concentrate on culinary images, drawing on the relevant analyses of visual semiotics in this field, and then propose a mediological reading of the phenomenon.

Jacques Fontanille (2006), analyzing from a semiological point of view the photos of the dishes made by the French chef Michel Bras, argues that the images share the perspective of the diner and that therefore they mimic the real spatial location of the dish and its consumer, making the observer of the photograph assume the same point of view. In this way, the man who looks and the one who eats have the same position with respect to the meal, and there is a convergence between the consumption of the image and that of the dish. For Fontanille, the two objects cannot be superimposed by virtue of the fact that the photographic version is a "second-degree artifact" that makes a "reduction" with respect to its referent because it simplifies the sensorial complexity, proposing it only in the visual dimension. What the semiotician tries to say is that photography is a representation of reality, the visual reproduction of a real prototype. What we are trying to show, instead, is that the photographic lens prolongs and strengthens the faculties of the gaze, consequently modifying the object observed. The manipulation of the image does not occur only after the shot, in a phase of postproduction; the author's intervention is inscribed in the shot, that is, in the particular way in which time is stopped to extract an image. Thanks to photography, the observer can grasp aspects that live enjoyment keeps hidden, that can change perspective, that can make immortal in his memory the original shape of a dish. The object photographed does not suffer any reduction; on the contrary, it is swollen through the juxtaposition of a second level that enhances the expressive possibilities. Vision and sight have a strong cultural connotation, that is to say that they are modelled by the medium that provides them with the symbolic technical apparatus through which to see. Each technology thus inaugurates its own mode of vision[37] and, at the same time, introduces an imaginary towards which the viewer converges.

While the photographic image celebrates the technological power of the modern gaze, on the other hand it shows its structural limit, that of the fixity of the point of observation. In front of the plate, this fixity is somehow outdated; indeed, it has never been the usual posture of the consumer, and it is certainly less so since television began to explore the plate through its filming, assuming the many observation points allowed by the mobility of its technologies, the direction and editing, which free the plate from the yoke of the photographic frame, transforming it into a process, a fluid substance: prepared, served, and consumed.

When dissimilar, photography and television converge in the effect of pushing food towards its culturalization. Each medium does this according to its own functioning, applying a specific filter and returning, consequently, deeply different images. In both cases, however, the image is not a falsification of food; it is rather to be understood as a veil that creates an erotic indistinguishability between the visual sign and its referent. The image of a food has the same function that, according to McLuhan, covers the silk sock on a woman's leg: "the silk net sock is much more

[37] On the relationship between the media and the eye, one of the first and most important studies is that of Wolfang Schivelbusch, *Storia dei viaggi in ferrovia*, Einaudi, Turin, 1988. The German scholar formulates the idea that train journeys produce a "panoramic view," that is, a view modeled on a vision experience hitherto unknown: the rapid succession of the panorama and the consequent sensory stimulation make the vision in motion a novelty to which the humus must adapt, producing a view that tends to fly over the details to concentrate on the overall view.

sensual than the nylon, because the eye must cooperate to fill and complete the image, just like in the mosaic of the television image" (1964, p. 39). In the same way, the photograph of a plate is its addition, a surplus that leads us to turn to it with greater attention. The photo acts on the plate as a showcase for the goods: it creates a phantasmagoria in which the goods acquire their own dignity and a life, regardless of the use that will be made of it. The showcase isolates the goods in a dream world (Codeluppi 2007) that provides a double access: the one linked to the visual connection of the passer-by with the goods on display and that of a more concrete act of consumption. In the same way, photography brings the user closer to the plate—as Fontanille intends to demonstrate—but it can also be an autonomous object of consumption, which does not need a reference point to seduce the viewer.

This means that food products and all the elements that compose them have the value of enunciative visual signs, that is, they acquire expressive capacity. In this way, the social process that Baudrillard has brilliantly described in his analyses takes shape, namely that of the "system of objects": in the consumer society, a food product and any other object acquire the faculty to express themselves, to communicate, regardless of their consumer. This is not only the case with works of art, such as those by Michel Bras, in which there is a clear expressive intentionality; each object has a communicative power disseminated in the visual traces that compose it. The use of the word "trace" in the analysis of visual semiotics is not random: the aim is to refer to an existing sign, which must be identified through an effort of understanding. There is nothing immediate, extraordinarily evident in the image of a food; everything is hidden and must be sought through a conjunction between the individual consumer and the goods. It is the characteristic of the "plastic" taste that is the possibility of a visual sign to create meanings not immediately referable to grids of previous knowledge.[38]

The expressive power of the object is not directed against the consumer; it does not overpower him—on the contrary, it is a function of its centrality in the production process. If the semiotic approach had the merit of shifting the focus on food and its composition to read about social processes, through the media studies of Henry Jenkins (2006) it is possible to integrate this perspective. In particular, the idea of "transmedia product" questions the distinction between the object and its image. Taken together, the various declensions of a dish form a "transmedia storytelling," or a dissemination of the different fragments of a narrative on multiple platforms, which focuses on the new participatory ethics of networks that has pervaded contemporary users. Here, transmedia storytelling is not a simple expansion of the object, a way to make it overflow beyond its original place of conception with the intention of colonizing new territories and new audiences. No, here we are speaking more of a

---

[38] On the distinction between figurativeness and plasticity of taste in gastronomy, see G. Marrone, "Levelli di senso: dal gustoso al saporito", E/C, 2013, n. 17.

complexification of the narrative object that finds in the articulation of the contemporary media landscape and in the polyphony of its languages the privileged place to create a "narrative world" (Herman, 2002)—in other words, an interpenetration of expressive forms that is configured as a magmatic movement and not as a closed and definitive form.

The concept of the narrative world indicates the potential of a text, its endless capacity to regenerate and expand its boundaries, thanks to which history finds the possibility to be reborn and expand. Today, every fan can contribute to writing or creating a piece of the narrative, thus expanding the narrative world of reference. This is what happens to food and its images: in the last decade, food has become a spreadable content because on it, it is possible to "baste a discussion", not a simple chat for its own sake, but a flow of words that creates a second level—and often also the main level—of the original image. A conversation triggered by a narrative handhold and, by virtue of its strong participatory nature, sets in motion a huge amount of conversations. The photo of shared food immediately becomes the property of the community that manages to coagulate and that, through its narrative work, re-creates the same photo, triggering an incessant mechanism of sharing. The second aspect is the public agency. Their role is different from that of the old broadcasting media: there is no greater or lesser "participation" than in the past but a real production that concerns both the content and its speakers (Jenkins, Ford, Green, 2013).

Therefore, the sharing of culinary images cannot be reduced only to a mere spectacularization, as the expression "food porn" could suggest—which, as it has been written, mainly evokes the exposure of food. Food, today, has undergone a process of mediatization that has changed its form, giving it a new complexity that can be fully described with the expression "food media" (Rousseau 2012)—a term that summarizes the processes of culturalization and mitigation that have involved modern food, while at the same time highlighting the communicative and participatory vocation of food.

References

Barthes, R. (1957). *Mythologies*. Paris: Editions du Seuil.
——— (1961). "Pour une psychosociologie de l'alimentation contemporaine." *Annales ESC XVI* 5, pp. 977–986.
——— (1980). *La Chambre claire*. Paris: Gallimard-Seuil.
Baudrillard, J. (1974). *La société de consommation. Ses mytes ses structures*. Paris: Éditions Gallimard.
——— (1979). *De la séduction*. Paris: Galilée.
——— (1995). *Le crime parfait*. Paris: Galilée.
Benjamin, W. (1936). *Das Kunstwerk im Zeitalter seiner technischen Reproduzierbarkeit*. Frankfurt am Main: Suhrkamp Verlag.
Codeluppi, V. (2015). *Mi metto in vetrina*. Udine: Mimesis.
——— (2007). *La vetrinizzazione sociale*. Torino: Bollati Boringhieri.
Coward, R. (1984). *Female Desire: Women's Sexuality*. London: HarperCollins.

Elsaesser, T., and M. Hagener (2007). *Teoria del film*. Torino: Einaudi.
Finkelstein, J. (2014). *Fashioning Appetite: Restaurants and the Making of Modern Identity*. London–New York: I.B. Tauris & Co.
Flichy, P. (2011). *Le Sacre de l'amateur*. Paris: Le Seuil.
Fontanille, J. (2006). "À deguster des yeux. Notes sémiotiques sur la 'mise en assiette'. À propos de la cuisine de Michel Bras." *Visible* 1, 195–216.
Gumbrecht, H. G. (2006). *In Praise of Athletic Beauty*. Cambridge, MA: Harvard University Press.
Han, B. C. (2012). *La società della trasparenza*. Roma: Nottetempo.
Herman, D. (2002). *Story Logic. Problems and Possibilities of Narrative*. Lincoln: University of Nebraska Press.
Lindenfeld, L., and P. Parasecoli (2016). *Feasting Our Eyes: Food Films and Cultural Identity in the United States*. New York: Columbia University Press.
Mangano, D. (2014). *Che cos'è il food design*. Roma: Carocci.
Manovich, L. (2016). "Instagram and Contemporary Image." Manovich. Manovich.net.
Marrone, G. (2013). "Livelli di senso: dal gustoso al saporito." E/C, n. 17.
——— (2014). *Gastromania*. Bologna: Bompiani.
——— (2014). *Gastromania*. Bologna: Bompiani.
(2015). *Buono da pensare. Cultura e comunicazione del gusto*. Roma: Mimesis.
McBride, A. E. (2010). "Food Porn." *Gastronomica* 1, 38–46.
McLuhan, M. (1964). *Understanding Media: The Extensions of Man*. New York: McGraw-Hill.
Mejova, Y., S. Abbar, and H. Haddadi (2016). "Fetishizing Food in Digital Age: #foodporn Around the World." Cornell University. http://arxiv.org/abs/1603.00229.
Jenkins, H. (2006). *Convergence Culture*. New York and London: New York University Press.
Jenkins, H., S. Ford, and J. Green. (2013). *Spreadable Media*. NY: New York University Press.
Kozinets, R. (2016). *Il culto di Star Trek*. Milano: Franco Angeli.
Perullo, N. (2016). *Il gusto come esperienza*. Roma: Giunti.
Pinotti, A., and A. Somaini (2016). *Cultura visuale*. Torino: Einaudi.
Rousseau, S. (2012). *Food media: Celebrity Chefs and the Politics of Everyday Interference*. Oxford: Berg Publishers.
Schivelbusch, W. (1977). *Geschichte der Eisenbabnreise*. Frankfurt am Main.
Sontag, S. (1977). *On Photography*. New York: Farrar, Straus and Giroux.
Stagi, L. (2004). *Food porn*. Milano: Egea.
Teti, V. (2015). *Fine pasto. Il cibo che verrà*. Torino: Einaudi.
Vagni, T. (2015). "EXPO 2015: La decrescita da utopia a stile di vita," in A. Abruzzese and L. Massidda, (eds.), *EXPO 1851–2015: storie e immagini delle Grandi Esposizioni*, pp. 469–477. Torino: UTET.
——— "Gastronomia e industria culturale in Italia." In M. Polesana, *La società italiana. Cambiamento sociale, consumi e media*, pp. 93–106. Milano: Guerini Next.
——— (2017). *Abitare la TV. Teorie, immaginari, reality show*. Milano: Franco Angeli.
Vattimo, G. (1989). *La società della trasparenza*. Milano: Garzanti.

DISAPPEARING *WOMAN*
A VOICE-OVER AS *SHE* PIXELATES

GWENDOLYN SPRING KURTZ

We've known for a while now that *woman* isn't something that individuals are, but rather something that individuals do. Since West and Zimmerman's landmark notion of gender as "undertaken by women and men whose competence as members of society is hostage to its production," we've been doing gender.[39] Gender exists in our expectations and actions, a performativity that structures the arguably private, and quite often public, lives that make for Western culture. A sightseeing, showcasing, and souveniring culture, we've captured those gendered moments on camera. We've eyed the category of individuals called *women* through film theorist Laura Mulvey's "male gaze," watching as the camera stylized *women* as objects of desire and for consumption.[40] In the shift from big screen to device screens, however, that style loses coherence, as does the camera's object. *Woman* falls apart in pixels.

In an increasingly digital age, gender is pictured in pixels, those small, controllable samples of an original image represented on a digital screen. Pixels are pieces that make for a recognizable whole. But the social category *woman* loses meaning when pictured in small samples, as *woman*'s ex/samples are too many and too much at odds with one another to make collective sense. To evidence this discursive conflict, I look to a few samples of *woman*: Marie Antoinette's breasts, Tatyana Fazlalizadeh's street art, and an individual's Facebook profile picture. In these pixelating sites, *woman* disorganizes into uncontrollable samples that do not reference a clear original image. At this point or, perhaps better, from these points forward, then, the word *woman* marks a cultural artifact. *Woman* as we knew *her* is slipping into the stuff of legend, soon an example of archaic language—because the pixels, and their gendered trope, are coming apart. In a digital take on the category that has been *woman*, I challenge the singleness and stability of this waning not-quite selfhood and instead call for a personhood that is newly seen, a person in disorganizing pieces.

Admittedly, feminist theorists have been looking at *women* for a while now. Laura Mulvey teaches us that in film and in life, we see *women* from culture's dominant heterosexual male perspective, as "the determining male gaze projects its fantasy onto the female figure, which is styled ac-

---

[39] Candace West and Don. H. Zimmerman, "Doing Gender," *Gender and Society* 1 (2), 1987, p. 126.

[40] I'm indebted here and throughout to Laura Mulvey, "Visual Pleasure and Narrative Cinema," in *Visual and Other Pleasures* (New York: Palgrave, 1989), p. 19.

cordingly. In their traditional exhibitionist role women are simultaneously looked at and displayed, with their appearance coded for strong visual and erotic impact so that they can be said to connote to-be-looked-at-ness."[41] Per this male gaze, *woman* acts as a visual pleasure rather than a person. Donna Haraway denies this kind of socially coded existence: "there is nothing about being female that naturally binds women together into a unified category. There is not even such a state as 'being' female, itself a highly complex category constructed in contested sexual scientific discourses and other social practices."[43] Judith Butler, too, sees the end of *woman* coming. Like Haraway, Butler argues that there are no shared conditions for satisfying the concept of *woman*. Butler reimagines a gender in the making, "a complexity whose totality is permanently deferred, never fully what it is at any given juncture in time."[44] Gender is continually at work, an always scripting and revising activity.

Theorists have argued for some time now that there is no still, signifying *woman*, but they've done so with limited audience. The digital image, however, breaks it down before our eyes so that we might more popularly see and believe *woman*'s disappearance. This visual deconstruction, then, follows the work of these and other prescient thinkers as it watches *woman* fall to irreconcilable pieces in hand, on the street, and at home. As sociologist Barbara Risman argues, "we can honor West and Zimmerman" and, I would argue, Mulvey, Haraway, Butler, and others, "no more than by moving beyond our reliance on a doing gender framework, because the very existence of that language has helped change the gender structure itself."[45] I want to look again at gender, then, as something we might be done with rather than do. I want to watch *woman* as *she* pixelates. To illustrate, I'll offer a few small samples, a few erratic pixels, of *woman*.

PIXEL: MARIE ANTOINETTE'S BREASTS

Never mind that the seventeenth-century saucer-shaped champagne coupe was not modeled on Marie Antoinette's eighteenth-century breasts. It's storied stuff and a story we continue to tell. As is the case with mythologies in general, there's something to this story. *Huffington Post*'s Joseph Edos reports that

---

[41] Mulvey, "Visual Pleasure," p. 19.

[43] We owe a certain consciousness to the call for affinity rather than identity in Donna Haraway, "A Cyborg Manifesto: Science, Technology, and Socialist Feminism in the Late Twentieth Century," in *Simians, Cyborgs and Women: The Reinvention of Nature* (New York: Routledge, 1991), p. 295.

[44] If we can imagine doing away with gender, it's because we've been shown what gender does in Judith Butler, *Gender Trouble: Feminism and the Subversion of Identity* (New York: Routledge, 2006), 22.

[45] Barbara J. Risman, "From Doing To Undoing: Gender As We Know It," *Gender and Society* 23 (1), 2009, p. 84, http://journals.sagepub.com.proxy.lib.utc.edu/doi/abs/10.1177/0891243208326874.

> history does show that in fact, Marie Antoinette had porcelain bowls molded from her breast. They were designed for drinking milk as part of her 'Pleasure Dairy' where the queen and her ladies-in-waiting would dress up as milkmaids and frolic, milking and churning butter all day in her rustically designed hamlet at Versailles.[46]

But history doesn't show, in fact, that Marie Antoinette made bowls from her breasts. Art historian Meredith Martin notes that Louis XIV's building director the Comte D'Angiviller commissioned the legendary pink-nippled porcelain bowl for a dairy the king built for the queen at Rambouillet. Busy outfitting her own hameau at Petit Trianon, Marie Antoinette was unaware of the king's secret work at Rambouillet, and unaware of the *jatte teton*. The milk bowl by Sèvres, history shows, is modeled not on Marie Antoinette but rather on "the ancient Greek *maestos*, which was shaped like a woman's breast and used to consume wine during Dionysian drinking rituals."[47] Here, Greek legend lends a misplaced mythos to the already maligned milkmaid queen, with her breast but without her consent.

The legend continues in 2008, when Karl Lagerfeld designs a glass modeled on Claudia Schiffer's breast for Moët & Chandon. Per *Wine Spectator*, the glass sells "along with a bottle of 1995 Dom Pérignon Oenothèque, but some still might find it a tad bit too expensive at $3,150 a pop, considering you don't get to take the real Schiffer home to share the bottle with you."[48] In this troublingly misogynistic quip, breasts and their likenesses are objects of someone else's appetite and desire. Not to be outmodeled, Kate Moss debuts a champagne glass in 2014 molded from Moss's left breast and designed by British artist (and Sigmund Freud's great-grandchild) Jane McAdam Freud. The problematic quips continue: "anyone hoping to cop a feel," jokes the *NY Post*, can do so at the several stylish restaurants serving celebration in Moss's coupes.[49] And anyone hoping to see, can. For the recent coupe's internet

---

[46] You can read some almost-factual history in Joseph Erdos, "The Glass Shaped Like Marie Antoinette's Anatomy," *The Huffington Post*, May 27, 2014, https://www.huffingtonpost.com/2012/04/13/marie-antoinette-champagne-coupe_n_1424686.html.

[47] You can read the real stuff by Meredith Martin, *Dairy Queens: The Politics of Pastoral Architecture From Catherine De' Medici to Marie-Antoinette* (Harvard: Cambridge, 2011), p. 240.

[48] Evidently, *Wine Spectator* rates *women* as well as wine. "Unfiltered: Dom Pérignon Pays Homage to a Top Model's Chest," *Wine Spectator*, December 4, 2008, http://www.winespectator.com/webfeature/show/id/Unfiltered-Dom-Perignon-Pays-Homage-to-a-Top-Models-Chest_4486.

[49] This joke belongs to Lindsay Putnam, "Kate Moss' Breast Serves As Mold for Champagne Glass," *New York Post*, October 9, 2014. https://nypost.com/2014/10/09/sip-champagne-in-a-glass-molded-from-kate-moss-breast. But everyone wants in on the joke. After parroting that "cop a feel" line, another style writer goes for size: "Thank goodness Moss's breast is barely a B-cup or patrons would be feeling one hell of a hangover." See Allyson Rees, "Champagne

hype, Moss poses in signature waif-like style, topless but with arms crossed over the chest as if to hide from Mulvey's male gaze.[50] But hiding is hardly possible in an intentionally infinite online exposure.

In this lineage of breasts, there's a mythology rich with beauty and pleasure layered over *women*'s bodies. But the stylized celebration served in these stories makes a literal object of *woman*: a champagne coupe. A champagne coupe no longer signifies a piece of the arguably original *woman*. Champagne and its vessel signal celebration, the stuff of New Year's Eve and after-parties. Here, the referent, embodied *woman* is lost in a crowd and to the crowd, as she loses agency. Per those too-familiar quips, *she* can be taken home for a price and copped without consent. *She* is the object of someone else's action, a vessel to hold and consume like the coupe and celebration *she* inspires. These are too many layers of desire and consumption acted upon a person. Mulvey helps us look away from the theorized screen and at the material being: "although none of these interacting layers is intrinsic to film, it is only in the film form that they can reach a perfect and beautiful contradiction."[51] In lived experience, the word *woman* writes a gendered vigilance and a social violence onto the body. The word and its work evidence real consequence in which both the oppressor and the oppressed share a single-minded perspective. These iconic breasts and their likenesses tell us that our culture permits—rises to, even raises in toast—sexist jokes and sexual assault, and that *women* have internalized male fantasy to the point that they are drinking, quite literally, from the same cup. This picture moves quickly from Marie Antoinette's faux-rustic hamlet to contemporary rape culture.

Mythos and marketing at odds fetishize breasts in layers of visual pleasure and violence and keep them in a comparatively staid domestic space. Whether champagne coupe or milk glass, glassware is kept in the kitchen. But hypersexualization doesn't really fit in a kitchen cupboard, so we move to the still domestic space of the bedroom—ineffectively. Champagne glasses are intentionally untouchable, designed for the drinker to hold the stem, so as not to affect the champagne's temperature and taste. Not unlike Cinderella's legendary glass slippers, these glass breasts are at hand, but the body they represent is not in hand. Mulvey describes this Freudian "fetishistic scopophilia" as to control *women*, the male unconscious "builds up the physical beauty of the object, transforming it into something satisfying in itself."[52] In this scene, the prince sighs with at once longing and contentedness as he raises the slipper like a champagne glass. But Cinderella doesn't go to the ball for

---

Served in Kate Moss' Breasts? We'll Take Two, Please.," *Ravishly*, August 22, 2014. https://ravishly.com/2014/08/22/champagne-served-kate-moss'-breasts-we'll-take-two-please.

[50] If so inclined, you can imagine "the Moss effect" with Scarlett Kilcooley-O'Halloran, "The Kate Moss Coupe," *Vogue*, August 21, 2014. http://www.vogue.co.uk/article/kate-moss-champagne-glass-coupe-34.

[51] Mulvey, "Visual Pleasure," p. 25.

[52] Mulvey, "Visual Pleasure," p. 21.

the prince to take pleasure in or purchase a piece of her. The fetish carries over when we look away from the fairy-tale scene (and screen) because the glass (and looking glass) are flawed: *bon vivants* know that the wide coupe isn't right for champagne, as the glass's surface area lets aromatic bubbles escape too quickly. This premature pleasure is over before it's had. In the breast-shaped coupe, celebratory sparkle goes flat, a midnight magic lost as it's poured. These made-of-glass stories are fragile, always poised to break in eager celebration or rough handling. Once *she*'s analyzed, once *she*'s broken into pieces for a closer look at the whole, *woman*'s pixels don't fit together again in an original picture that is desirable off screen, if at all.

If not for champagne coupe or milk glass molds, American consumer culture arranges breasts in molded bra cups. The shopping mall and internet ad fixture Victoria's Secret markets cleavage as always conscious of the "to-be-looked-at-ness" of Mulvey's camera. In stores, Victoria's Secret employees mill about with nipple-pink measuring tapes draped around their necks, *women* hung with rubrics, ready to measure other *women*. If we're all drinking from the same cup, we're also all measuring one another with the same gendered norms. From posters hung high on store walls, cleavage-conscious celebrity "Angels" watch over the norming event, promising heterosex appeal and cisnormative style to the brand faithful. Like the champagne coupe, this gender self-policing is almost enough to make a misnomer of Mulvey's so-called male gaze. But again, these ex/samples of what might have once been an original picture—breasts from Marie Antoinette's to supermodels' to those measured by salespersons and sales—don't make for a reasonable picture. Breasts can't be at once untouchably mythological and readily available for purchase in a mass market.

It might help to remember here that baby bottle nipples, too, are modeled after breasts, if anonymous breasts.[53] In her engaging *Breasts: A Natural and Unnatural History*, Florence Williams relies on zoologists, anthropologists, evolutionary biologists, and an array of other scientists and doctors for insights into breasts. Like this one: lactation expert Peter Hartmann tells us that breast-feeding "represents 30% of a woman's energy output."[54] Williams translates: "what he meant by that energy bit is that while a woman is lactating, the metabolic energy required to feed

---

[53] There's an infantilization at play among these breasts, bras, and babies: Joan Jacobs Brumberg reminds us that even the bra, from the French *brassière*, "actually means an infant's undergarment or harness." In an uncanny homage to all things patriarchy, the bra harnesses girls, women, and sexuality, and brings in postwar profits, in Joan Jacobs Brumberg, "Breast Buds and the 'Training Bra,'" in *Women's Voices, Feminist Visions*, eds. Susan M. Shaw and Janet Lee (New York: McGraw Hill, 2015), p. 205.

[54] Ever wonder where "paint thinners, dry-cleaning fluids, wood preservatives, toilet deodorizers, cosmetic additives, gasoline by-products, rocket fuel, termite poisons, [and] fungicides" go after we (think we) dispose of them? See Florence Williams, *Breasts: A Natural and Unnatural History* (New York: Norton, 2012), p. 174.

her infant is 30% of her total output—or the energy equivalent of walking seven miles—every day. Looked at another way, a male baby requires 1,000 megajoules of energy the first year of life.

That is the equivalent of one thousand light trucks moving one hundred miles per hour."[55] Again, there's something to the mythology as *woman*'s face launches a thousand ships and *her* breasts move a thousand trucks. Both the mythologies and the bodies are impressive. Objectifying breasts, however—valuing the "to-be-looked-at-ness" of champagne coupes or lacy lingerie over the lived experience of individuals—makes a spectacle of *woman* as an object of someone else's celebrated pleasure, a pleasure so far removed from *her* that *she* is left out of the picture. That's the way of a social construction: we collectively hold one another in place and out of privileged places. But the individuals facing the conceptual boundary called *woman* resist place. They move. They're on one of the thousand ships they launch, and drive one of the thousand trucks they power.

As if the miracles of childbirth and childcare weren't enough, there's breast milk. Even Marie Antoinette followed Jean-Jacques Rousseau's campaign encouraging *women* to live idyllic pastoral lives and breastfeed their own children. As goes the saying, breast is best, for infants and maybe even more. Williams reports that breast milk boasts as many as eight hundred species of bacteria and two hundred oligosaccharides that we don't yet understand, that "human milk inhibits the transmission of HIV," and that one of our milk's proteins, alpha-lactalbumin, a.k.a. HAMLET,[56] "kills forty different types of cancer cells in a dish, including those of the bladder, lymphoma, skin, and brain."[57] It seems that breasts continue to promise great stories. But even storied breasts can't at once nurse children and cure cancer and exist exclusively for their desirable, sellable "to-be-looked-at-ness." These loosely associated pieces of a person don't make for a stable picture of a single, original subject. Seductive though the legend of *woman* might be, it's not a legend individuals can, or necessarily should try, to live up to.

PIXEL: A POSTER

Director Spike Lee recounts to *NPR*'s "Fresh Air" host Terry Gross: "I was walking down downtown Brooklyn. I saw these things—Stop Telling Me To Smile. And I put it up on my Instagram because I was intrigued by it. And I said, who is this person?" Lee's story is about Brooklyn artist Tatyana Fazlalizadeh, whose street art campaign "Stop Telling Women to Smile" inspires the "My Name Isn't" campaign in Lee's *Netflix* series

---

[55] Williams, *Breasts*, p. 175.

[56] Curious that the breast milk protein alpha-lactalbumin, a.k.a. HAMLET (human alpha-lactalbumin made lethal to cancer cells), is acronymed after the bard's tragic misogynist. Donna Haraway might remind us here that the sciences are sexist. Think I'm reaching too far back into history with Haraway? In 2015, Nobel laureate Tim Hunt made news with this gloss on female scientists: "'Three things happen when they are in the lab,' he said, 'you fall in love with them, they fall in love with you, and when you criticize them they cry.'" What's more important and much more fun than Hunt's gaffe is viewing "Female Scientists Post 'Distractingly Sexy' Photos," *BBC News*, June 11, 2015, http://www.bbc.com/news/blogs-trending-33099289.

[57] Williams, *Breasts*, pp. 178–189.

STOP TELLING
WOMEN
TO SMILE

"She's Gotta Have It." In the rebooted series, protagonist Nola Darling paints portraits (which are really Fazlalizadeh's portraits) that respond to gender-based street harassment. There's a doubling and then some to this resistance discourse, as two artists throw up two campaigns of two titles, making for a dual-consciousness artist who potentially speaks for themselves. But when the "She's Gotta Have It" audience sees the posters, they see them through Fazlalizadeh's real-life eye, through Darling's character's eye, through Lee's camera's eye, through Netflix's digital stream. The individuals who pose for and create the art are lost in "to-be-looked-at-ness," subject to too many layers of expectant gaze. The *woman* represented on the screen is ex/samples, sampled by several artists, a derivative with an unclear original. The pixels do not make for a recognizable whole, leaving *woman* more confusion than coherence.

Like their breasts, *women*'s movement through public spaces is storied. Their walks have inspired the likes of the Brazilian jazz classic "The Girl from Ipanema" (although the *girl* heard "a cacophony of wolf-whistles," not bossa nova, as the composers watched her walk by[58]) and less lyrical, less lauded catcalls. Fazlalizadeh's art speaks back to the calls and whistles, and seeks to rewrite the story of *woman*'s "to-be-looked-at-ness." As one of Fazlalizadeh's poster captions reads, "Women Are Not Outside For Your Entertainment." Of course, individuals don't enter into public spaces for passerby pleasure or purchase any more than Cinderella goes to the ball for the prince's pleasure or purchase. But American culture suggests that *woman*, in public spaces, does exist at the public's disposal. Street harassment is more accepted than street art. Fazlalizadeh's postering is an illegal act, and posters often end within days if not hours of their installation, pulled down or covered over by supposed authorities and supposed vandals. Catcalls, wolf whistles, and worse, however, are not policed like public art. Street harassment is a long-standing, legal violence that solicits *women* in public spaces as if they were so-called street walkers rather than just walking down the street, like Spike Lee can, without catcalls, at the beginning of this pixel.

Fazlalizadeh frames activism with this street-turned-digital art and, in doing so, participates in what Chela Sandoval and Guisela Latorre coin "digital artivism," an artistic sensibility that seeks social justice for people of color through digital media.[59] Fazlalizadeh admits an activist aim: "this project takes women's voices, and faces, and puts them in the street—creating a bold presence for women in an environment where they are so often made to feel uncomfortable and unsafe."[60] The artist's posters confront harassers and challenge *woman*'s social function as an

---

[58] Ruy Castro, *Bossa Nova: The Story of the Brazilian Music That Seduced the World* (Chicago: Chicago Review Press, 2000), p. 240.

[59] Chela Sandoval and Guisela Latorre, "Chicana/o Artivism: Judy Baca's Digital Work with Youth of Color," in *Learning Race and Ethnicity: Youth and Digital Media*, ed. Anna Everett (Cambridge, MA: The MIT Press, 2008), p. 81.

[60] "Stop Telling Women to Smile," Tatyana Fazlalizadeh, accessed December 20, 2017, http://www.tlynnfaz.com/Stop-Telling-Women-to-Smile.

object of desire and for consumption. Or, as Fazlalizadeh captions a self-portrait, "I am not here for you." A poster featuring a Latinx face and captioned "*Yo merezco ser respetada*" nods again towards Sandoval and Latorre's digital artivism that "advances the expression of a mode of liberatory consciousness that Chicana feminist philosopher Gloria Anzaldua calls *la conciencia de la mestiza*, that is, the consciousness of the mixed-race woman."[61] Whether or not Fazlalizadeh identifies as *mestiza*, the curated voice that speaks from the campaign's captions does. That collective voice also speaks to Kimberlé Crenshaw's theory of "intersectionality," the kind of oppression that occurs when identities overlap or intersect.[62] Fazlalizadeh puts the theory into practice, explaining: "because I'm black, the street harassment and the sexual harassment that I receive is often overlapped with racism."[63] The catcalls Fazlalizadeh and others hear act as raced, gendered placeholders, objectifying remarks that discriminate on at least two intersecting lines: Fazlalizadeh is targeted for blackness and *woman*hood. But unlike identity markers like race and ethnicity, gender is a concept without a clear presentation. In this presentation, *woman* is one of any number of targets attached to a person. Like a champagne coupe, a gendered target for sexual violence no longer references an original image. Nor does a gendered target for sexual violence reference a desirable image.

Fazlalizadeh captions another portrait, "My worth extends far beyond my body." I would argue beyond social categories, too. Fazlalizadeh (and others) and their likenesses are more than *women*. They are, as the street

[61] Sandoval and Guisela, "Chicana/o Artivism," p. 82.

[62] See feminist and antiracist Kimberlé Crenshaw, "The Urgency of Intersectionality," filmed October 2016 at *TEDWomen 2016*, San Francisco, CA, video, 18:37, https://www.ted.com/talks/kimberle_crenshaw_the_urgency_of_intersectionality.

[63] Hannah Pikaart, "Meet the Brooklyn Artist Who Lent Art-World Cred to Netflix 'She's Gotta Have It' Reboot," *Art Net News* (December 11, 2017), https://news.artnet.com/art-world/brooklyn-artist-shes-gotta-have-it-1163225.

art campaign insists, nuanced individuals, more than an object of someone else's racism or desire, more than a sight or site in which others take pleasure. But this sight and site is what *women* are and cannot, in embodied experience, unbecome. An individual is, as Haraway caveats, "a creature of social reality as well as a creature of fiction" or here, of theory.[64] In Fazlalizadeh's campaign, experience and expression try to refuse *woman*—because *woman* suffers intersectional oppression and is *herself* an intersection, a space in which others take pleasure, held in place by a gaze, a stem, a hand, or a word. Whether in a hamlet or restaurant, in a kitchen or bedroom, on a screen, or on the street, *women* are always on display, poised to be enjoyed or interrupted or talked over or even taken against their will. Again, *woman* is too many pictures of forcibly fulfilled and unfulfillable desire to make for a desirable picture. Fazlalizadeh's captions deny these irreconcilable functions: "I am not here for you." Neither the model, nor the artist, nor their likenesses are someone's *woman*.

All this aside, "She's Gotta Have It" is a show, shown through Spike Lee's lens, shown on *Netflix*. Lee tells Terry Gross, "this Netflix series could not be strictly a male gaze." Lee continues, "there was a huge presence of black women in the room. It was the most truthful thing to do."[65] *The New York Times*'s James Poniewozik, too, promises *women*'s agency, writing of protagonist Nola Darling: "Art is her means of self-assertion and self-defense."[66] Fazlalizadeh, however, remembers, "Spike had the final say in a lot of things. I found that to be frustrating at moments, but I am creating these pieces for the show."[67] Once Fazlalizadeh works with Lee, the art is for the sake of the film, so the camera (and the show's creator, producer, and distributor) takes precedence over the *woman*'s art. Fazlalizadeh continues:

> So being an artist who works with street harassment, who has a real-life art series about this, who is a painter of women, who feels really close to the character of Nola, I wanted the work to be as realistic as possible, but there were some moments and some things that Spike interjected and said, "This is what we need for the show."[68]

Resistance narrative notwithstanding, Fazlalizadeh's creative experience is subject to Lee's privilege. Even as he defends against the male gaze,

---

64 Haraway, "A Cyborg Manifesto," p. 291.

65 Terry Gross, "31 Years Later, Spike Lee Puts A New Spin On 'She's Gotta Have It,'" NPR, December 14, 2017, https://www.npr.org/2017/12/14/570761380/31-years-later-spike-lees-got-a-fresh-take-on-shes-gotta-have-it.

66 James Poniewozik, "'She's Gotta Have It' From Netflix Is a Bold Reboot from Spike Lee," *The New York Times*, November 22, 2017, https://www.nytimes.com/2017/11/22/arts/television/shes-gotta-have-it-netflix-tv-review.html.

67 Pikaart, "Meet the Brooklyn Artist."

68 Pikaart, "Meet the Brooklyn Artist."

Lee privileges that gaze: he references "a huge presence of black women in the room," but in comparing the independent film he directs at age 29 and the rebooted series he directs at age 60, Lee says, "it's a universe from those two different Spike Lees."[69] Fazlalizadeh's art and ideas are subordinate to Lee's camera because the cultural script has *woman* subordinate to *man*. *Woman* is not a desirable role. Let's no longer script, cast, or take the role.

69 Gross, "31 Years Later."

PIXEL: MY PICTURE[70]

In "The Feminine Mystique on Facebook," journalist Katie Roiphe critiques the loss of *women*'s identities when they post pictures of their children to social media. The fourth wall breaks when I realize that I'm one of the people Roiphe targets. I've selected an image of myself with my children Maia and Eve for my Facebook profile picture. Roiphe pantomimes my social networking: "here is my pretty family, she seems to be saying, I don't matter anymore."[71] Roiphe's taunt catches my self-consciousness, and I argue: parenthood makes me matter all the more. I ask your patience, reader, as I claim that you and I matter to this cultural study, too. Feminism has long tied life stories and larger narratives. The personal is political, after all. In my story, my children watch me. They watch as I share ideas in university classrooms, enjoy friendships and introduce myself to new people, delight in books, talk myself and them through difficulties, and serve my community. Here, my "to-be-looked-at-ness" is about agency, not objectivity. I'm shadowed by two beautiful people (despite herself, Roiphe's kind of right when she calls my family "pretty"[72]) who will learn to live as timidly or as fully as I do. I matter.

Roiphe's essay also appears as online by another title, "Disappearing Mothers."[73] While I'm watching *woman* disappear in these pixels, I'm not one of *los desaparecidos* as Roiphe's title suggests. I've not been disappeared by parenthood. Instead, I've become something more than my self, something more than singular, even: my youngest Eve recently forgot her usual "mama" and began calling me "mimi." When calling for me, then, Eve names us both: Me, Me. Not unlike Tatyana Fazlalizadeh and Nola Darling, we are two as one, a dual consciousness speaking for ourselves. This heightened sense of selves doesn't mean that I've forgotten my feminism, as Roiphe worries. Parenthood inspires, not dulls, my other roles. If I've wondered how to break that notorious binary theorists critically eye, I find hope in the revelation of parenthood. I'm not parent *or* child, but rather both at once, "me, me," literally from the mouth of babes. I am a plural subjectivity, shifting as does my lived experience, learning as I go. This kind of fluidity is simple enough to see in my pixels.

Roiphe opens her essay: "If from beyond the grave Betty Friedan were to review the Facebook habits of the over-thirty set, I am afraid she would be very disappointed in us."[74] But my profile picture isn't a "problem that

---

[70] A version of this pixel originally appeared as a *Love and Biscuits* blog post. See Gwendolyn Spring Kurtz, "My Mystique," *Love And Biscuits*, April 26, 2013, https://loveandbiscuitsblog.wordpress.com/2013/04/.

[71] I argue here, but I'm an admirer of the smart, incisive essays in Katie Roiphe, *In Praise of Messy Lives: Essays* (New York: Dial Press, 2012), p. 207.

[72] I'll qualify my retort here, as I don't mean to suggest that I take words like *pretty* (or *woman*) lightly. If you haven't already, watch Katie Makkai, "Pretty," November 3, 2007, YouTube video, 3:28, https://www.youtube.com/watch?v=M6wJl37N9C0.

[73] Katie Roiphe, "Disappearing Mothers," *Financial Times*, August 31, 2012, https://www.ft.com/content/0bf95f3c-f234-11e1-bba3-00144feabdc0.

[74] Roiphe, "In Praise," p. 205.

has no name."[75] It's an image of me, as I am in this moment: me, me. For Roiphe to suggest I'm a problem smacks of victim-blaming. I'm not a problem, and I'm not disappearing.

The cultural category I didn't ask to join, however, is. Friedan focuses on young, unhappy, and, perhaps most importantly, socially prescribed *woman*hoods.

This is all problematic, even more so than Friedan intends as those *woman*hoods, admittedly not unlike the mythological breasts in this piece, include relatively young, white, wealthy models at the exclusion of minority images and experiences. And Friedan's critique speaks to another age and age group. I gave birth in my thirties after years of coffee drinking and book reading, backpacking through mountainscapes and cityscapes, sampling California's academic trajectories and wines, and following *Zapatistas* and community ideals in southeastern Mexico and southeastern Tennessee. I'm not, as Roiphe suggests, Friedan's disappointed, optionless housewife. Not unlike Roiphe, I'm an educator. I read intriguingly controversial feminists on popular culture. Sometimes, I argue with them.

Katie Roiphe helps me rethink my self-image and the digital media that makes up my appearance. Whether my face or my children's faces, we're captured in as many layers of social expectation as pixels. So maybe part of me is disappearing. Or, at least, an unreasonable name I've been bound by is. But what Roiphe's critical aim grazes is that losing *woman* isn't a loss. Neither my corporeality nor my virtuality suffers *woman*lessness. I do, however, like Marie Antoinette and Fazlalizadeh and others, wear *woman*hood like a target, and *woman*hood too often makes an object of me. So let the pixels fall apart. *Woman* is not going to win the so-called cultural wars, just like *she* couldn't win Susan Faludi's "undeclared war."[76] It's time, then, to let *her* go.

LIGHTS, CAMERA

Feminism can go on without *woman*. Self-described feminist Chimamanda Ngozi Adichie puts this perfectly: "for centuries, the world divided human beings into two groups and then proceeded to exclude and oppress one group. It is only fair that the solution to the problem should acknowledge that."[77] Feminism seeks social justice and will do so as long as it needs to. But not only is *woman* not prerequisite to social change; *she* stalls change. In problematizing *woman*, I'm admittedly arguing over a word. But this word has shaped, and reshaped, and reshaped us all until we are no longer made in a recognizable image. I can

---

[75] Roiphe and I respond here to the 1963 classic: Betty Friedan's *The Feminine Mystique* (New York: Norton, 2013).

[76] Susan Faludi, *Backlash: the Undeclared War against American Women* (New York: Crown, 1991).

[77] Chimamanda Ngozi Adichie, *We Should All Be Feminists* (New York: Anchor, 2015), p. 41.

no longer see *women* for their so many discordant pieces. And importantly, I can no longer bear *woman* on my back. *Woman* is a discursive commitment and also an embodied commitment with too many expectations and limitations to manage.

An awareness of gender performativity doesn't mean we necessarily stop performing and posing, even if we can point to our posturing. Add another layer to the lens, digitize the image, and we're too many points that matter in too many ways. We are literally too much matter, material reproduced through devices until we're an overcrowded instability. In the digital age, *woman* has gone supernova. A new mode of reproduction has made for a momentary new *woman*. She's been a star—see muses Helen of Troy, Marie Antoinette, Claudia Schiffer, and Kate Moss above. But *she*'s a dying star. Cue the lights, camera,

ACTION

The move from silver screen to device screen doesn't clean up *woman*'s image. *She* is contested subjectivities in contested spaces, too many and too mediated to be pictured clearly. And individuals placed in *her* category share *her* fated role to embody pleasure and consumption, and suffer systemic violence. Admittedly, we've seen some change. Whereas Marie Antoinette's *jatte teton* was commissioned without her knowledge, much less her consent, Claudia Schiffer and Kate Moss participate in their coupe and internet ad campaigns, and Victoria's Secret models find financial success in displaying their breasts in bras. The internet has amplified voices that mainstream American audiences have chosen traditionally not to hear. Fazlalizadeh finds audience through Instagram, and others do so through Facebook. But markers like race, place, and price haven't disappeared from devices. Rather, they're displayed in the same pixels that show us *woman*'s irreconcilable pieces.

A gendered cliché might work here: *woman* is a vision—a shifting vision, too many times remade in our own unmanageable cultural image. Feminists have kept Laura Mulvey in mind so as to resist the male gaze that is our own, too. But we can't look away from culture, and we can't look in from the outside of culture. We're our own captive audience, and we can't get a hold on ourselves because there's too much movement, and too much conflict, to hold onto. I claim this shifting subjectivity while it's moving. I call for a personhood that is self-selected, a personhood in pieces, not for disorder's sake but rather with an appreciation of *woman*'s mythos in mind. *Woman* might once have been an original picture, and *women* have certainly made for memorable images. And the individuals who would be limited by the category still do. But *woman*'s samples are too storied and too many stories to fit into a singular creature category. As Haraway muses, this is still "a mythic time, we are all chimeras."[78] Disappearing *woman* into memory is a chance for us all to participate

[78] Haraway, "Cyborg Manifesto," p. 292.

fully in our ongoing selves and communities. *Her* own telos, *woman* is finished because *she* cannot be a finished category.

This almost glitch feminist[79] pixelation challenges the coherence of *woman* and, in the spirit of Audre Lorde, calls *her* social category a "master's tool [that] will never dismantle the master's house."[80] It's time for new tools, or tools new to hegemonic white feminism, like Sandoval's "differential consciousness," which "operates like the clutch of an automobile: the mechanism that permits the driver to select, engage, and disengage gears in a system for the transmission of power."[81] Sandoval's switching among oppositional sites might help us all better navigate Crenshaw's intersection. The conversation during this ride might rely on gender-neutral pronouns to talk about individuals' experiences, so that we are not limited by a singular linguistic and lived marker like *woman*. Gender fluidity gets at the mobility I'm after here, but with *man* and *woman* still anchoring the gender spectrum, nonbinary remains something of a third wheel to a cisnormative couple. To free us all, we must leave behind, not stand between, the iconic binary pair. My intention here follows Haraway's, "to build an ironic political myth faithful to feminism, socialism, and materialism. Perhaps more faithful as blasphemy is faithful, than a reverent worship and identification."[82]

I want to make, like Judith Butler, gender trouble.

To analyze like Laura Mulvey, who admits, "it is said that analysing pleasure, or beauty" or *woman*, I'll add, "destroys it. That is the intention of this article."[83]

References

Adichie, Chimamanda Ngozi (2015). *We Should All Be Feminists.* New York: Anchor.

Butler, Judith (2006). *Gender Trouble: Feminism and the Subversion of Identity*. New York: Routledge.

Castro, Ruy (2000). *Bossa Nova: The Story of the Brazilian Music That Seduced the World*. Chicago: Chicago Review Press.

---

[79] See Legacy Russell on glitch feminism, which "embraces the causality of 'error', and turns the gloomy implication of *glitch* on its ear by acknowledging that an error in a social system that has already been disturbed by economic, racial, social, sexual, and cultural stratification and the imperialist wrecking-ball of globalization—processes that continue to enact violence on all bodies—may not, in fact, be an *error* at all, but rather a much-needed erratum. This glitch is a correction to the 'machine,' and, in turn, a positive departure." Legacy Russell, "Digital Dualism and the Glitch Feminism Manifesto," *The Society Pages*, December 10, 2012. https://thesocietypages.org/cyborgology/2012/12/10/digital-dualism-and-the-glitch-feminism-manifesto/.

[80] Audre Lorde, *Sister Outsider: Essays and Speeches* (Trumansburg, NY: Crossing Press, 1984), p. 27.

[81] Chela Sandoval, "U. S. Third World Feminism: The Theory and Method of Oppositional Consciousness in the Postmodern World," *Genders* 10 (Spring 1991), p. 14.

[82] Haraway, "Cyborg Manifesto," p. 291.

[83] Mulvey, "Visual Pleasure," p. 16.

Crenshaw, Kimberlé (2016). “The Urgency of Intersectionality.” Filmed October 2016 at *TEDWomen 2016*, San Francisco, CA. Video, 18:37. https://www.ted.com/talks/kimberle_crenshaw_the_urgency_of_intersectionality.

Erdos, Joseph (2014). “The Glass Shaped Like Marie Antoinette’s Anatomy.” *Huffington Post*, May 27, 2014. https://www.huffingtonpost.com/2012/04/13/marie-antoinette-champagne-coupe_n_1424686.html.

BBC News (2015). “Female Scientists Post 'Distractingly Sexy' Photos.” June 11, 2015. http://www.bbc.com/news/blogs-trending-33099289.

Friedan, Betty (2013). *The Feminine Mystique*. New York: Norton.

Gross, Terry (2017). “31 Years Later, Spike Lee Puts A New Spin On ‘She's Gotta Have It.’” NPR, December 14, 2017. https://www.npr.org/2017/12/14/570761380/31-years-later-spike-lees-got-a-fresh-take-on-shes-gotta-have-it.

Haraway, Donna (1991). “A Cyborg Manifesto: Science, Technology, and Socialist Feminism in the Late Twentieth Century.” In *Simians, Cyborgs and Women: The Reinvention of Nature*, pp. 149–181. New York: Routledge.

Kilcooley-O’Halloran, Scarlett (2014). “The Kate Moss Coupe.” *Vogue*, August 21, 2014. http://www.vogue.co.uk/article/kate-moss-champagne-glass-coupe-34.

Kurtz, Gwendolyn Spring (2013). “My Mystique.” *Love And Biscuits*, April 26, 2013. https://loveandbiscuitsblog.wordpress.com/2013/04/.

Lorde, Audre (1984). *Sister Outsider: Essays and Speeches*. Trumansburg, NY: Crossing Press.

Makkai, Katie (2007). “Pretty.” YouTube video, 3:28. November 3, 2007. https://www.youtube.com/watch?v=M6wJl37N9C0.

Martin, Meredith (2011). *Dairy Queens: The Politics of Pastoral Architecture From Catherine De’ Medici to Marie-Antoinette*. Harvard: Cambridge.

Mulvey, Laura (1989). “Visual Pleasure and Narrative Cinema.” In *Visual and Other Pleasures*, pp. 14–26. New York: Palgrave.

Peterson, Dean (2014). “Stop Telling Women to Smile,” filmed 2014, video, 6:05. http://stoptellingwomentosmile.com.

Pikaart, Hannah (2017). “Meet the Brooklyn Artist Who Lent Art-World Cred to Netflix ‘She’s Gotta Have It’ Reboot.” *Art Net News*, December 11, 2017. https://news.artnet.com/art-world/brooklyn-artist-shes-gotta-have-it-1163225.

Poniewozik, James (2017). “‘She’s Gotta Have It’ From Netflix Is a Bold Reboot from Spike Lee.” *New York Times*, November 22, 2017. https://www.nytimes.com/2017/11/22/arts/television/shes-gotta-have-it-netflix-tv-review.html.

Putnam, Lindsay (2014). “Kate Moss’ Breast Serves As Mold for Champagne Glass.” *New York Post*, October 9, 2014. https://nypost.com/2014/10/09/sip-champagne-in-a-glass-molded-from-kate-moss-breast.

Rees, Allyson (2014). “Champagne Served in Kate Moss’ Breasts? We’ll Take Two, Please.” *Ravishly*, August 22, 2014. https://ravishly.com/2014/08/22/champagne-served-kate-moss’-breasts-we’ll-take-two-please.

Risman, Barbara J (2009). “From Doing to Undoing: Gender as We Know It.” *Gender and Society* 23 (1), pp. 81–84. Accessed March 26, 2018. http://journals.sagepub.com.proxy.lib.utc.edu/doi/abs/10.1177/0891243208326874.

Roiphe, Katie (2012). “Disappearing Mothers.” *Financial Times*, August 31, 2012. https://www.ft.com/content/0bf95f3c-f234-11e1-bba3-00144feabdc0.

——— (2012). *In Praise of Messy Lives: Essays*. New York: Dial Press, 2012.

Russell, Legacy (2012). “Digital Dualism and the Glitch Feminism Manifesto.” *The Society Pages*, December 10, 2012. https://thesocietypages.org/cyborgology/2012/12/10/digital-dualism-and-the-glitch-feminism-manifesto/.

Sandoval, Chela (1991). "U. S. Third World Feminism: The Theory and Method of Oppositional Consciousness in the Postmodern World." *Genders* 10 (Spring 1991), pp. 1–24.

Sandoval, Chela, and Guisela Latorre (2008). "Chicana/o Artivism: Judy Baca's Digital Work with Youth of Color." In Anna Everett, ed., *Learning Race and Ethnicity: Youth and Digital Media*, pp. 81–108. Cambridge, MA: The MIT Press, 2008.

Tatyana Fazlalizadeh (2017). "Stop Telling Women to Smile." Accessed December 20, 2017. http://www.tlynnfaz.com/Stop-Telling-Women-to-Smile.

Wine Spectator (2008). "Unfiltered: Dom Pérignon Pays Homage to a Top Model's Chest." December 4, 2008. http://www.winespectator.com/webfeature/show/id/Unfiltered-Dom-Perignon-Pays-Homage-to-a-Top-Models-Chest_4486.

West, Candace, and Don. H. Zimmerman (1987). "Doing Gender." In *Gender and Society* 1 (2), pp. 125–151.

Williams, Florence (2012). *Breasts: A Natural and Unnatural History*. New York: Norton.

BUILD BRIDGES, PAINT WALLS
STREET ART AS FEEDBACK FOR CULTURAL STUDIES AND BEYOND

KRISTAL BIVONA

Academia and street art appear to be opposing sides of cultural production.[84] Whereas academic production is dependent upon the gatekeepers of a given field to accept proposals and anonymous reviewers to determine the merits of research, the street artist produces their work in public spaces, oftentimes without permission. While academia runs on protocols, permissions, submission guidelines, deadlines, selection committees, registration fees, institutional affiliations, letters of recommendation, and restricted funds, street art can happen anywhere at any time in a variety of media, styles, and techniques, and the artist can be anyone. Academia struggles to engage the community whereas street art infiltrates public spaces. Thus, academics can learn a lot from street artists, not only in the burgeoning and interdisciplinary academic field of inquiry known as street art studies, but also through collaborative projects.

Street art is not the kind of art that is quarantined within the aseptic white walls of a museum or that accumulates monetary value in a vault; rather, it is part of quotidian life and daily experience. Street art provides vital insight into the community in which it is found; it spreads ideas and inspires further interventions in a myriad of forms that range from tagging to photography to criticism. As such, street art is a point of departure for countless projects that contribute to social change. Cultural studies scholars can work with street artists from the former's culture of scholarly inquiry to create engaging projects that, at best, can prompt pleasurable experiences for those involved. This essay proposes a framework of interpreting street art through the analogy of feedback and then examines a project where research, community engagement, and teaching converge to produce an opportunity for art making, the humanistic interpretation of art, and mutual cultural understanding—it emerges as a model for cultural studies departments seeking stronger connections between the communities in which we live and work and those that we research.[85] Indeed, street art's capacity for contributing to movements for social change has a history, within which the Latin American context is

---

[84] My sincere thanks to the Clandestine Writer's Group, Darrah Lustig, D. Emily Hicks, Athina Kontos, and the Build Bridges, Paint Walls artists and photographers for their feedback and support during the writing process.

[85] Doris Sommer's case for civic agency and defense of the public humanities has been foundational for my understanding of my work as a teacher, scholar, and cultural agent. I first encountered Sommer's work on the role of art making, humanistic interpretation, and the public humanities at Dartmouth College in 2011 when she gave the Annual Zantop Memorial Lecture, "Welcome Back: The Humanities as Civic Education." See the book, Sommer 2014.

exemplary. The 2019 UCLA Arts Initiative project, Build Bridges, Paint Walls, was a theoretical and practical experiment involving Colombian street artists DJLU, Erre, and Toxicómano with Latin Americanist scholars, students of Spanish, and other community partners.[86] The weeklong series of events included mural painting, workshops, public lectures, a documentary screening, and an art exhibition. Street art proved to be a uniquely engaging medium in part because of its accessibility as art in public as well as its cultural capital. An analysis of a selection of the artworks produced during the project will exemplify what street art studies can add to our understanding of aesthetics, public space, and the role of art making in our society.

Despite its growing popularity, street art is an artistic medium that defies definition and can mean a number of artistic interventions in public. In the field of street art studies, there is dissensus about how to define its object of study. In his 2016 introduction to *The Routledge Handbook of Graffiti and Street Art*, Jeffrey Ian Ross describes the problem of defining street art and the tendency for scholars to attempt to do so in light of questions concerning legality, sanctioning of artwork, content, and who "perpetrates" the artwork.[87] In the same *Handbook*, Maia Morgan Wells explores graffiti and street art in the context of the "art world," noting that, "as the stylistic innovations of *the streets* have become increasingly established inside the formal art world, labels of Graffiti and Street Art tend to refer not only to the concrete elements of *what* and *who* and *where* of how art is produced, but also to an *ethos* of youth, rebellion and outsiderism."[88] Oftentimes, scholars writing on street art attempt to define it in relation to graffiti, which itself is sometimes included as a form of street art and other times is set apart as something else. Wells, for example, argues that "there may never be a full integration of Graffiti into the formal art world, because the authenticity so craved within the genre—its main location of value—must continue to be derived from the illegal placement of works in public places and the reputation that results for the artists."[89] Stephen Pritchard also posits that its acceptance into the art world sets graffiti and street art apart:

> Street art moved from being an activist practice with radical hopes and dreams for a radically different future to an artistic practice that is at once integral to today's art world, property world, economic

---

[86] Thank you to the UCLA Arts Initiative and the UCLA Department of Spanish and Portuguese for supporting the Build Bridges, Paint Walls project. Thank you to the community partners, the Las Fotos Project, the Rendon Gallery, Cartwheel Art, Graphaids Art Supply, Smile South Central, Los Angeles K-Town Wallz Project, the Art Department of the Sandra Cisneros Learning Academy, Deseos Primitivos, Andrés y la Música, Picture Perfect Music, DJ Skwirl Sauce, Angel City Brewery, Topo Chico, Sus Arepas, and Memije's Tacos and Catering. Special thanks to DJLU, Elcarebarbie, Erre, Toxicómano, Isaac Giménez, Maite Zubiaurre, Ralph Ziman, Maria Greenshields-Ziman, Espy, María Flores, and Adam Ayala.

[87] Ross 2016, pp. 1–10.

[88] Wells 2016, p. 466.

[89] Ibid., 472–473.

> world; in short, part of the Creative City and Creative Class narratives that merge nostalgic pasts with shiny new futures—an integral part of the global neoliberal capitalist hegemony. [...] Street art, unlike graffiti, is part of the art world. It is part of neoliberalism.[90]

For Pritchard, street art's complicity in all stages of gentrification renders it a form of artwashing.[91] Lindsey Mancini approaches the difference not in terms how these two forms of expression function in capitalism, but how "street art functions as a gift: where graffiti was a reclamation of space for what is 'mine,' street art is an acknowledgement of the 'us' [...] street art should be defined by its sense of duty: vandalism with purpose, whether political, aesthetic, or otherwise."[92] Mancini posits, "it was not until graffiti became an aggressive target of law enforcement in the 1980's that street art emerged as a substantively new entity, and began to develop on its own aesthetic and historically-informed trajectory," while also noting that the two are not mutually exclusive.[93]

This chapter does not attempt to distinguish the difference between street art and graffiti, but for the sake of clarity, what I mean by street art here is aligned with Olivier Dabène's definition in his book, *Street Art and Democracy in Latin America*, as "a broad category that includes all forms of expression on city walls, including graffiti."[94] Considering murals, paste-ups, stickers, and graffiti among such forms of expression is a convenient way to highlight what they have in common as artistic interventions in the street even though each has its own relationship to activism and artistic movements across time, as well as its own history with criminalization and social acceptance.

Many contemporary histories of street art begin their narratives with graffiti artists in US cities, especially New York, citing the increase in tagging on trains and walls in the 1970s and 1980s as well as graffiti as an essential element of the hip hop movement.[95] However, street art in Latin America is not a recent import. While US graffiti culture has certainly influenced Latin American street artists, other movements are also precursors to contemporary street art movements in the region, such as Mexican *muralismo*, early twentieth-century communist murals in the Southern Cone, and political poster and stencil art during the military dictatorships that terrorized several Latin American countries in the 1960s through the 1980s.[96] After all, street art is a traditional means of mass communication in Latin America through which, as Chaffee argues,

---

90 Pritchard 2019, p. 19.

91 Ibid., p. 11.

92 Mancini 2019, p. 31.

93 Ibid.

94 Dabène 2020, p. 6.

95 See Cooper and Chalfant 1984. Also pertinent are Pabón 2016, p. 78, and Wells 2016, p. 464.

96 For more on Mexican *muralismo*, see Azpéitia 1981, p. 61. For an analysis of the struggle for public space for political murals in Chile, see Palmer 2016, pp. 258–271. For information about interventions in the street in the contexts of the South American dictatorships, see Indij 2018, p. 26., and Dabiène 2020, pp. 77–78, 114.

collectives have used the street as a space to convey messages to the public regarding protests, cultural identity, electoral campaigns, political intimidation, territorial demarcation, and special events.[97] All of these messages compete with commercial advertising and mainstream media communications in the public space.

Politicized street art has a long history in Latin America. Themes of social inequality, urban and rural workers' rights, Indigenous cultural influences, political repression, and other critiques are common in contemporary Latin American street art, and these themes have roots in early twentieth-century public art, such as the *muralismo* movement dating back to postrevolution Mexico. In 1923, the "Manifesto of the Union of Workers, Technicians, Painters, and Sculptors" proclaimed that Mexican creators of beauty must socialize art and make public art that celebrates Indigenous culture and expresses solidarity with popular struggles.[98] Their rebellion against the Mexican ruling class and association with communist ideology, the working class, and revolution "has been a major factor in changing muralism's rank within the hierarchy of the 'fine arts' from the highest to the lowest."[99] Yet, as muralist Eva Sperling Cockcroft and art historian Holly Barnet-Sánchez argue, despite the repression from national elites, *muralismo* became internationally renowned and the three most influential artists of the movement—Diego Rivera, José Clemente Orozco, and David Alfaro Siqueiros—painted celebrated and controversial murals all over the world.[100] They write, "The stylistic innovations of the Mexicans have provided the basis for a modern mural language and most contemporary muralism is based to some extent or another on the Mexican model."[101] Therefore, Latin American street art is not an imported product of cultural imperialism from the United States, but rather an autochthonous movement with a history that is often overlooked by global street art fans.

As an art in public space, street art is part of quotidian life and everyday experience. Considering the conditions of production and consumption, its disruptive potential, and its role in social transformation, it is useful to consider street art as analogous to feedback. Street art functions as feedback in all three senses of the word. According to the Oxford's online dictionary Lexico, feedback is: 1. information about reactions to a product, a person's performance of a task, etc. which is used as a basis for improvement; 2. the modification or control of a process or system by its results or effects, e.g., in a biochemical pathway or behavioral response; and 3. the return of a fraction of the output signal from an amplifier, microphone, or other device to the input of the same device; sound distortion produced by this.[102] Thinking of street art as feedback can shed light

---

[97] Chaffee 1993, pp. 8–20.

[98] Siqueiros et al. 1923.

[99] Cockcroft and Barnet-Sánchez 1990, p. 6.

[100] Ibid.

[101] Ibid., p. 7.

[102] Lexico, s.v. "feedback," accessed December 10, 2019,

on this unique form of expression, and in this chapter, I will work my way backwards through the definitions of feedback and how each can function as an analogy.

The third definition describes audio feedback from a public address system, which when unintentional assaults the ears with loud screeches and booms. Feedback is traditionally considered a mistake that happens when a microphone is too close to the speaker, and sound engineers work hard to avoid unintentional feedback. While feedback is used intentionally sometimes, notably by Jimi Hendrix and in rock subgenres like punk rock, metal, noise, and hardcore, when concertgoers of other genres hear feedback, some immediately cover their ears for protection. In an analogous way, street art is seen by some as an anomaly from which people must protect their eyes, vandalism that must be endured because it exists. Others find street art and graffiti fascinating and enjoy it, not unlike the fans who enjoy the genres and songs that use feedback creatively. It is, in the end, a matter of taste.

The framework for thinking through art and politics that Jacques Rancière puts forth in *The Politics of Aesthetics* is useful for understanding the visibility and invisibility of street art in society. I am particularly interested in how Rancière's notion of the distribution of the sensible, which his translator, Gabriel Rockhill has defined as the "system of divisions and boundaries that define, among other things, what is visible and audible within a particular aesthetic-political regime"[103] and his understanding of the political contributes to the discussion of street art as an aesthetic form that is also political. In Rancière's work, politics is a dissensus in the distribution that occurs when the invisible, the inaudible, and the unsayable subject must suddenly be taken into account, thereby shifting the coordinates of the whole of the system. This political act, which Rancière calls "subjectivization," renders visible, audible, and sayable that which was previously unaccounted for or excluded, and such changes in the distribution of the sensible are only possible because of the democratic premise that we are all equal.[104] Rancière's framework contributes productively to my understanding of street art as an act of dissensus with the distribution of the sensible. As the force that constitutes and maintains the regime of visibility and intelligibility, the police aim to keep unauthorized artistic interventions invisible and outside of the distribution of the sensible, while the people (the *demos*) are unconsciously trained not to pay it any heed.

Street art is in public and accessible to anyone who wants to notice it. I emphasize the desire of the art beholder because in the city, it is easy to overlook street art—the city inundates us with information through

https://www.lexico.com/en/definition/feedback.

103 Gabriel Rockhill, "Editor's Introduction: Jacques Rancière's Politics of Perception," in Rancière 2018, p. xii.

104 Rancière 2018.

signs, advertising, sounds, traffic, smells, and people. One poignant example of this phenomenon is in Gregory J. Snyder's analysis of graffiti subculture, where he narrates his first meaningful encounter with a piece: "Although surrounded by graffiti in New York City, it had finally penetrated my indifference. This was the first time I had really looked at it as something to be seen, instead of just white noise, something to be overlooked and avoided."[105] Snyder lived among graffiti without ever really looking at it—avoiding it even—and this avoidance is the social norm.

In cities like Los Angeles, painting in the street is associated with gangs and criminality, which is a difficult association to overcome and impacts the way that city residents interact with it. Graffiti, in particular, has been reduced to a purely criminal activity that enables and begets further criminal activity, signaling the destruction of the very fabric of our communities. To be sure, the website of the Los Angeles Police Department still correlates graffiti to gang violence clearly: "A community's first step to taking back its streets is getting rid of graffiti immediately."

In the logic of the LAPD, graffiti removal is the key to enabling residents to exert agency over their neighborhood and foster community. This logic erroneously points to graffiti as a cause of social ills rather than a response to the inequality and social exclusion that undermine efforts to foster stability in underresourced neighborhoods. It also presupposes that graffiti writers are not themselves part of the community and that they have no interests or rights to participate in community building and strengthening.

The logic of the LAPD is to marginalize graffiti writers and to keep graffiti and other forms of unauthorized street art from becoming socially unacceptable, effectively determining that graffiti is feedback that no one would want to hear. The criminalization of graffiti and other forms of street art may also contribute to its exclusion from museums despite its status as "a global commercial juggernaut with a diverse audience" and "the world's most globally accessible form of contemporary art" according to the associate editor of *Artnet*, Naomi Rea.[106] However, in the few instances in which prestigious museums such as Los Angeles's Museum of Contemporary Art and the Brooklyn Museum organized exhibitions of street art, local law enforcement and government created a narrative that was leaked to the media that connected the museum shows to increased incidences of vandalism, calling for their respective cancellations.[107] The 2012 exhibition planned for the Brooklyn Museum titled "Art in the Streets" was indeed eventually cancelled. Even influential cultural institutions are suppressed when they attempt to make street art more visible or frame the art form as one that merits a place within a museum. Despite the recent opening of Miami's Museum of Graffiti and the work of

---

105 Snyder 2016, p. 204.
106 Rea 2019.
107 Ibid.

street artists selling at art auctions, the art form is largely still associated with criminality. This, too, contributes to the phenomenon of people seeing graffiti and street art without actually looking at it.

Today, musicians employ feedback in songs that are accepted by the mainstream, played on the radio, and otherwise exist as lucrative commercial commodities. Screechy feedback, once a sound associated with underground clubs and dives, now fills stadiums and arenas as bands implement the technique to play to large crowds. Street art, too, is increasingly revered in the mainstream, which is made evident by the high value of the artwork of street artists in auctions, such as Banksy's "Devolved Parliament," which sold for $12.2 million in 2019.[108] Despite mainstream acceptance and commercial success, musical genres that employ feedback are rooted in transgression and counterculture. Street art, too, can be transgressive and countercultural even as it attracts more admirers and becomes more acceptable around the world.

Street artworks, as public interventions, are feedback in the second sense of the word in that they modify or attempt to control each other. Unlike the artists who paint the works of so-called fine art that will grace the walls of a museum or a collector's home, street artists paint over each other, and this is an essential feature of the medium. For example, it is common practice in cities around the world for property owners and street art festivals to commission murals as a deterrent to unauthorized graffiti, attempting to inoculate their walls against the tags and pieces that are illegible and unintelligible to the public outside of the graffiti community. Meanwhile, graffiti writers cross each other out, street artists paint over graffiti pieces, and taggers modify or deface extant artwork. This sort of feedback is a characteristic of public interventions in that they are vulnerable to modification and erasure as they belong to the public.

The ephemeral nature of street art and graffiti is part of the experience, both for the artist and for the spectator. However, preservation efforts intervene in the way that we can access and experience street art. Legal scholar Enrico Bonadio observes that, "preserving a piece, either *in situ* with protective glass or *ex situ* (for example, via a surgical removal of the mural from the wall), or even through photographs, has always a negative impact on its authenticity (albeit, with different degrees of intensity)."[109] *In situ* and *ex situ* forms of preservation essentially deny street art an essential part of the medium, which is its vulnerability by virtue of being in the street (see Figure 1).

108 Dixon and Linford 2019.

109 Bonadio 2019, p. 39.

Figure 1. Artwork by Banksy preserved in situ.
Photo by Athina Kontos. Used with permission.

Nonetheless, they also operate as a form of feedback as they attempt to consecrate and preserve the artwork for posterity as something worth saving. Street art photography documents and preserves works that will inevitably be erased, modified, or replaced. However, unlike the other aforementioned preservation tactics, street art photography is itself a form of creative expression. Digital media like Tumblr, Instagram, and Flickr have become archives for street art photography, presenting it to viewers who are otherwise removed from the street artwork in time and space. Viewing street art through photography is not unlike gazing at the stars in that we can see the murals that may be long extinguished without knowing that they are gone. The photography, books, and digital media through which street art circulates operate as a kind of feedback that can consecrate works of street art and make them "go viral." While a rival artist or a disgruntled neighbor can destroy a work of street art, photography and digital media can prolong its impact beyond its physical existence. Both respond to the inherent ephemerality and susceptibility of street art to change.

In keeping with the first sense of the word, street art is feedback in that it is both a reaction to and interaction with society in general, and the immediate community more specifically, that can be used to inform efforts towards improvement. Even when the artist has no intent to produce a socially engaged work of art, street art responds to social conditions by virtue of existing in the street. Whether or not the artist intentionally assumes the task of social critique, street art is political in Rancière's sense. Street art intervenes in public space and conveys information to the viewer who happens upon the artwork. In one of his well-known stencil pieces, Banksy says that "society gets the kind of vandalism it deserves," echoing here the notion that street art functions as feedback through social commentary or simply by existing in a particular place.

Street art reveals who has been there, as the artist leaves physical evidence of both the act of art and themselves, their own trace, on the streets with works that are *in situ*. This is just as true for the tagging crews who mark territory as it is for a commissioned artist paid to paint a mural on a renovated building in a neighborhood undergoing artwashing and gentrification. By virtue of existing, the work of street art conveys information to the viewer. Further, the content of graffiti and street art is also feedback, as it can convey a message to the viewer using text and image. In some cases, this information is meant for a specific audience like in graffiti culture, where pieces may not be legible to the uninitiated. Other times, a work of art in the street contains an explicit message accessible to the masses, such as social critiques, homages, or memorials. Some of these messages found within the content may be explicit. Hate messages such as neo-Nazi tags also reveal the presence and boldness of these ideologies in a given location. Other times, the message is implicit, as in when murals challenge white supremacist mainstream beauty standards by spotlighting people of color as subjects for murals. The content of street art, even when it appears to be apolitical or art for art's sake, is conveying information about the world to the viewer and disrupting the distribution of the sensible—in other words, enacting communication as feedback.

If street art is like feedback, then what does the art that resulted from Build Bridges, Paint Walls tell an Angeleno audience? An analysis of the murals painted with Smile South Central and the multimedia artwork at the Build Bridges, Paint Walls exhibition show the conceptual sophistication of the invited street artists and their self-awareness of the power of their work to impact spectators. The subjects of the murals show that the artists dedicate distinct themes to different audiences. Their audiences are different in part because the gallery patrons went to see their work deliberately, while the passers-by are interpellated by their work in the street, perhaps completely by surprise. By examining their murals, we encounter a version of street art acting as praxis (even when the works of street artists end up in a gallery)—as feedback opening up further possibilities for social critique.

The murals on the street in South LA present people of color as subjects, showing Black and Latinx folks as subjects of works of art in a neighborhood that has been red-lined, criminalized, and stereotyped.[110] South Central is infamous for 1990s' gang warfare and the LA rebellions of 1992 in response to police brutality, and it has figured prominently in hip hop, film, and other depictions of gang culture in Los Angeles. The history of violence in the neighborhood and its artistic representation repeatedly through several manifestations of pop culture have solidified the reputation of South Central as a "rough neighborhood" with most discourse about South Central focusing on it as a site for police brutality and gangs rather than presenting it as home to many working-class families from diverse backgrounds. Indeed, the neighborhood's infamy in the media eclipses the history of discrimination, negligence, and even public health emergencies that the state has enabled and perpetrated.[111]

Smile South Central is a one-man operation run by Adam Ayala, a street art aficionado raised in South Central. As a street curator, Ayala reaches out to local business owners around the neighborhood and gets their permission for artists to paint murals on their property to give residents art to enjoy. He then arranges for world-renowned artists to go to his

---

[110] Reibel shows that mortgage lenders disproportionately rejected loan applications from Black applicants in Los Angeles in 1990, and concludes that not only is redlining occurring in the city, but that such lending practices also encourage segregation, with Black applicants more likely to be approved for home loans in Black neighborhoods and Latinx applicants more likely to approved for home loans in Latinx neighborhoods. See Reibel 2000, pp. 45–60.

[111] A 1996 study showed that the lower the income, the higher the lead levels found in the blood of children in South Central Los Angeles, showing the high risk level for lead poisoning there. See Rothenberg et al. 1996. Still today efforts for environmental and food justice are consistently undermined. See Irazábal and Punja 2009.

neighborhood to paint in the spaces where he has secured permission. From the wall space, to the materials, to the artists' time and work, everything is donated, including the meals that feed the artists.

It is important to note that Smile South Central is not a project of so-called artwashing, in which artists and art spaces move into working-class neighborhoods, usually drawn by cheaper rent and consequently initiate a process of gentrification—first come the galleries and the studios, next come boutique coffee shops and restaurants, and soon enough the rent skyrockets. In their book *Gentrifier*, John Joe Schlictman, Jason Patch, and Marc Lamont Hill propose that "the conceptualization of gentrification is a complex mixture of migration, transformation, and reinvestment; forced migration and displacement; class, racial, and ethnic transformation; and investment for new residents to the exclusion of old residents."[112] Artwashing may seem to be a phenomenon driven by individual "starving artist" types migrating to low-income neighborhoods for economic reasons or for the adventure of living somewhere gritty, but it can also be a coordinated effort between arts communities and developers to ~~revitalize~~ gentrify a neighborhood, raise property values, and push out long-time low-income residents to make way for tenants who are able to pay much more. Jillian Billard of *Artspace* notes that processes of artwashing clash with the existing community:

> When artists and galleries move into what is branded as a "newly established art community," they generally don't think of themselves as gentrifiers so much as they think of themselves as pioneers of a "new community," (as opposed to new members of the pre-existing, already culturally-rich community). So it's not just that these art galleries attract developers like ants to a picnic; it's also that they often display a blatant disregard for the rich history of the community they are overtaking.[113]

Artwashing, as a feature of gentrification and social cleansing, contributes to white supremacist oppression and privilege, as the residents priced out of their homes are oftentimes people of color, while the young, affluent, and typically white newcomers replace the businesses and institutions of the neighborhood with new ones that cater to white commercial and consumer interests.[114] Of what he calls Creative Class gentrification, Pritchard notes how

> [t]he process wipes clean entire areas, communities, classes, ethnicities, announcing their erasure with a heady mix of glass-fronted luxury apartments, repurposed ex-public service buildings, community festivals, flamboyant art spectaculars, and huge new citadels for the celebration of white, middle-class culture: art galleries, opera

112 Schlichtman, Patch, and Hill 2017, p. 4.
113 Billard 2017.
114 Donnely 2018, pp. 374–393.

> houses, theatres. Many of these hugely expensive artistic behemoths wryly wink at their neighbourhoods' working-class heritage.[115]

Smile South Central is not a movement to establish a new arts community, nor is it feeding into a gentrification effort. Rather, Smile South Central provides art for residents of the neighborhood to enjoy while simultaneously investing in existing businesses by painting murals on them. The work that Smile South Central does fosters stronger community connection rather than supporting a parallel community of gentrifiers that is at odds with long-time residents. Looking closely at the content of the murals painted for Smile South Central will further illustrate how street art, under certain conditions, can still function to support and inspire the community that is already there.

The three murals painted for Smile South Central are located side by side, each on its own roller shutter on a market. Each artist was responsible for their own piece, and while each artist used their own style and color palette, the three murals go well together (see Figure 2).

Figure 2. Murals by Toxicómano (left), Erre (center), and DJLU (right) painted for Smile South Central. Photo by Adam Ayala. Used with permission.

---

115 Pritchard 2019, p. 18.

The mural in the center by Erre utilizes her signature color palette of bold reds and cyan with an illustration of three girls in black and white with their fists in the air. Their eyes are covered by masks and sunglasses, their hair is curly and natural, and their shirts each have a letter. Together the shirts read "P-W-R." The message of their shirts and their fists in the air allude to the Black Power movement, and the sky-blue sunburst emanating from behind the three figures centers them in the mural, while also anchoring them as the source of light, of power, and of beauty. To the left, Toxicómano's mural shows a young Black girl as a protagonist and prompts the beholder to consider what the future can be like for her, raising questions of opportunity, access and the progress that is always yet to come. Her face is painted yellow with black shading, but her eyes are white, which makes them stand out, adding an intimate tone to the high-contrast image. Her face is framed by jagged shapes painted in orange, black, and white, and the question, "[w]here is the future?" Juegasiempre's mural was part of his series *Orgullo de calle/Street Pride*, which takes the portraits of anonymous people that he typically snaps himself and transforms them into large-scale stencils, simultaneously paying homage to them as subjects and prompting the spectator to recognize their faces as beautiful works of art. In his mural for Smile South Central, he creates a design featuring two local children, transforming them into neighborhood celebrities.

These murals for Smile South Central are significant as feedback in the neighborhood because they not only engage with the community by portraying the actual people who live there, but also because they provoke the beholder to think about histories of racism and resistance and how they impact the youth in the neighborhood. By painting faces and characters with whom people in the neighborhood can identify, the artists show solidarity with residents and also reject white supremacist beauty standards, which Tressie McMillan Cottom posits have systemically excluded darker-skinned people from the category of beauty and depend upon that exclusion in order to reinforce the value and privilege of whiteness. Her conclusions here function as a sort of epiphany:

> Big Beauty—the structure of who can be beautiful, the stories we tell about beauty, the value we assign beauty, the power given to those with beauty, the disciplining effect of the fear of losing beauty you might possess—definitionally excludes the kind of blackness I carry in my history and in my bones. [...] But if I believe I can become beautiful, I become an economic subject. My desire becomes a market. And my faith becomes a salve for white women who want to have the right politics while keeping the privilege of never having to live them.[116]

The murals may seem inconsequential as the kind of feedback that some people drown out, or that may be quickly painted over or defaced, but

[116] Cottom 2019, p. 65. For analyses of the impact of white supremacist beauty standards on women of color's identities and experiences, see also Jah 2015 and Robinson-Moore 2008.

they have significance because of what they depict. The beauty depicted in the murals is not reinforcing the white beauty norms of advertising, nor is it trying to sell a lifestyle of whiteness; instead, it centers on Black and Brown faces as subjects of artwork and connects them to a history of resistance and a better future.[117] The murals reconfigure the coordinates of the distribution of the sensible by making visible, thinkable, and sayable a history of anti-racist activism. They suggest that one need not be a star in nearby Hollywood to have a large-scale portrait of their face in the street, and most importantly, that there is a future in play. As feedback, the murals aim to prop up a neighborhood and bring joy to those who behold them, and as such they offer a form of support to the community. These murals produced a lot of pleasure during their creation as neighbors stopped to talk to the artists and with Ayala, thereby fostering human connection in a climate where paranoia, fear of the other, and the destruction of public interactions is encouraged from the highest ranks of government.[118]

The murals for Smile South Central contrast greatly with the murals on display at the Build Bridges, Paint Walls exhibition at the Rendon Gallery. The objective of the latter exhibition was to make space for people to view Bogotá street art, see the incredible work of the nonprofit Las Fotos Project, contemplate the importance of cultural exchange to Los Angeles and the imperative to support Latinx and immigrant communities facing systemic prejudice and repression, and to, lastly, experience the pleasure of visual arts, music, food, and bilingualism. The project was an experiment in pleasure activism, in adrienne maree brown's sense of "the work we do to reclaim our whole, happy, and satisfiable selves from the impacts, delusions, and limitations of oppression and/or supremacy."[119] Indeed every aspect of the project had pleasure in mind as an antidote to the political climate—from the pleasure of the youth at Las Fotos Project honing their skills in self-expression through stencil making and crafting college essays at Build Bridges, Paint Walls workshops, to the delicious conversations about art on the university campus, to the delight of neighbors discussing the new murals in their streets. Along with the pleasure came the work of confronting heavy themes and topics of life and death, white nationalism, ICE raids, the migration crisis, and poverty among other social problems that afflict both Los Angeles and Bogotá. Ultimately, Build Bridges, Paint Walls addressed pain through pleasure and left its mark on the city, both figuratively through the community ties that were forged and literally on the painted walls.

---

[117] Cottom also cogently notes that the "lifestyle" put forth in magazines is made for "white western women of a certain status, class, profession, and disposition," which ultimately creates a lifestyle that is unattainable to women of color, whose pursuit of the lifestyle and exclusion from it reinforce the hegemonic beauty structure of white femininity. Ibid, p. 64.

[118] The work of adrienne maree brown on pleasure as a principle that can bring about liberation and justice informs my understanding of how mural painting can foster social transformation. For example, she insists that we "center pleasure as an organizing principle" in our work. See brown 2019.

[119] Ibid., p. 13.

The murals at the Rendon Gallery were appropriate to the theme of Build Bridges, Paint Walls—the title of the project itself evokes resistance to the Trump administration's cruel policies toward immigrants, Trump's racist rhetoric that demonizes and criminalizes immigrants, and his well-known plan of building a reinforced and militarized wall along the border with Mexico. The street artists painted murals on the walls of the gallery themselves, and the beholders were not unassuming passers-by on the street, but rather gallery patrons who chose to attend the pop-up to see their work. As such, the artists had no responsibility to the neighborhood and they could choose what they wanted gallery patrons to see. Furthermore, unlike typical art shows for street artists where there are canvases on display that may be purchased, by painting directly on the wall, the works at the Rendon were not subject to a possible market of collectors. As such, the artists painted whatever they wanted without any pressure to consider sales. The artists' work for the exhibition appears to have taken on a twofold purpose: they taught their audience about Bogotá street art and different media through which artists in Colombia express themselves in the street, and they also painted murals, which as far as feedback goes, offered a searing critique of the political conjuncture of US politics to prompt thought and discussion among gallery patrons.

Upon entering the gallery space, the patrons were confronted with a collage of paste-ups—posters designed to be "pasted up" in the street—and stickers from various Latin American street artists. For practical reasons, paste-ups are an important medium in street art, especially politicized street art. They are easily printed in mass, they are relatively inexpensive to reproduce, they can be affixed to walls with homemade wheat paste, and they go up in a matter of seconds, whereas painting can take minutes to days depending on the size of the piece. Chaffee notes that "the poster's greatest asset is its mass duplication and, thus, visibility. One can disseminate a message by massively papering a city or town overnight. Posters are designed to be ephemeral, for quick consumption and short duration, not to constitute artistic treasures to be preserved."[120] Another advantage of the paste-up is that they are a legal grey area, if not entirely legal, in many cities.[121] In the street, paste-ups and stickers compete with advertisements and bills for the attention of passers-by. Paste-ups appropriate the medium of the poster and interpolate them into the space of advertising. In Bogotá, for example, paste-ups exist among election flyers and promotional posters, and they borrow the language of advertising to intervene in the space and medium that aims to influence the electoral or consumer behavior of the beholder, uncannily transforming it into a space and medium that can provoke interpretation of art and critical thought. As such, paste-ups always engage in politics, in keeping

---

[120] Chaffee 1993, p. 7.

[121] Police officers in Bogotá, Colombia told the author, who was interpellated *in flagrante delicto*, that while aerosol art on private property was illegal, paste-ups were not a crime.

with Rancière's sense of the political in that it ruptures the aesthetic division by making art visible in a space and through a medium where advertising is normalized and where art *per se* would not exist. Rather than convincing the viewer to consume a particular product or vote for a political candidate, the paste-up engages the viewer with artwork and prompts a process of critical interpretation and evaluation. This intervention in public space interrupts the mode of thinking about advertisements, which create desire or foster the rejection of products, candidates, and policies, and instead prompts humanistic interpretation. The viewer considers the artwork, judges its beauty, and considers what its message might be and why it exists where it does.

The "Collage Arte Urbano Latinoamericano" (Latin American Urban Art Collage) represented 24 artists—mostly from Colombia, but also from other Latin American countries—and as such displayed a diverse range of styles and of discourses (see Figure 3).

Figure 3. "Collage Arte Urbano Latinoamericano" at the Build Bridges, Paint Walls exhibition at the Rendon Gallery, May 2019. Photo by Erre. Used with permission.

Many of the works that composed the collage were critical of politics, raising awareness of several political issues in Latin America, such as the peace process in Colombia, enforced disappearance, and corruption scandals. In particular, the posters in the "Collage Arte Urbano Latinoamericano" demand the whereabouts of Colombia's estimated sixty thousand disappeared people; they denounce the policies of former-president Álvaro Uribe (2002–2010) and his "uribista" successors, Juan Manuel Santos (2010–2018) and Ivan Duque (2018–today); and they call for the abolition of the Esquadrón Móvil Antidisturbios (ESMAD), a riot squad infamous for brutality and extrajudicial killing.[122] Other posters employ irony to prompt the spectator to think about social problems in new ways, such as in Elcarebarbie's poster "Copa Odebrecht," which portrays Marcelo Odebrecht winning the championship of corruption, suggesting that kickbacks and graft are as quintessential to Latin American culture as soccer, or DJLU's "War Bugs/Fastidiosa Guerra" that presents war as an annoying yet deadly plague with the wings of insects transformed into assault weapons. The collage prepared the gallery patrons for what was to come: politicized artwork from which no public figure would be safe. Additionally, the collage took on a pedagogical function, teaching the beholder about the political conjuncture in Latin America, particularly in Colombia.

The "Collage Arte Urbano Latinoamericano" was, arguably, the only installation that dealt exclusively with a critique of politics and society in a Latin American context. This is in part because the paste-ups and stickers were not destined for the gallery, but meant for the street and intended for an audience that would immediately understand the references and recognize the faces of the politicians. The other artwork engaged with themes to which people from a variety of national political contexts could relate, while some pieces could be read as acute condemnations of Donald Trump.

One striking and acute response to Trump's immigration policies was a 4 x 8 foot mixed media piece by Colombian-born immigrant and Los Angeles–based street artist Elcarebarbie. "The KKK Took My Baby Away" is a collage of plywood with acrylic paint, aerosol paint, paper, barbed wire, and chicken wire. It refers specifically to the 69,550 traumatized children who were detained by US immigration authorities in 2019 (see Figure 4).[123]

---

122 The ESMAD made recent headlines when they were deployed to repress National Strike, which began November 21, 2019. After an eighteen-year-old student, Dilan Cruz, died from injuries from a bean bag shot by an ESMAD officer, demands to abolish the ESMAD rang out in the city streets and on social media. For more, see Quintero 2018. See also Fiorella 2019.
123 Christopher Sherman, Martha Mendoza and Garance Burke, "US Held Record Number of Migrant Children in Custody in 2019," *AP News*, November 12, 2019, https://apnews.com/015702afdb4d4fbf85cf5070cd2c6824.

Figure 4. "The KKK Took My Baby Away," mixed media collage by Elcarebarbie for Build Bridges, Paint Walls. Photo by Erre. Used with permission.

The collage contains images of a boy in distress behind chicken wire, Trump holding crying babies, both hired and "undocumented" Border Patrol agents surveilling with binoculars, surveillance cameras, and the apparatus of the Department of Homeland Security represented by the logo and the ICE agent. If taken as a recreation of the US flag, which the red and barbed-wire stripes suggest, then in the top left corner, the blue and starry canton is replaced with John Hood's iconography of the migrant family that any commuter of the borderland freeways in California in the last thirty years would recognize.[124] Whereas Hood's original iconography depicts a family running, presumably for their lives, Elcarebarbie's remix of the image features the parents running after their daughter who has been abducted by the eagle in the seal of the office of the Presidency.

Elcarebarbie's collage depicts the horror show of Trump's inhumane and cruel immigration policy, which uses child abuse to deter asylum seekers and other migrants fleeing violence and crises in their homelands from attempting to immigrate to the United States. The distress of children, the popular support for Trump and his border policy as embodied by the "Undocumented Border Patrol Agent," and both the government infrastructure that executes these policies and the physical and technological infrastructure that enforce them all figure into the visual representation

---

[124] For the history of the Caltrans sign, see Carcamo 2017.

of Trump's immigration policy. "The KKK Took My Baby Away" confronts the Angeleno gallery patron with this reality, enabling them to engage in humanistic interpretation of art with regard to the humanitarian crisis happening at the border and in an undisclosed number of concentration camps throughout the country.

Toxicómano's "Richie Trump" takes the iconic image of Richie Rich from the comic book series of the same name, which debuted in 1960, and transforms him into a sinister necropolitician that wields control over commodities and telecommunications.[125] Richie Rich's slicked, middle-parted hairstyle is replaced with a "Trumpadour," Trump's signature style that appears to combine a comb-over with a pompadour (see Figure 5).[126]

Figure 5. "Richie Trump" (left) by Toxicómano for Build Bridges, Paint Walls. Photo by Erre. Used with permission.

125 While Richie Rich as a character first appeared in 1953 in the Harvey Comic *Little Dot*, the series *Richie Rich* was launched seven years later. See ComicVine.com, "Richie Rich (Harvey comic book) Page 1," http://www.comicvine.com/richie-rich/49-11660.
126 Urban Dictionary, s.v. "Trumpadour," accessed February 2, 2020, https://www.urbandictionary.com/define.php?term=Trumpadour.

His arched eyebrows and black eyes suggest demonic mischief. Richie Trump holds a marionette, but instead of playing with a puppet, he holds an atom, a telecommunications antenna, and a barrel of oil. Beholding Richie Trump's manipulations of the atom and recalling the President's access to nuclear access codes, it becomes clear that Richie Trump is a necropolitician, one who executes laws that "[subjugate] life to the power of death."[127] The satellite can signal the necropolitics of the United States' leading role in numerous conflicts and the militarization of space's impact on contemporary and future wars through Trump's signature Space Force program as much as it can symbolize telecommunications and propaganda proliferation.[128] The barrel reminds the spectator of the resources at stake in the context of the US-fought wars and US-aided conflicts, for wars abroad can potentially bring about unfettered access to the natural resources of other sovereign lands. All the while, Richie Trump's old school aesthetic nostalgically harks back to the mythical midcentury period when America was allegedly great. This artwork's feedback points to the inextricability of the nostalgia for the past with the death and violence of war.

The murals painted on the walls of the Rendon Gallery offer both nuanced and brash political critiques that, arguably, prompt interpretations that engage with the US political context, or the Colombian, or numerpous others. DJLU's "Political Animal" mural conveys a critique of the political class, infusing them with the symbols and allegories of the pigs, rats, and wolves from the fairy tales and stories of our collective imaginary (see Figures 6 & 7).

Figure 6. "Political Animals" by DJLU/JuegaSiempre for Build Bridges, Paint Walls. Photo by Erre. Used with permission.

[127] Mbembe 2003.

[128] See the Official Space Force website, https://www.spaceforce.mil/.

Figure 7. “Political Animals” by DJLU/JuegaSiempre for Build Bridges, Paint Walls. Photo by Erre. Used with permission.

The pig-headed politician assumes an iconic stance of Trump, pointing up with his stubby finger, although anyone familiar with Colombian political cartoons and jokes might read the pig as President Ivan Duque, who is often portrayed with a snout. Behind the politicians is a backdrop featuring DJLU’s “War Bugs” in formation, linking the politicians with the deadly contest of war. “Political Animal” depicts a structure of war even though the casualties of war are omitted, particular nations are not named, and the issues over which wars are fought are not made explicit. As such, the mural is open to the interpretation of the gallery patron, who may think first of Colombia’s roughly sixty years of civil war, or might immediately consider the American history of fighting wars and intervening in conflict, especially the ongoing wars in Afghanistan and Iraq, or perhaps the war that Colombia and the United States share: the so-called War on Drugs. Regardless of the specifics of a particular war, DJLU’s work engages with how war is structured, which Elaine Scarry so thoroughly elucidates in *The Body in Pain*:

> The essential structure of war, its juxtaposition of the extreme facts of body and voice, resides in the *relation* between its own largest parts, the relation between the collective casualties that occur *within* war, and the verbal issues (freedom, national sovereignty, the right to disputed ground, the extra-territorial authority of a particular ideology) that stand *outside* of war, that are *before* the act of war begins and *after* it ends, that are understood by warring populations as the motive and justification and will again be recognized after the war as the thing substantiated or (if one is on the losing side) not substantiated by war's activity.[129]

"Political Animal" deals precisely with the relation between the casualties and verbal issues that foreground and survive the war precisely by depicting such moments of enunciation, justification, and rallying for popular support. While we do not see casualties *per se*, the weapons on the wings of the war bugs symbolize the violence that injures bodies. The gallery patron faces this feedback, which is a reminder that politicians make war, and that we do, too, by virtue of the democratic process.

Another political question that Build Bridges, Paint Walls raises was the phenomenon of so-called fake news and the acute manipulations of democratic processes in order to control the outcomes of elections and therefore to wrest power away from the people (the *demos*).[130] Erre's mural "Fake News" contains several images that address the systematic undermining of democracy through falsehoods that are disseminated through a variety of media spectacles, including sensationalist television news, dramatic press conferences, and, as the playful depiction of shit-headed Rock 'Em Sock 'Em Robots suggests, political debates (see Figure 8, next page).

The politician, his head replaced with a gnarled pile of shit and his thumb and index finger pinching in a Trumpian fashion, gives his statement to Shit TV. Meanwhile, the general public consumes lies as represented by the woman enthusiastically eating from her bucket of shit. Another woman, masked and yelling into a loudspeaker, appears to speak out against the farce in protest. The central images of the mural, however, suggest that the fake news phenomenon will only lead to death, destruction, division, and perhaps detention. The skull, the rat skeleton, and the bear trap point to death and destruction as inherent to contemporary manipulations of the media and the erosion of democracy. The chains, the chain-link fence, and the barbed wire denote the prison-

---

[129] Scarry 1985, p. 63.

[130] Collins Dictionary officially added fake news as an entry in 2017 and defined the term as "false, often sensational, information disseminated under the guise of news reporting."

Figure 8. "Fake News" by Erre for Build Bridges, Paint Walls.
Photo by Erre. Used with permission.

industrial complex, including within it the immigrant concentration camps, and can signify the border fence itself. All of these symbols function to divide the population into binary groups, from legal residents to illegal aliens, from citizens to criminals, from us on this side to those on the other side. Indeed, the broadcasted political drama enables these very policies.

As street art covers increasingly more square footage of the earth's wall space and street art studies grows as a scholarly field, the impact of art in the streets of our societies will also grow. Cultural studies scholars can partner with street artists to carry out meaningful projects through which university departments can engage with their communities off campus. Projects such as the Build Bridges, Paint Walls case examined here have the potential to educate the public about the cultures that we research and to prompt pleasurable collective experiences through art making and art interpretation. For those of us who work with international cultural traditions, we are uniquely positioned to engage with local immigrant communities and art projects that include educational events and workshops that create networks of support.

Academics in cultural studies can learn another important lesson from street artists as we face increasing precarity and crises in our fields and in our universities. Street artists continue to make meaningful interventions in the public space knowing that their work is ephemeral. Scholars, too, can intervene in the public space even if the impact is unknown because we have something to contribute beyond our scholarly writing and classroom teaching. We are capable of bringing communities together across time and space with our work, which includes promoting a deep, humanistic interpretation of cultural production as well as cultural and historical analyses, and contemplating a diversity of ideas, thereby fostering pleasure and enriching the human experience. Like street artists, we must also do it ourselves.

References

Azpéitia, Rafael Carrillo (1981). Pintura Mural en México. Mexico City: Panorama Editorial.

Billard, Jillian (2017). "Art & Gentrification: What Is Artwashing and What Are Galleries Doing to Resist It?" In Artspace, November 20, 2017. https://www.artspace.com/magazine/art_101/in_depth/art-gentrification-what-is-artwashing-and-what-are-galleries-doing-to-resist-it-55124.

Bonadio, Enrico (2019). "Does Preserving Street Art Destroy Its Authenticity?" In Nuart Journal 1, no. 2 (2019), pp. 36–40.

brown, adrienne maree (2019). Pleasure Activism: The Politics of Feeling Good. Chico: AK Press.

Carcamo, Cindy (2017). "With Only One Left, Iconic Yellow Road Sign Showing Running Immigrants Now Borders on Extinction." In Los Angeles Times, July 7, 2017. https://www.latimes.com/local/california/la-me-immigrants-running-road-sign-20170614-htmlstory.html.

Chaffee, Lyman G. (1993). Political Protest and Street Art: Popular Tools for Democratization in Hispanic Countries. Westport: Greenwood Press.

Cockcroft, Eva Sperling and Holly Barnet-Sánchez (1990). Signs from the Heart: California Chicano Murals. Venice: Social and Public Art Resource Center.

Cooper, Martha and Henry Chalfant (1984). Subway Art. London: Thames and Hudson.

Cottom, Tressie McMillan (2019). "In the Name of Beauty." In Thick and Other Essays. New York: The New Press, 2019.

Dabène, Olivier (2020). Street Art and Democracy in Latin America. Cham: Springer Nature.

Dixon, Emily, and Maisie Linford (2019). "Banksy's 'Devolved Parliament' Sells at Auction for $12.2 million." CNN, October 3, 2019. https://www.cnn.com/style/article/banksy-devolved-parliament-auction-final-sale-price/index.html.

Donnely, Kathleen (2018). "A Gentrifier's Dilemma: Narrative Strategies and Self-Justifications of Incoming Residents in Bedford-Stuyvesant, Brooklyn." In City and Community 17, no. 2 (2018), pp. 374–393. https://doi.org/10.1111/cico.12296.

Fiorella, Giancarlo (2019). "The Dilan Cruz Shooting: Tracking Officer 003478." Bellingcat, December 3, 2019. https://www.bellingcat.com/news/2019/12/03/the-dilan-cruz-shooting-tracking-officer-003478.

Indij, Guido (2018). Hasta la victoria, stencil! Buenos Aires: La Marca Editora.

Irazábal, Clara and Anita Punja (2009). "Cultivating Just Planning and Legal Institutions: A Critical Assessment of the South Central Farm Struggle in Los Angeles." In Journal of Urban Affairs 31, no. 1 (2009), pp. 1–23. DOI: 10.1111/j.1467-9906.2008.00426.x.

Jah, Meeta (2015). The Global Beauty Industry: Colorism, Racism, and the National Body. New York: Routledge.
Mancini, Lindsey (2019). "Graffiti as Gift: Street Art's Conceptual Emergence." In Nuart Journal 1, no. 2 (2019), pp. 30–35.
Mbembe, Achille (2003). "Necropolitics." Translated by Libby Meintjes. In Public Culture 15, no. 1 (Winter 2003), pp. 11–40.
Pabón, Jessica N. (2016). "Ways of Being Seen: Gender and the Writing on the Wall." In Jeffrey Ian Ross, ed., The Routledge Handbook of Graffiti and Street Art. London: Routledge.
Palmer, Rodney (2016). "The Battle for Public Space along the Mapocho River, Santiage de Chile, 1964–2014." In Jeffrey Ian Ross, ed., The Routledge Handbook of Graffiti and Street Art, pp. 258–271. London: Routledge.
Pritchard, Stephen (2019). "More Today than Yesterday (But Less than There'll Be Tomorrow)." In Nuart Journal 2, no. 1 (2019), pp. 10–20.
Quintero, Mateo (2018). "Esmad y Uso De La Fuerza." In Fundación Paz y Reconciliación, Redacción Pares, December 29, 2018. pares.com.co/2018/11/17/esmad-y-uso-de-la-fuerza.
Rancière, Jacques (2018). The Politics of Aesthetics. London: Bloomsbury.
Rea, Naomi (2019). "Street Art Is a Global Commercial Juggernaut with a Diverse Audience. Why Don't Museums Know What to Do With It?" Artnet, August 7, 2019. https://news.artnet.com/art-world/street-art-museums-1617037.
Reibel, Michael (2000). "Geographic Variation in Mortgage Discrimination: Evidence from Los Angeles." In Urban Geography 21, no. 1 (2000), pp. 45–60. DOI: 10.2747/0272-3638.21.1.45.
Robinson-Moore, Cynthia L. (2008). "Beauty Standards Reflect Eurocentric Paradigms- So What? Skin Color, Identity, and Black Female Beauty." In Journal of Race & Policy 4 , no. 1 (May 2008), pp. 66–85.
Ross, Jeffrey Ian (2016). "Introduction: Sorting It All Out." In Jeffrey Ian Ross, ed., The Routledge Handbook of Graffiti and Street Art, pp. 1–10. London: Routledge.
Rothenberg, Stephen J., Freddie A. Williams Jr., Sandra Delrahim, Fuad Khan, Michael Kraft, Minhui Lu, Mario Manalo, Margarita Sanchez, and Daniel J. Wooten (1996). "Blood Lead Levels in Children in South Central Los Angeles." In Archives of Environmental Health: An International Journal 55, no. 5 (1996), pp. 383–388.
Scarry, Elaine (1985). The Body in Pain: The Making and Unmaking of the World. Oxford: Oxford University Press.
Schlichtman, John Joe, Jason Patch, and Marc Lamont Hill (2017). Gentrifier. Toronto: University of Toronto Press.
Sherman, Christopher, Martha Mendoza and Garance Burke (2019). "US Held Record Number of Migrant Children in Custody in 2019." AP News, November 12, 2019. https://apnews.com/015702afdb4d4fbf85cf5070cd2c6824.
Siqueiros, David Alfaro, Diego Rivera, Xavier Guerrero, Fermín Revueltas, José Clemente Orozco, Ramón Alva Guadarrama, Germán Cueto and Carlos Mérida (1923). "Manifiesto del Sindicato de Obreros Técnicos Pintores y Escultores." Mexico City: Sala de Arte Público Siqueiros.
Snyder, Gregory J. (2016). "Graffiti and the Subculture Career." In Jeffrey Ian Moss, ed., The Routledge Handbook of Graffiti and Street Art, pp. 204–213. London: Routledge.
Sommer, Doris (2014). The Work of Art in the World: Civic Agency and Public Humanities. Durham: Duke University Press.
Wells, Maia Morgan (2016). "Graffiti, Street Art, and the Evolution of the Art Market." In Jeffrey Ian Moss, ed., The Routledge Handbook of Graffiti and Street Art, pp. 464–474. London: Routledge.

THE SPECTACLE OF REENACTMENT AND THE CRITICAL TIME OF TESTIMONY IN IÑARRITU'S *CARNE Y ARENA*

LUCA ACQUARELLI

SHOES AS TRACE

Shoes are at the same time an extension of the human body and a body in itself: more than the clothes or objects we usually use, they carry our traces, our shapes, even the traits of our character; they age with us and for us.[131] Exhibiting these "metonymic prostheses" to indicate memory and evoke the presence of victims of oppression, of injustices—in essence of real crimes, often gone unpunished—and therefore to remind us that we have contracted a witness debt to these victims and help combat the oblivion that falls on events, is something that generally belongs to the strategies of museums of memory, starting from the place of memory par excellence, Auschwitz.

So, we are led to say, it happens in the second room of *Carne y Arena* by Alejandro González Iñárritu, an installation that aims to tell us, through different narrative and experiential strategies, the stories of some migrants from various countries of Central America, driven by different reasons but all having to do with poverty (with "corruption and impunity" adds González Iñárritu), having experienced the inhuman desert crossing to enter the United States. A frontier, that of southern North America, meticulously controlled by practices of subjection in which the abuse of power ends up being the norm and where, in all probability, according to the policies of the current American administration, a much more "technological" concrete partition wall will shortly replace the current one of metal.

In this room, we said, the shoes are blown away. Of different sizes, more or less worn or deteriorated, some calcified in unnatural torsions: trainers, sandals, wedges, lace-up boots, slippers—all that belonged to men, women, children who lost them during the crossing. Many of the lives embodied by these objects, have inevitably died out in the inertia of the path or under the insistence of an overzealous order. Others, such as those that we will discover to be the protagonists of the installation, survived, by chance or stubbornness, the violence of the journey. As itinerant spectators, we adhere to the emotional game of entering this room which, together with the two that follow it, will set the rhythm of a reconstitution in which we will be actors and spectators at the same time.

---

[131] Photography opposite, "Untitled," found migrant boot—by Michael Wells; part of the *Undocumented Migration Project* by Jason de Leon & Michael Wells.

The shoes, among which there are also makeshift water bottles, are collected in the desert by an association that tries, with this and other actions, to keep alive the memory of the victims fallen in the Arizona desert, a place that a recent article in *The Guardian* calls "America's secret graveyard."[133]

The shoes are arranged in bulk at the base of the walls of the room and under the benches of polished aluminum that are the only furniture of what we discover to be the reconstruction of a "freezer," a *hielera*: a room of "waiting" where, in a perverse extreme logic—since these rooms, as the name suggests, are notoriously cold—people are placed. People are taken into custody by the security forces that manage the mechanisms and, after all, the very nature of the border between Mexico and the United States—a desert border, a place of low political density, where the political itself offers no resistance to the law of the strongest and best equipped.

But this room, the one reconstructed by González Iñárritu, is also an exhibition room, the benches so clean (something beyond their hygienic connotation as a detention center) they become frames; the walls so ordered—exhibition supports, of course, but also something more.

This is the counterpoint of this section of the work: on the one hand, a theatrical architecture strongly pushes us to relive the same experience as the migrant ("being in their shoes," we might say) and thus think of this aesthetic experience in the context of experiences of memorial reenactment; on the other hand, a system of exhibition, more or less disguised, tends to provide the frame that gives sense and strength to the objects. The testimonial value of the objects is inserted in this counterpoint.

Stated briefly, we can speak of the installation as a path that takes place in six rooms. The first presents the work with explanatory signs, the second is the one just evoked, and in the third we see an immersive environment using virtual reality technologies. In the fourth room, you enter again in a room reminiscent of the *hielera*, and, after passing through a corridor bordered by a barrier, you find yourself in a fifth room, filled with video portraits. The sixth room is completely dark. In this article, we propose to examine some analytical approaches and to test them with a disciplinary theory between aesthetics, semiotics, the study of devices, and the political value of the image.

As one follows the installation path, the migrants's image evolve: from an anonymous multitude of subjects conceivable in empathic and inevitably nonpolitical terms, they will progressively take shape in specific personal stories, those already announced in the first "didactic" room by the words of González Iñárritu himself, quoted in an explanatory panel: "There are

---

[133] A. Hannaford, "Missing in the US Desert: Finding the Migrants Dying on the Trail North," in *The Guardian*, August 20, 2017.

no actors here. These are true stories re-enacted by the people who experienced them. Even some of the clothes they wear are pieces they wore while crossing the border."[134] The value of the trace and the relic is then unfolded in the plot of more or less spectacular reenactments, that here take a risk: the fetishization of the victim. We will come back later to the question of these reenactments.

The writing on the walls of the second room are instructions for the discipline of the spectator: we, visitors, have to take off our shoes and for a moment, before putting them in a room, even though they appear and are inscribed in the symmetry of an unviolated body, they share the same space as the others. An alarm and a flashing light signal us the possibility of advancing into the next room, where, barefoot, on a fine gravel ground we will move into the darkness to wear a virtual reality headset and once again receive instructions, this time verbal, on things to do and not to do. So begins the spectacle of an immersive environment, lasting about seven minutes, which projects us into another darkness, that of the desert.

IMMERSIVE SPECTACLE, MOTION CAPTURE, AND REENACTMENT PERFORMANCE

If in this virtual immersive environment, the background, a sidereal desert landscape, filmed during the dawn, is cinematographic in its nature, the subjects of the action (migrants, *coyotes*, and policemen) are the result of synthetic images. In particular, the movements and expressions of migrant subjects were captured through *motion capture* devices that render gestures and expressions re-enacted by survivors under the direction of González Iñárritu and, based on the data produced, "photorealistic" avatars were created.[135] Therefore, two main aspects come together: this just mentioned, which concerns the *aesthetic* dimension of the reenactment, and the *aesthesic* one, linked to the vision device that provides the spectator with the possibility to walk (within the physical limits of the room) through the immersive space.

The term "photorealistic," used by González Iñárritu himself ("photo-realistic avatars"[136]), can be deceptive, because the effect of photographic reality is made of a completely different weaving of the iconic material. The characters of the environment evoke rather the unrealism of the subjects of the latest generation of video games. If motion capture tends to capture the articulation of movement, gesture and expression and then transfer it to animation, in the final rendering there is a decisive gap between the human gesture and that of animation. What is the relationship between the fluidity of human movement and that of the animated character on

[134] *Quaderno Fondazione Prada # 12*, Fondazione Prada Publisher, Milano, 2017, p. 41.
[135] For the moment we don't have precise information about the other characters, the coyote and the guards: it's reasonable to think that they too are built on the basis of motion capture but with the use of professional actors.
[136] *Quaderno Fondazione Prada # 12*, p. 41.

the basis of motion capture? Can we think of it in the same way as the relationship between the movements of the dancer and those of the puppet imagined by von Kleist in his famous *On the Marionette Theatre*?[137]

In fact, the body of the actor—who in this case is also a survivor, that is, not a professional actor but witness who accepts the staging of the reenactment under the direction of González Iñárritu—is segmented through markers placed along the structure of the bones and its articulations, building a veritable diagram of the body in motion. Moreover facial expressions are also captured. A complex diagrammatic testimonial base is then used as the skeleton of avatars trying to imitate photography, but taking, by default, only the appearance of the fragmentation of the world, the realistic rendering of the boundaries between things, subjects, figures, and backgrounds—a realistic effect by default that is nourished, precisely, by a "computational record" (to use the terms of Sutil[138]), a trace of human gestures and their memory.

Are the places of these markers comparable to the "centres of gravity" that animate the limbs described by the interlocutor of Kleist's imaginary dialogue? Partly yes, because the first are the points that "direct"[140] the other parts; partly no, because they, being so numerous, do not give practically space to the mechanical movement of the hanging limbs moved only by the gravitational force (as von Kleist wrote: "therefore all the other limbs are what they should be—dead, pure pendulums following the simple law

[137] H. von Kleist, "On the Marionette Theatre," in *The Drama Review: TDR* 16 (3), 1972, pp. 22–26, (Über das Marionetten Theater, 1812).

[138] N.S. Sutil, Motion and Representation: The Language of Human Movement, MIT Press, Cambridge and London, 2015, p. 5.

[140] Von Kleist, "On the Marionette Theatre," p. 22.

of gravity, an outstanding quality that we look for in vain in most dancers"[141]), which was the very essence of the grace of the puppets evoked in Kleist's text. According to the German writer, man had lost, because of his knowledge, this grace, and only after reaching an infinite knowledge (like the one that a god can have), could he return to measure himself with this grace.

There is therefore a tension between the grace of the puppet, the movement on the human gestural trace of the avatar on the basis of motion capture, and the actor (or dancer) in flesh and blood.

We are concerned here with the temporal aspect of this tension. As Lyotard writes in the case of the grace of puppets, there is a suspension of time: "affranchissement de l'esprit de toute diachronie, de toute tâche de synthèse ... suspension de la tâche d'actualiser et de ré-actualiser les passés et les futurs."[143] This suspension of time for lack of intentionality and affectation established a dialectic with the testimonial trace of motion capture, the diagrammatic soul that moves the three-dimensional avatar. This new form of realism that interweaves these two temporal polarities, timeless grace and gestural trace full of memory, could be very fruitful for a new testimonial impact on the viewer. A larger study could provide us with important aesthetic answers.

The other problem lies more in the dimension of reenactment by these subjects who, although witnesses of their own misfortune, are not professionals in the management of gesture and expression: what gestures and what expressions to choose by the survivors in the "theatricalization" of the event? To which anthropological and gestural repertoire does the memory of these subjects appeal, and, above all, through what forms (we could say, for good reason, pathetic Warburgian formulas) is this repertoire staged under the direction of a Hollywood director like González Iñárritu? Which dialectics go through the gestures and expressions of a physical and psychic suffering that nevertheless looks with hope to a future different from the abandoned one? What mimicry inspires what will then be the digitally derived figuration of the representation of migrants in the middle of the desert? These questions remain difficult to answer for the moment, largely because the immersive experience (as well as the entire installation) cannot be easily reviewed today, and all we write is on the memory of a single experience, on the frantic notes that follow it and on the press kit (in our case the one provided by the Fondazione Prada), poor in images and without any immersive "sequence" available.

As for the aesthesic question, we must emphasize that this is not just a film shot with a 360-degree camera, an experience of immersive space that we can do in our daily lives through the many applications that offer such filmed captures. Here we are faced with a walkable environment: it is not

[141] Von Kleist, "On the Marionette Theatre," p. 24.

[143] F. Lyotard, *L'inhumain. Causeries sur le temps*, Galilée, Paris, 1988, p. 175.

only the movements of the head that are localized but also those of the body in space. Thanks to this walkability, we can experience a wide range of points of view (and not only of reframes) *vis-à-vis* the entire immersive environment. We can approach the characters, try to touch them, or push ourselves out to gain an overview of the event staged. In short, a range of possibilities is opening up in order to take our place as a spectator-actor in this new device that does not yet have such a long practice of use as to provide us with a standard of observation.

Seen from this global perspective, the immersive room is the place of a representation but at the same time also of a performance of reenactment. In fact, if we look at the history of immersion, pushing for example to the case of nineteenth-century *panoramas*, we can find as a preeminent theme that of the historical-documental reenactment. The immersive will has therefore long been linked to the tradition of making historical events.[144] In our case, the reenactors are, by mediation of gestural trace, also the witnesses of this phenomenon. It happens that unlike the battles that have become historical (also thanks to their reenactment), migration is often relegated to a wide-ranging phenomenon but whose protagonists are largely anonymous, a mass of "nameless" lost in the flow of the repetition of the same catastrophic event at several geographical points of migration (and González Iñárritu does not fail to point this out with the surreal appearance of a small ship, the object of connotation for migrants crossing the Mediterranean).

The few minutes of VR tell a short story, and the grammar of this immersive story, translated into cinematic terms is very poor: three long takes, a first editing cut with transition and a second one without. The laborious pace of migrants behind a coyote struggling with an excited phone call,

[144] See Olivier Grau, *Virtual Art: From Illusion to Immersion*, MIT Press, Cambridge, 2003 and in particular pp. 91–139 on the panorama of the Battle of Sedan.

then the sudden arrival of guards and soldiers. After, we'll experience a sequence that sees migrants around a table, probably after moments of beating by the guards, and, finally, a last short sequence: the sun has now risen, in the desert there is no one left, there remains only an envelope that rolls in the air—a drapery harbinger of a condensation of ancient anthropological gestures if you think back to the lesson of Aby Warburg—and a shoe, an object that repeats one of the themes of the whole installation, entangled between the rocks and shrubs of the desert.

From a narrative point of view, we could summarize the unfolding of events in a rather simple way. There are some vulnerable people in obvious difficulties who are approaching, the only presence in the barren and hostile desert of the border: we already know who they are from what we have seen in the previous rooms, but before really understanding what the specific problems of each are, a sudden and imposing volume of light, wind, and noise bursts into the immersive environment. The lights come from the helicopter above and the cars at our height—they illuminate the scene and focus us on it. Armed men with equipped uniforms threaten the vulnerable migrants to stop, to declare their identity in an agitated and aggressive rhythm. It's a Manichaean scene, where we immediately understand who the victims are and who the executioners: there is no time to deepen the characters of each actor. This scene is endowed with another negative subject, at the same time a non-disinterested helper and probably a persecutor in collusion with the most ruthless guards, one of the figures par excellence of the traitor of our times: the *coyote* (corresponding to the Mediterranean "people trafficker"), who in the representation risks to be overshadowed because of the Manichaeanism that we mentioned above.

In any case, the border area is thus staged as an area of confrontation between victims and persecutors, characterized by belonging to different communities (indicated mainly through the language, Spanish for the victims—excluding the *coyote*—English for persecutors). The deconstruction of the Manichaeanism of this scene will take place in the penultimate room, that of the video-portraits with testimonies, where, among other things, we will see a porosity of the actantial roles of victim/persecutor: a sort of "catharsis" of an American policeman and of his passage from the group of persecutors to those of helpers and denouncers, and an emphasis on the fact that among the persecutors there were U.S. Homeland Security agents with "Mexican accents."[145]

As spectators, we "throw" ourselves on the scene because we want to become "subjects" of it at all costs: in reality we are excluded as if an invisible and intangible edge was always keeping us out of this environment that, at the same time, does everything to make us feel inside (and sometimes

---

145 "There was an officer with a Mexican accent who was very cruel," in *Quaderno Fondazione Prada # 12*, p. 11.

we feel the inadequacy of the device with respect to its spectacular goal[146]) *Sucked away by something that excludes us, we might say*. The off-screens selected from the very thin edge of the frame of our viewer are at hand, or rather "at head," since the headset recognizes the movements of our head, and the spatial movement inside the room. The environment simulated by González Iñárritu seems to insufflate its viewer a new agency, a new capacity to act on the environment, and then deny it.

To stop this irreconcilable dialectic, present since the subtitle of the installation (Virtually Present, Physically Invisible), the immersive bubble seems then to resort to the old strategy of the reflexive enunciation: a shadow that appears fleeting on the texture of the images and simulating "our" shadow, or the look and the rifle of the policeman pointing in our direction, echo of a real *trouvaille* of the cinema of attraction of the early days. When we try to experience our own proprioception trying to "touch" a virtual subject, our sensory conflict is resolved in what can be seen as a digital glitch: if we try to touch one of the migrants or if one of them crosses us, the image of a pulsing heart suddenly appears, as a dissonance of reality planes, an error that shows us the dense materiality of the digital image despite its apparent superficiality.

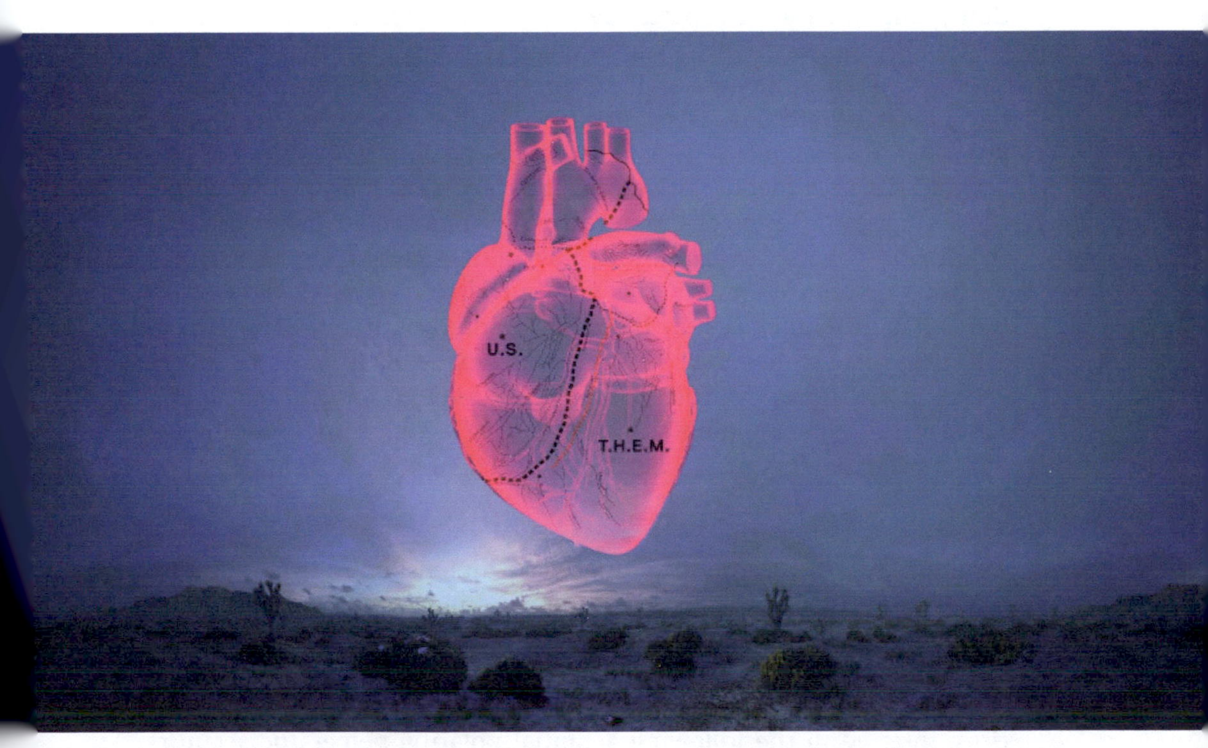

In this virtual scene, a phenomenon present in the very nature of the spectator seems to intensify: the split between the immersed spectator and the spectator who looks at himself while he is building himself as such. The

[146]González Iñarrítu writes about this: "There are infinite limitations and possibilities, multiple tricks where we had to use the limits of technology to hide the technological deficiencies, so that faces and things don't look so plastic or digital," *Quaderno Fondazione Prada # 12*, p. 41.

two dimensions, the contemplative and the sensorimotor, presented phenomenologically in every gaze, seem here to amplify their polarity, their distance, while remaining obviously co-implicated. In some way, then, we think of ourselves, at the same time, as spectators of a scene and as spectators of an agent inside the scene (invisible but tangible in its sensorimotoriety that makes the immersive vault flow). This amplification, in addition to the more sociological data that concerns the mere confrontation with a device made recently popular, is given by the shift of the pregnancy of the "frame" in the pictorial or cinematographic sense to that of the body installation of the spectator (from the revolving chair to the platforms of simulation of the movement such as the omnidirectional treadmills up to the walkable environments) and the tracking abilities inserted in the viewers and in the locators.

This last dimension seems therefore to be the new place of the "frame" of the immersive environment, and I believe that it will mark in a more and more remarkable way the future studies on the virtual reality of the phenomenological-aesthetic approach.

Not secondary is the fact that it is not the effect of reality that creates credulity, but rather it is the partial response of the image to our movements that makes the experience meaningful at the sensorimotor level: the gradient of presence, namely, that adapts the image to our body, making it partially adhere to our senses. In this regard, once again, the real focal point of virtual simulations is the tracking system, that is, in a nutshell, the membrane of mediation between the texture of body movements and the responses in terms of the portion of immersive environment shown through the headset.[147] If the immersive environment moves with us, it perfectly takes into account much of our kinetic activity; as soon as we touch it or generally raise the threshold of interactivity, it reminds us brutally of its phantasmagorical nature. And then the barefoot solicited by the gravel in the immersive room of *Carne y Arena* no longer contributes to the effect of immersion but rather opens gaps connecting with the room lost in the darkness that we were experiencing before wearing a headset.

These aesthesic hypotheses, however, must not make us forget their coimplication with the aesthetic and narrative dimension. In his book on suffering at a distance, Luc Boltanski points out a central node of political theory on the basis of the writings of Hannah Arendt, namely the introduction of piety in the political dimension, whose tensions are modalized starting from the subjects that can or cannot be profiled next to the victims, inserted in the following actantial polarity: helpers/persecutors.[148] The immersive environment of *Carne y Arena* can in fact produce a feeling

---

[147] These reflections are in line with some of the positions of a work of few years ago which placed the kinesthetic and proprioceptive dimension at the center of its study on virtual reality and other digital experiences—see Mark B.N. Hansen, *Bodies in Code, Interfaces with Digital Media*, Routledge, New York and London, 2006.

[148] Luc Boltanski, Distant Suffering, Morality, Media and Politics, Cambridge University Press, 2004.

like that of sympathy for migrants given their condition of difficulty in the crossing: they are accompanied by a wicked person, the coyote, but that seems however not to differ so much from the group of migrants. Subsequently, the arrival of the police generates a second state, prevalent, of antipathy towards the police and sympathy towards the resentment of the victims as well as indignation *tout court*. If the first state seems to tend towards a politics of piety, the second seems rather to turn to a politics of justice. Sympathy, resentment, and indignation, however, are destined to remain without a directly explicit action and therefore remain "in reserve" in the spectator as a political position that can potentially turn into a political act. The immersive environment intensifies the scope of these feelings through a strong experiential charge and, by contrast, through narrative and character simplicity.

VIDEO PORTRAITS, THE INTIMATE PLACE OF TESTIMONY

The penultimate room, before a completely obscure final vestibule, offers an additional element of contrast. On the back wall of small cavities open on the wall at the height of our head—niches similar to small *cameras obscuras*—are projected video portraits, silent faces of individual migrants looking into the room in a pose almost (and this "almost" is key) fixed. The strength of the moving portrait consists of the affection of the close-up associated with slowed micromovements that return all the *pathos* of the face—small shifts in the facial muscles, temporary eye movements, distortion or stretching of the mouth, more or less accentuated movement of the eyelids. On these portraits, on a background always blurred to accentuate the presence of the face, in superimposition, is the written testimony of the stories associated with these people. These short stories, adapted from interviews with the survivors, are generally structured around key moments: the difficult life in the country of origin, the passage of the river (critical moment of the crossing), the violence suffered owing to the guards, the memory of trauma, and the desire for redemption in American life.

Some phrases translate the figurativity of the installation path made so far by the spectator. By way of example, here are some significant extracts.

> "Amaru: "We were suddenly surrounded by an helicopter, quadbikes, horses and dogs. They put some of us in dog cages and let the *coyote* go free."[149] Carmen: "My son was so dehydrated that I feared he would die. They told us to take our shoes off and hand-cuffed us."[150] Manuel: "One night, an helicopter and three patrol cars caught us. ... They took us to the holding cells known as 'the freezers' because of how cold they kept them."[151] Lina: "In the distance, we saw a highway and ran. I lost a shoe in the sand."

---

[149] Quaderno Fondazione Prada # 12, p. 13.
[150] Quaderno Fondazione Prada # 12, p. 11.
[151] Quaderno Fondazione Prada # 12, p. 7.

The story of the catharsis caused by the experience of the extreme suffering of the other flows on the only portrait of a (white) police officer as a guard of the border, John:

> When you deal with somebody who is dying of heat exhaustion, they are scared, they moan, like life is just coming out of them. To watch that and not be able to do anything, it's heartbreaking. ... So, anyone who doesn't have empathy for an alien who is trying to come to the US for a better life, I don't want to talk to you—that's just the way I felt.[152]

This new room supports, then, with features of "faciality" the previous aesthetic experiences and sutures them temporarily, avoiding the dissipation of empathy and engagement, with an apparently contrary movement, the openness to the willingness to meet with the other.[153] They are meetings that *reterritorialize* the *derritorializing* experience of the desert, in a new possibility of memory. One survives the physical and symbolic desert that erases the faces of the victims of the crossing, and takes up the oral testimony, and uses it for a written memory, which now has a place, that of the face, a space for intercepting political lines.

From being simple actants of a rather banal clash between victims and executioners, the migrants then become actors. It is true that the dimension of confession in which we are exposed to sustain this *face-to-face* with the close-ups of the protagonists is supported by the narrative curiosity of wanting to know what happens after the vicissitude that has united them all, and in which we too, in a consciously abusive way, feel part of: the crossing of the desert. The immersive sensorimotor experience that precedes these "encounters," diametrically in contrast with their intimacy and sobriety, then seems to act as an "intensifier," impressing a drive to want to listen, to that willingness to enter into communion with those migrants who finally have a name. But it is the affectivity of the close-up, which therefore returns features of faciality rather than significant faces, that undermines the mere narrative thread by introducing the full "intensive" dimension of the images.

Pietro Montani, today one of the first to have written about *Carne y Arena* in a theoretical work,[154] inserting his analysis in a very broad discourse, speaks in particular of the experience of the "virtual room" as one of the

---

[152] *Quaderno Fondazione Prada # 12*, p. 41.

[153] Cf. G. Deleuze and F. Guattari, *A Thousand Plateaus. Capitalism and Schizophrenia*, University of Minnesota Press, Minneapolis and London, p. 167.

[154] See Pietro Montani, Tre forme di creatività*: tecnica arte politica*, Cronopio, Napoli, 2017, pp. 132–138. Another recent work is that of Adriano D'Aloia, "Virtualmente presente, fisicamente invisibile. Immersività ed emersività nella realtà virtuale a partire da 'Carne y Arena' di Alejandro G. Iñárritu" in C. Dalpozzo, F. Negri, A. Novaga, eds., *La realtà virtuale. Dispositivi, Estetiche, Immagini*, Mimesis, Milano, 2018.

attempts of "perceptual and patemic disautomatization," that is, a new relationship between images and language in order to dissolve "the blind automatisms of repetition" in the reading of the world provided by some media practices.[157] Indeed, as we said, the nature of virtual experience is the trigger of a disorienting dimension: between "disautomatization" and intensification, it creates a state of sensory excitement that shakes us.

I wonder, however, and this seems to me the central point, whether what "triggers" is the spectacle of the show (i.e., the surprising event of the immersion precisely because not yet widely shared by a "common feeling") or the very nature of the immersive environment that presents us in a way still to be studied, these "photosynthetic avatars" of victims and executioners. What is certain is that if a critical effectiveness lies in González Iñárritu's work, it is the one that considers the "montage" among the various experiences and among the various devices that the installation includes.

CRITICAL SPACE OF *MONTAGE* BETWEEN DEVICES AND ENVIRONMENTS

To show the pain of others is to run the risk of showing it off and thus of "derealizing" the substance of the injustices in question. As Boltanski asked in the introduction to his essay: "On what conditions is the spectacle of distant suffering brought to us by the media morally acceptable?"[158]

Montani, in another text of his a few years ago, which I think is more pertinent for the analysis of *Carne y Arena*, talks about processes of "authentication" of images obtainable through *intermedial* aesthetics.[160] The term "intermedial," that has, over the years, developed an important literature, is used here as a philosophical option summarized in the first pages of the book:

> only by starting from an active comparison between different technical formats of the image (optical and digital, for example) and between its different discursive forms (fictional and documentary, for example), can one do justice to the irreducible otherness of the real

---

157 To give us an idea of this disautomatization, Montani travels a long and fruitful path, which here we cannot reproduce, between the Kantian imagination (and in particular that "without concept") and the imaginative regression of the dream according to Freud, to then come to new paradigms of technoaesthetics, the subject at the center of the latest studies of the Roman philosopher— Montani, *Tre forme*, p. 50.

158 Luc Boltanski, *Distant Suffering*, p. 15.

160 The term "authentication" can be problematic because it resonates with the semantic area of authenticity and therefore of truth. Montani immediately wished to disambiguate the term, distancing it from this risk of confusion: "Authenticating, from this point of view, is akin to 're-realizing', rehabilitating the image to the relationship with its irreducible other, with its radical and inappropriable off-screen." Pietro Montani, *L'immaginazione intermediale*, Laterza, Roma and Bari, 2010, p. 14.

> world and the testimony of facts, (media or not media facts), that happen there.[161]

It is actually a question of reflecting on an extended meaning of the concept of montage, present in the *aesthetic* and *aesthesic* dimension of discursive forms. Montani, in his book, analyzes forms of intermedial montage internal to the cinema, but I believe that his theses can also be effective in "environmental" installation contexts such as that of *Carne y Arena*.

In fact, we could think of González Iñárritu's entire installation as an intermedial path that produces meaning precisely because it intertwines different aesthetics that resonate with the various media, or more precisely, the various apparatus (*dispositifs*) used. If the not-insignificant theoretical problem of evaluating the dialectic between aesthetic and aesthesic dimensions interior to this montage remains, I guess that thinking of intermediality as a possible process of critical mediation can be quite productive.

The encounter with the "irreducible otherness" of the migrant, and even before that, with the symbolic order that supports the representation of the "migrant," is proposed in the various rooms in different gradations of simulation and contemplation. These aesthetic and media variations on the same "subject" trigger a process of empathic solicitation and a path of political subjectivity that attempts to avoid cliché. It is not, therefore, as one might expect, the virtual environment that best embodies the presence of migrants, but it is only its location within this montage of environments that creates a critical shift on the subject and therefore potentially opens a space for testimonial elaboration.

As the "shoe room" seems to trigger the theme of presence and simulation that will be typical of the room of the immersive environment, in the same way, the room of portraits seems to disactivate the spectacular dimension of virtual reality. And again, as we said, the immersive environment seems to trigger the shock that leads us to an intensified attention to the stories proposed in the form of portrait-videos. Moreover, the corridor from the immersive room to the portrait room, until now only briefly evoked, next to a metal wall which we discover to be the authentic metal wall of the boundary barrier, resonates the aesthetics of the trace that crosses the entire installation. This continuous aesthetic and medial counterpoint works on the shifting and holding of the critical spectoriality in front of the same subject, detaching it from some possible connotative and derealizing automatisms. We could speak of a theater of empathic gradients lived in the spectators' solitude, destined to make a work of internal remontage of the experience.

---

161 Montani, L'immaginazione intermediale, p. 13.

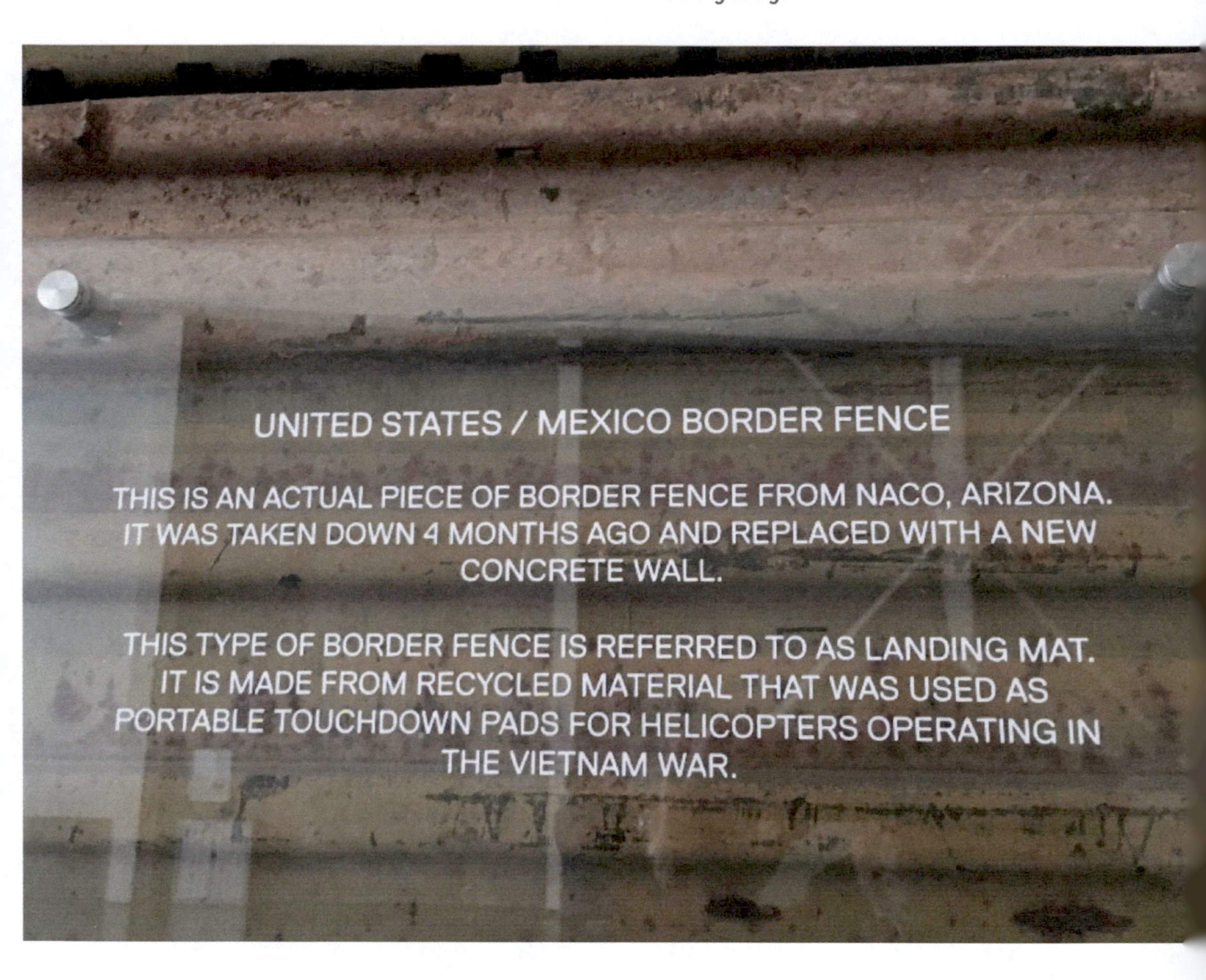

So a space of critical engagement is created where, and I quote Montani again, "images can once again be exposed, through authentication, to a debt of witness."[163] Nevertheless, the montage of the rooms seems to be built suggesting a preeminence to the immersive room compared to the others, seen almost as peritextual elements. As in a sacred architecture that conveys the moments of prayer, ritual, or conversion, the silence of the "peritextual" rooms seems to mark them as antechamber, on the one hand, and decompression room, on the other, of a central event where sounds and voices burst onto the spherical bubble of the images. This rhetoric of intensity could make us think of a path instructed by the work itself and that would go in the direction of a strongly heterodirect montage.

If, therefore, an opportunity of critical subjectivity can be experienced by visiting the installation, it is precisely in breaking down the instruction of the path: reinterpreting the spatiality of the montage in a place of *telescopage* without fixed spatiotemporal hierarchies—the only way, it seems to us, to think of *Carne y Arena* as a political place.

---

[163] Montani, L'immaginazione intermediale, p. 16.

Six Shape Shifting Surrealists Walk Into a Bar

Ruminations on Chris Ware, Nahui Olin (aka Carmen Mondregón), Helmut Newton, Meret Oppenheim, Man Ray, and Tina Modotti

William Anthony Nericcio

PARSING CHRIS WARE

Verisimilitude in the arts is common enough. From Cervantes's Don Quixote chasing down his counterfeit others in Book II of the Spanish masterwork, to Borges weaving his circular ruins, or, shifting from literature to cinema, to Spike Jonze picturing John Malkovich surf the subterranean intestines (psychic entrails?) of his own unconscious in *Being John Malkovich*, to master printmaker Escher crafting pre-*Inception* uncanny mirrors in epic black and white, storytellers across various media have launched narratives wherein the telling of the tale is key, but where the telling of the telling of the tale is even more immediate and in the foreground—noviciates confuse this mirror-impulse with the postmodern, but it's been around since the Assyrians cut friezes onto stone. Cue Chris Ware, America's sequential art James Joyce, our own United Statesian graphic narrative Cortázar. A child of the Silver Age of Comics, Ware's spare, clockwork micro and macro tragedies are awash in self-awareness, a formalist mirror for the painful, autobiographically-laced stories that populate his panels: Big Tex is a mentally-challenged cowboy subject to the abusive loathings of his parents.

Rocket Sam, a space castaway, fashions robots as company and then, wracked by self-hate, loneliness, and petty narcissism, destroys them in orgasmic fits of anger that leave him more lonely and Ware's readers awash in shock and awesome sadness.

But perhaps Ware's piece de resistance is Rusty Brown, seen "playing" with dolls on the next page, a comic-collecting, overweight schlub, who is like some sort of obese Peter Pan (as played by the late Philip Seymour Hoffman and raised by E.T.A. Hoffman's the Sandman)—a pathetic lunatic and a self-destructive and abhorrent loaner subject to acts of hate and pathological ugliness that make Dostoyevski's personages seem tame, Ballard's psychopaths seem like pussycats.

Chris Ware is not *like* James Joyce—a sublime seer and narrative architect whose *Ulysses* changed the course of the novel in the 20th century. He *is* Joyce's successor, but with this difference: he is both a writer *and* an artist, a storyteller and an illustrator, whose canon of works is changing the history of literature, comic books, and, even, yes, high art, simultaneously.

THE VORACIOUS EYES OF CARMEN MONDREGÓN / NAHUI OLIN

On 8, July 1893, Nahui Olin, barrels onto the planet as María del Carmen Mondragón Valseca —eyes, in Europe and Latin America and across the planet, will never be the same again.

An artist, artists' model, painter, poet, and all-around Mexican bohemian, Olin is born into a Mexican industrialist family of privilege (la familia Mondrágon manufactured rifles and ammo). The money does some good, as Olin is educated in Mexico and France with an intellectual and arts regime focused on the arts.

Soon her circle includes a Euro arts league of usual suspects including Jean Cocteau, Pablo Picasso, and Henri Matisse; the same goes down on her return to Mexico, with Olin falling in with the crème de la crème of the Mexican burgeoning arts scene. Soon she's modeling for Diego Rivera and Tina Modotti, hanging with José Vasconcelos, father of "La Raza Cosmica," and composing poetry with Gerardo Murillo, the one and only Dr. Atl, a writer and painter whose Mexican circle (he's sort of DF-style Gertrude Stein) fired the palettes of 20th century Mexican muralists.

Olin is Madonna, pre-Madonna, a pre-cursor herself to celebrity/artist fusions like Lady Gaga—no meat frock for Nahui Olin, but she did wear the first miniskirt in Mexico!

Olin is yet another American original, a bon vivant artist-lover whose work, predictably, as a model, as an object of art, at times overshadows her efforts as a poet, painter, and photographer—see Meret Oppenheim, below. In her wake, Mexico reacts as most of the world does to strong sexy shamanic female forces shameless in speech and actions, with the Mexican popular press scandalously framing her as a witch/mad-woman—these are trite, predictable responses to an aesthetic hurricane prone to nudity and random sexual co-conspirators but they did their damage as Olin's legacy is largely anonymous!

Olin's work as a poet is worth a second look, her two major works being *Óptica cerebral, poemas dinámicos* (*Brainy Optics, Dynamic Poems*) and *Calinement je suis dedans* (*Tender, I am inside*). Her paintings, raw, bursting with color, are worth a second peek as well, with Olin revealed as the seeming mother of Big-Eye paintings (someone tell Tim Burton or Margaret and Walter Keane).

As your own eyes wander from her work as a painter (*Autorretrato en los jardines de versalles, Self-portrait in the Versailles Gardens*) to that of a model (here in a haunting capture by Edward Weston), one is struck by the dance of optics at work in Olin's work. As if infected by retinally-conveyed viruses borne of the cameras and canvases that captured her unique power, Olin's own aesthetic destiny moves to the rhythm of this optic beat: eyes themselves grow larger than life in her haunting, uncanny paintings.

INTO THE UNCANNY VALLEY WITH HELMUT NEWTON

Imagine yourself waking up in a veritable *Land of the Giants*, but all its denizens are female, and not *just* female, but aggresively dominant—women like Leopold Von Sacher Masoch dreamt of in the night.

Imagine further: you are in a silent noir world (even the colors are noir), where *haute* couture monstresses rule the land with a steely eye and a firmer grip. Now, then, you have stumbled into the erotically-charged, surrealistic photographic universe of Helmut Newton.

Berlin-born October 30, 1920 as "Helmut Neustädter," the Jewish photographer escaped Nazi persecution (surviving the *Kristallnacht*, no less) for the wilds of Australia where he honed his craft via portraiture.

His work emerges from the world of high fashion photography with *French Vogue* and *Harper's Bazaar* where his special semiotic kink emerges. Before Baudrillard and Laura Mulvey patented their lurid meditations on the gaze, simulacra and more, Newton was out there, outing photographer's secrets and crafting tour de force erotic mise-en-scènes that performed meditations on the nature of representation, on female/male objectification, and more—his unambiguous fetish? A deific contemplation of woman's body as sexual source and spur, is everywhere.

Newton authors many a masterpiece including his infamous "Naked" – above, Sie Kommen, Paris, 1981)—a diptych featuring four monumental female models, shot low-angle (so that they tower more) once fully clothed, the next wholly nude.

And my favorite, pictured here, from a series shot with actress Nastassja Kinski, posed by director James Toback, and, as an utterly uncany touch, a child-sized Marlene Dietrich doll (Newton is buried next to Dietrich in Berlin!).

Helmut Newton dies January 2004 when he wrecks his car at the Chateau Marmont Hotel, off Sunset Boulevard. Newton looms large in the universe as an inverted Surrealist Man Ray: Man Ray, the famous surrealist artist/innovator whose life in fashion photography is shadowed by his high art career vs. Helmut Newton, the famous fashion photographer who in my view is our late 20th century surrealist extraordinaire.

BETWEEN THE MIRROR AND THE PRINTING PRESS WITH MERET OPPENHEIM

She is a shaman—an avatar given to feats of mesmerizing transubstantiation so memorable that they become the defining moments of a movement in art (photography, performance, painting, what have you) called surrealism.

And whilst Man Ray, Luis Buñuel, and Salvador Dali are the proper names usually associated with surrealism, it was Méret Oppenheim for me that defines the movement. Is she a writhing winsome collaborative nude model at times? Yes.

A medium-defying, inventive sculptor? Yes.

A risk-taking celebrity artist whose antics fueled the fishwrap in the days before TMZ? Oh yes!

We cannot blame the heternormative gaze for having imprisoned much of our memory of Méret as Man Ray's lover/model, cavorting nude in surprising, uncanny ways for his camera.

This was Meret's destiny.

Born 6, October 1913, in Berlin, she is allegedly named after "Meretlein" a singular, untamed, child in Gottfried Keller's bizarre autobiographical bildüngsroman *Der grüne Heinrich*—as a result she comes to be the sexy poster child for the Surrealist clan, a "*femme-enfant*," a young, unpredictable woman with a direct subterranean pathway to dreams and unconsciousness.

From the start she is immersed in and expresses herself through the arts, palling around with Picasso and André

Breton, and blasting onto the Parisian arts scene at 22 with "Object," also known as *Le Déjeuner en fourrure* (thanks to Breton), the fur-lined teacup piece that goes on to become "the quintessential Surrealist object" at MOMA in New York (1936-37).

Whilst this sexy, furry place setting comes to define Oppenheim, and, some say, limit her reach, (thanks MOMA!), Méret goes on to enjoy a long, dynamic career. Whether it was her youthful, edgy performance art collaborations with Man Ray, her surrealist experimentations with sculpture, or her late-career confabulations of fashion and high art— "Glove" here, is typical, echoing the work of Frida Kahlo in an evocative fashion, the biological dimensions of aesthetics laid bare for all to see—Oppenheim leaves her mark on 20th century art in indelible fashion.

MAN RAY MEET EMMANUEL RADNITZKY

Take it from me, intellectuals are not knee-jerk lovers of comedy—or, for that matter, anything funny.

'Funny ain't smart' is all when it comes to the Ivory Tower poindexters of academe and that's too bad because funny is hard, funny is harder than tragedy, and the comedic is a vein best left to the experts.

Enter Jewish-American "funny man" Emmanuel Radnitzky, better known as Man Ray to students of surrealism, Dada, photography, sculpture, and experimental cinema. Born with a Semitic moniker, Radnitzky became Ray, Man Ray, to the masses, when his family ditched 'Radnitzky' when things got hot for the tribe in Philly where he grew up. So Manny Ray, the son a tailor (watch for sewing symbolism whilst perusing his oeurvre—irons, scissors, ironing boards, mannekins), goes on to be one of the ringleaders of Surrealism, hanging out and making art for decades with the likes of Salvador Dalí, Luis Buñuel, Frances Picabia, Max Ernst, and, my favorite, Meret Oppenheim.

Ray's cinematic experiments are the fundament for Warhol's later experiments, and while he thought of himself as a painter, it is Ray's experiments throughout his life with photography that left their mark not only on the world of Art but on that of fashion—much of what we associate with Steven Meisel and Ellen Van Unworth (and, of course, that old cad, Helmut Newton) would not have come to be without Ray's semiotic tinkering and experimentation.

My favorite Man Ray photographic series are the "Erotique Voilée" panels shot with Meret and Louis Marcoussis (featured above in Meret Oppenheim's section, shot in Marcoussis' printing workshop). This improvised photosession with Meret, nude ('*those* Surrealists'), covered in printer's ink and posing in front of and behind a huge press become an allegory of the Surrealist project itself.

Do we read the photographs as art? As a record of a staged, theatrical improvisation? As a snapshot of the war between word and image staged through a naked female form?

Yes. Yes. And Yes.

All of the above, and more.

But it was also pure schtick, noir Vaudeville, performance art before Performance Art—'is that a printing press handle or are you just glad to see me'—and another reminder that Ray's ludic legacy, his comedic, improvisational collaborations *on* film, *with* other artists, are here to stay.

TINA MODOTTI'S MIRROR TYPEWRITER

If you are seeking to fuse the traditions of the bohemian avant-garde in the twentieth century and do so with some sensitivity to both Europe and the Americas, you could do worse than to spend time pondering the peripatetic wonder that was the life of Tina Modotti.

Both model *and* photographer, radical lefty political activist and bon vivant, Modotti cascades across the oceans during the twentieth century working with Edward Weston as a model, below, defining the Mexican left (especially the muralists) with her documentarian's eye, fighting with the resistance in Spain, befriending Frida Kahlo and sharing a bed with philandering hubby Diego Rivera.

While her work as a model is notable, see Weston's *Nude on the Azotea* (1924), it is her photography I find more striking.

Modotti's legacy is not unlike Meret Oppenheim's whose modeling work with Man Ray is often foregrounded *before* her sculpture and art (something about female nudies have lasting high cultural value, but that's fodder for another book).

Modotti's work behind the camera deserves lasting critical scrutiny. *Mella's Typewriter,* one of her iconic renderings, pictures the tool of her soon-to-be-assassinated lover's writing machine. Aesthetically, Modotti was an inveterate modernist and her portraits of people and objects are both stark and moving—the symbolism is never subtle. *Mella's Typewriter* is colossally curious as it captures one medium pondering the nature of an other medium—here the camera 'eyes' the typewriter, a *semiotic* machine face to face with the grandfather (now) of *semantics*.

Mella had been writing of "una sintesis" when Modotti captured the lasting image of her lover's writing machine and it is now part of Modotti's legacy.

Modotti dies of congestive heart failure far too young in her forties—all that is left to us is the synthesis of word, image, art, and revolution immortalized by her lens and her remarkable imagination.

[A personal thank you to legendary Hermenaut, Joshua Glenn, in whose online wonder Hilobrow.com these pieces first appeared. WAN]

Appropriating *The Lord of the Rings*: Tracing the Movements of Medieval Architecture in Popular Culture

Katie Waltman

When Kevin J. Harty writes that "it seems that people like the Middle Ages"—he isn't kidding.[164]

Some of these people are J.R.R. Tolkien, Alan Lee, and Peter Jackson: the author, novel illustrator, and film director of *The Lord of the Rings*, respectively. Tolkien's *The Lord of the Rings (LoTR)* was published in the 1950's. This fantasy novel traces a young hobbit's quest to destroy an ancient ring with dark powers. Readers of the novel and viewers of the films (released worldwide in the early 2000's) follow the main character through Middle-earth as he treks through different geographies inhabited by different peoples. The story progresses through a quiet countryside of kind and gentle hobbits and then through different lands of other good creatures, such as elves, dwarves, and men, who ultimately triumph over evil, darker creatures. The novel and film trilogy clearly convey medieval themes and visuals. For instance, the weapons, battle armor, costumes, and even the romantic, medieval theme of "self-realization" evoke the Middle Ages.[165] *LoTR* also possesses different types of medieval architecture that compare to structures from different locations across Europe from the early to late Middle Ages.[166] The architecture evolves from Tolkien's descriptions and illustrations, into Lee's novel illustrations, and then finally into Jackson's final films. However, the medieval architecture in *LoTR* may provide more than just interesting visuals for audiences. Indeed, a closer look reveals how specific types of medieval architecture was adapted for these stories. For instance, a stereotypical Gothic church interior illustrates the depths of evil and darkness. *LoTR* has a major, worldwide public presence, so the story's use of medieval architecture is one way to understand how literature and film sustain specific stereotypes of the Middle Ages in popular culture.

---

164 Kevin J. Harty, *The Reel Middle Ages: Films About Medieval Europe* (Jefferson, NC: McFarland, 1999): 3. This quote originates in the work of 20th century literary critic and philosopher Umberto Eco.

165 W.A. Senior, "Medieval Literature and Modern Fantasy: Toward a Common Metaphysic," *Journal of the Fantastic in the Arts* 3, no. ¾ (1994): 38. Medieval scholar Raymond H. Thompson asserts that, "It is important to realize that the great theme of medieval romance is self-realization."

166 For the purpose of this paper, the Early Middle Ages are the fifth through tenth centuries, the High Middle Ages are the eleventh through thirteenth centuries, and the Late Middle Ages are the fourteenth through fifteenth centuries.

We have been studying the Middle Ages since just after it ended, and authors and filmmakers have been interpreting the Middle Ages for decades. Literature first illustrated concepts of "medievalism," and many popular Hollywood films have especially evoked it.[167] Contemporary characters are placed in a medieval world, where audiences can often visit chivalrous and "barbaric" themes.[168] The Middle Ages are distorted to conform to contemporary perceptions of what society considers medieval.[169] Tolkien, Lee, and Jackson have taken both stereotypical and less well-known medieval architecture found in medieval western Europe to depict the medieval world of *LoTR*. The comparisons of their medieval architecture to historical medieval structures demonstrate *LoTR* promotes a generalized yet particular perception of the Middle Ages.

The Tournament of Kings show at the Excalibur, Las Vegas, Nevada.

---

[167] Harty, *The Reel Middle Ages*, 3. Harty defines the term "medievalism" as "a continuing process of creating and recreating ideas of the medieval that began almost as soon as the Middle Ages had come to an end." Examples of major, worldwide released Hollywood films that evoke "medieval" themes are *First Knight* (Columbia Pictures, 1995), *King Arthur* (Touchstone Pictures, 2004), *Kingdom of Heaven* (20th Century Fox, 2005), and *King Arthur* (Warner Bros. Pictures, 2017).

[168] Harty, The Reel Middle Ages, 4.

[169] Harty, *The Reel Middle Ages*, 4. Eco theorized how the idea of popular culture participates in "messing up" the Middle Ages to conform to how different time periods viewed it.

Tolkien created *LoTR's* medieval realm through his theory of subcreation – that with fantasy, man becomes sub-creator by creating various forms and images.[170] But where do these new forms and images come from? Often they are formed from familiar things one knows.[171] Much of Tolkien's literary descriptions and sketches for the architecture found in *LoTR* novels are often structures one might see from medieval times. His medieval context for *LoTR* was partially influenced by his British and Celtic medieval studies research and by the Gothic architecture that surrounded him while he was a scholar at Oxford University.[172] He also was an aficionado and researcher of *Beowulf*, an Old English poem that many scholars used as a source of early medieval history. Logically, one may understand why Tolkien "subcreated" *LoTR's* middle-earth realm with medieval architecture and other medieval motifs. But it merits attention that Tolkien's *LoTR* encompassed a particular form of medievalism, "a reimagining of the Middle Ages that blends contemporary preoccupations with the historical realities of medieval Europe."[173]

Alan Lee also made medieval-themed choices when he illustrated Tolkien's novel. Lee admitted where he lived influenced his decisions. One of his residencies was Devon, located in southwest England. A popular attraction in Devon is the medieval structure Exeter Cathedral. Although there may not be any evidence from Lee that mentions he specifically used this cathedral as inspiration, it is worth considering, as Lee says he was inspired by the landscapes from places he had lived. Lee also drew inspiration from early medieval Irish and British illuminated manuscripts and metalwork as well as from Celtic and Germanic art.[174] Lee makes clear his wish to bring Tolkien's descriptions to life: "I was trying to summon up what Tolkien was describing." Lee, via Tolkein, begins to unintentionally awaken "an idealized medieval past," a continuation of the nineteenth-century movement of medieval nostalgia.[175]

Peter Jackson drew heavily on Lee's illustrations and hired him as a set designer for the *LoTR* films. Some argue that Peter Jackson wanted *LoTR* to appeal to audiences by exploiting notions of medievalism.[176] Since as early as 1912, some films have often reimagined the Middle Ages based on

---

170 J.R.R. Tolkien, *The Monsters and the Critics: And Other Essays* (HarperCollins, 2007), 122.
171 Tolkien, The Monsters and the Critics, 146.
172 Christopher Synder, *The Making of Middle-Earth: A New Look Inside the World of J.R.R. Tolkien* (New York: Sterling, 2013): 38. See also Steven Woodward and Kostis Kourelis, "Urban Legend: Architecture in The Lord of the Rings," in *From Hobbits to Hollywood: Essays on Peter Jackson's Lord of the Rings*," eds. Ernest Mathijs and Murray Pomerance (Amsterdam: Editions Rodopi, 2006): 204.
173 Snyder, Making of Middle-Earth, 39.
174 Snyder, Making of Middle-Earth, 239.
175 Woodward and Kourelis, "Urban Legend," 211.
176Carol A. Leibiger, review of *Picturing Tolkien: Essays on Peter Jackson's The Lord of the Rings Film Trilogy,* by Janice M. Bogstad and Philip E. Kaveny, *Journal of the Fantastic in the Arts* 23 (2012): 518. For instance, folklorist Dimitra Fimi thinks Jackson exploited "cultural medievalism in the 'Celtic' elves" to appeal to audiences.

both history and enduring contemporary perceptions about that time.[177] Over time, the Middle Ages began to be understood in terms of medieval stereotypes. Representing medievalism in film is often indeed "driven by the nostalgia of popular culture."[178] Filmmakers often use their films to project dominant medieval themes and visuals to satisfy popular interests in the Middle Ages. Jackson's films contain many different types of medieval architecture, and using those various types in film evokes a familiarity with the Middle Ages for an abundance of viewers. As commercial films with wide release often provide "something for everyone," sum, one comes to understand what inspired and influenced the author and illustrator and why they chose medieval descriptions and illustrations for architecture in *LoTR*.[179] And one can perceive how those images—rooted in Tolkien's subcreation—evolved over time, in Lee's book illustrations and Jackson's films. However, this is not an attempt to argue that they were all trying to satisfy popular ideas of the Middle Ages and medieval fandom.

There is no denying that medieval characteristics common in *LoTR* are well-known in popular culture. Much of the architecture in *LoTR* looks very "medieval." Placing this architecture outside its historical context can result in the recognition that the architecture is "medieval" while simultaneously overlooking cultural appropriation. As a result, (and no revelation here) viewers may not get an accurate understanding of western medieval history. The comparative analysis of medieval architecture in *LoTR* to historical medieval architecture demonstrates how *LoTR*—particularly the films—promote flattened, homogenized medieval themes that erase much of the Middle Ages' complexities.

To understand how these modern attitudes infiltrate depictions of the Middle Ages, we must move to a closer analysis of medieval architecture as it evolves in these filmed entertainments and compare these constructions to actual structures from medieval western Europe. For example, the Shire realm of the hobbit characters possesses medieval-inspired homes. For one hobbit house, Tolkien sketched an enclosed yard that has a gate, farm buildings on the side, and a round front door (Fig. 1).[180] Lee's illustration (Fig. 2) and Jackson's creation of a hobbit house (Fig. 3) stay loyal to Tolkien's description. Similar domestic structures exist from the early Middle Ages. For example, the hobbit houses resemble ninth- to eleventh-century Nordic medieval turf homes from Iceland (Fig. 4).[181]

---

177 Harty, *The Reel Middle Ages*, 384. The 1912 Italian film *Parsifal* details Parsifal "as the guardian of the Holy Grail."

178 Nickolas Haydock, *Movie Medievalism: The Imaginary Middle Ages* (Jefferson, N.C.: McFarland, 2008): 5.

179 Susanne Eichner, Lothar Mikos, and Michael Wedel, "'Apocalyse Now in Middle Earth': 'Genre' in the Critical Reception of *The Lord of the Rings in Germany*," in *The Lord of the Rings: Popular Culture in Global Context* by Ernest Mathijs. P. 143

180 Wayne G. Hammond and Christina Scull, *The Art of The Lord of the Rings by J.R.R. Tolkien* (New York: Houghton Mifflin Harcourt, 2015): 197-198.

181 Woodward and Kourelis, "Urban Legend," 200.

Additionally, tenth-century domestic palloza houses (Fig, 5) with rounded doors and earth-covered roofs from O Cebreiro, Spain also bear resemblance to the hobbit houses. Because of its architecture, the hobbit culture in *LoTR* may now appear to represent certain Nordic and Spanish cultures from the early Middle Ages.

Men live in the village of Edoras in the realm of Rohan, where Jackson models Edoras' domestic dwellings off of tenth-century Viking houses. In Jackson's film *The Two Towers*, the peasant houses in Edoras have simple wooden walls with a roof of dry vegetation (Fig. 6), which look like tenth-century Danish Viking houses (Fig, 7). The royal house in Edoras also resembles a tenth-century Viking fortress.[182] This Trelleborg Viking fortress from Denmark has a pediment at the top of the façade with a projecting, crossed structure at the point of the pediment (Fig. 8). It certainly evokes the pediment on Edoras's royal house (Fig. 9). The similarities are not a coincidence. Lee illustrates the Edoras royal house interior by using large columns crowned with Corinthian capitals (Fig. 10), which reflect the Romanesque style from the early Middle Ages. Jackson's depiction of the throne hall (Fig. 11), is also decorated with popular Romanesque motifs, like the connecting, arched columns down next to both sides of the aisle that lead to an apse behind the throne, as it would commonly appear in a Romanesque church (Fig. 12). Edoras has both Danish Viking and Western European Romanesque architecture. It is a hybrid of medieval cultures and styles, and like the Shire, the potential problem is that the audience likely does not recognize the cultural differences. Rather, they merely see what they believe is a medieval realm.

The city of Minas Tirith from the realm of Gondor also looks notably medieval (Fig. 13). It displays more stereotypical medieval architecture of fortresses and castles (Fig. 14). Tolkien wrote that Minas Tirith resembles a small mountain built on seven levels[183] with bright banners on top of turrets[184] and that its "white banners broke and fluttered from the battlements in the morning breeze."[185] It also draws inspiration from more than one type of medieval architectural style.[186]

182 Woodward and Kourelis, "Urban Legend," 204.
183 Hammond and Scull, Art of The Lord of the Rings, 140.
184 J.R.R. Tolkien, *The Lord of the Rings,* bk. II, ch. 10.
185 J.R.R. Tolkien, *The Lord of the Rings,* bk. V, ch. 1.
186 Snyder, Making of Middle-Earth, 155.

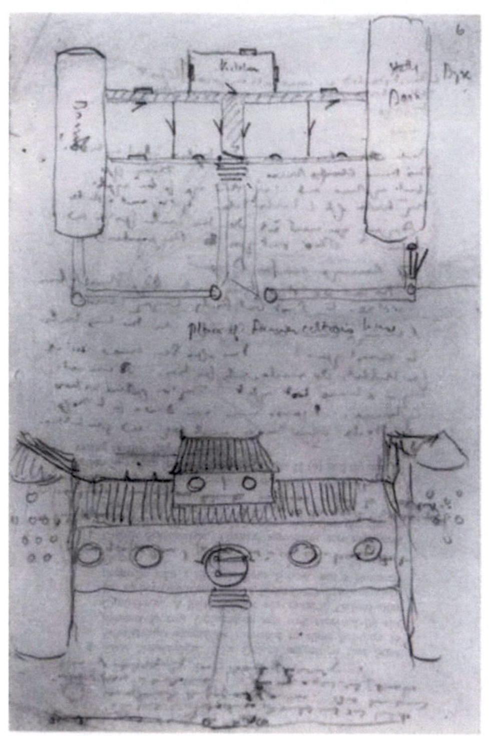

Fig. 1 Sketch of Farmer Cotton House (Tolkien).

Fig 2. Hobbit house (Lee).

Fig. 3 Hobbit house, *LoTR* film set, 1999 (Dailymail.com).

'ig.4 Medieval turf house, Iceland, 9th-11th c. nationalgeographic.com)

Fig. 5 Palloza, O Cebreiro, Spain, 10th c. (www.museos.xunta.gal)

g. 6 Edoras house, *The Two Towers*

Fig. 7 Danish Viking house, Hedeby, Germany, 10th c. (www.vikingdenmark.com)

:. 8 Trelleborg Viking Fortress, Denmark, 10th c. (National Museum of Denmark)

9 Edoras royal house, *The Two Towers*

Fig. 10 Edoras royal house interior (Lee)

Fig. 11 Edoras royal house interior, *The Two Towers*

Fig. 12 Basilica of Santa Maria, Rome, Italy, 3rd – 13th c.

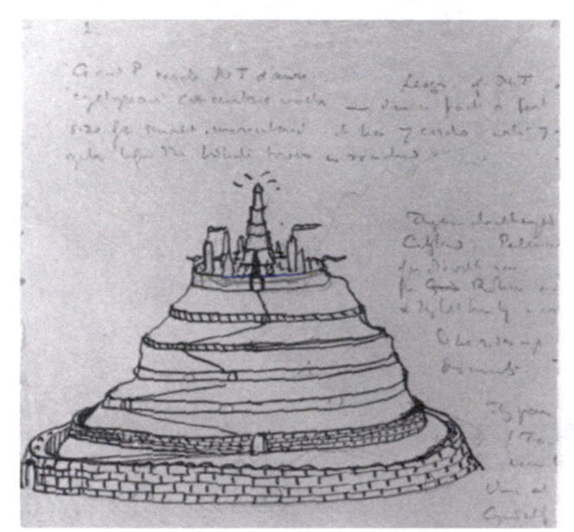

Fig. 13 Minas Tirith sketch (Tolkien)

Fig. 14 Coca castle, 15th c., Spain (castles.org)

Fig. 15 Minas Tirith, *The Return of the King*

Fig. 16 Minas Tirith, throne hall, *The Return of the* King

Fig. 17 Santa Maria Novella, begun 13th c., Florence, Italy (www.museumsinflorence.com)

Jackson depicts the first level of Minas Tirith with rounded watchtowers and high fortress wall (Fig. 15). Once again, Jackson borrows features from medieval church architecture for the interior of king's throne hall. Minas Tirith's main throne hall interior has an aisle lined on each side with a row of columns that have black and white striped arches connecting the columns (Fig. 16). This decorative feature is popular in medieval church and cathedral interiors (Fig. 17). While Jackson stays true to those motifs for the exterior of Minas Tirith, he incorporates medieval church architecture for the interior of the king's hall. Once more, another realm in *LoTR* has architecture that is characterized by more than one Western European medieval style. And once more, the audience may not see cultural differences. Instead, Minas Tirith is a city that is "medieval."

The residential evil settings in *LoTR* appear to contain architecture that resembles the Gothic style of the Middle Ages.[187] The evil wizard Saruman resides in Isengard's Orthanc tower, which has tiered towers ascending in size as they rise and culminate at the top with "teeth or "horns."[188] The tower conveys a harsh image when Tolkien describes it as "black and gleaming hard" and with "pinnacles sharp as the points of spears, keen-edged as knives."[189] Tolkien's first sketch of Orthanc tower reflects the tiered towers, while the other sketch reflects the harsh tower walls culminating in pointed spears (Fig. 18). The evil lord Sauron resides in Barad-dur, the other Dark Tower that Tolkien sketched like a castle or fortress structure (Fig. 19).

But his description is similar to that of Orthanc, as he describes Barad-dur as "rising black, blacker and darker...the cruel pinnacles and iron crown of the top-most tower."[190] Jackson stays loyal to Tolkien's description by depicting both towers as black and "cruel" structures that rise from the ground. Jackson's rendition of the Orthanc and Barad-dur towers (Fig. 20) look similar to the tall, pointed structures of medieval Gothic churches and cathedrals. Gothic cathedrals often have narrow, tall, pointed towers (Fig. 21).

---

187 Nicola Coldstream, *Medieval Architecture* (Oxford: Oxford University Press, 2002): 40. Gothic architecture dates rough from the mid-twelfth century to the early sixteenth century. Various descriptions of Gothic architecture include sculptured ornament, rib vaults, pointed arches, thin moldings, flame-like tracery patterns, tall, thin structures, and much surface decoration.

188 Hammond and Scull, *Art of The Lord of the Rings*, 105, 108.

189 J.R.R. Tolkien, *The Lord of the Rings,* bk. III, ch. 8.

190 J.R.R. Tolkien, *The Lord of the Rings,* bk. IV, ch. 3.

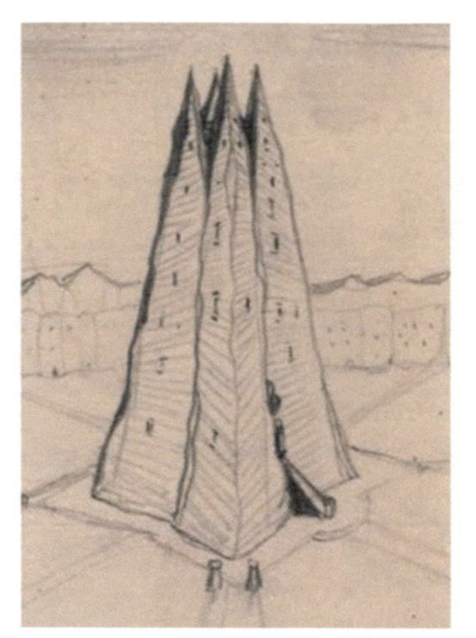

Fig. 18 Two Orthanc sketches (Tolkien)

Fig. 19 Barad-dur sketch (Tolkien)

Fig. 20 Orthanc, *The Two Towers* (left) and Barad-dur, *The Return of the King* (right)

Fig. 21 Ulm Minster, 14th c., Germany (wikipedia.org)

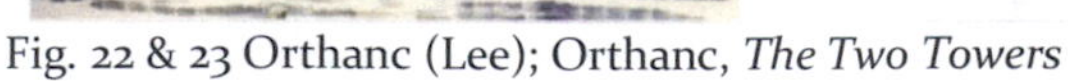

Fig. 22 & 23 Orthanc (Lee); Orthanc, *The Two Towers*

Lee's Orthanc illustration also looks Gothic. The exterior façade has repetitive, narrow, stone structures stacked closely next to each other (Fig. 22), and Jackson's rendition looks almost identical (Fig. 23). Their Orthanc exterior facades look similar to common facades from the portals on the exterior of some Gothic cathedrals from the High Middle Ages. For instance, the west portal from France's Chartres Cathedral has jambs, which are narrow statues of religious figures stacked closely next to each other (Fig. 24). The pattern of these jambs looks extremely similar to the pattern on Orthanc's exterior. Jackson further draws on Gothic architecture with the interior of Orthanc. Saruman's dwelling inside the tower has windows with iron designs that resemble window traceries from Western European Gothic churches and cathedrals from the High to Late Middle Ages. The Orthanc window designs have floral-looking patterns (Fig. 25), comparable to those seen in the windows of Amiens Cathedral in France (Fig. 26). The two evil towers noticeably reflect the Gothic style, which encourages the stereotype of Gothic as something dark and wicked.

Depictions for the mines of Moria – another dark and evil setting – also resemble religious Gothic architecture. As the Fellowship of the ring arrive at a large space in the mines, Tolkien writes, "Before them was another cavernous hall...down the centre stalked a double line of towering pillars."[191] This description could have been illustrated several different ways. But Lee depicts Tolkien's description with arched, ribbed vaulted ceilings supported by columns (Fig. 27), like those often found in Gothic cathedral and religious architecture. For instance, Moria's hall looks remarkably similar to the twelfth-century monastery of Santa Maria de Alcobaca in Portugal (Fig. 28). Italian church cathedral ceilings from the High Middle Ages also look like Moria's hall. Jackson's Moria hall is loyal Lee's painting when the audience sees the fellowship run through this hall in *The Fellowship of the Ring.*[192] Moria is supposed to be cold, dark, and made of solid, angular stone.[193] This type of description often characterizes Gothic architecture, as well as the "evil" architecture in *LoTR*. By exhibiting Gothic-looking architecture in its evil dwellings, *LoTR* seems to confirm the stereotype that Gothic is something dark and barbaric, or that the Middle Ages were the "dark ages."

---

[191] J.R.R. Tolkien, *The Lord of the Rings,* bk. II, ch. 5.
[192] J.R.R. Tolkien, *The Lord of the Rings,* bk. II, ch. 5.
[193] Snyder, Making of Middle-Earth, 250.

Fig. 24 West portal, Chartres Cathedral, France, 12th c. (wikipedia.org)

Fig. 25 Interior, Orthanc, *The Fellowship of the Ring*

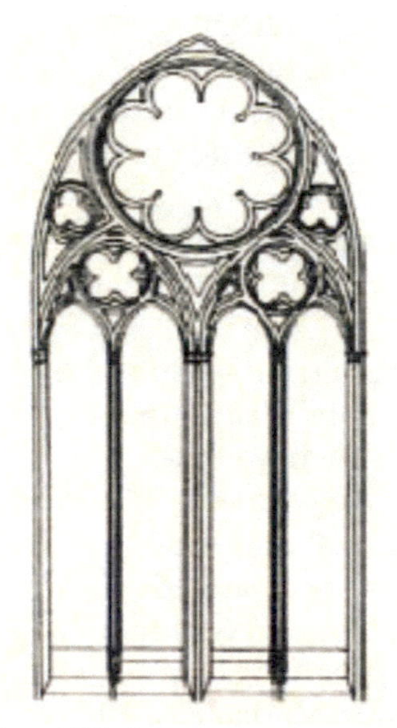

Fig. 26 Window tracery pattern, Amiens Cathedral, France, 13th c. (Coldstream p. 40)

Fig. 27 Moria interior (Lee)

Fig. 28 Monastery of Santa Maria de Alcobaca, Portugal, 12th c. (thoughtco.com)

An analysis of Tolkien's architecture sketches and novel descriptions demonstrates that his architecture in his subcreated world only subtly alludes to the medieval.[194] Out of the mentioned sketches and architectural descriptions, only his sketch and description of Minas Tirith hint at the medieval style, with the fortress wall, "pinnacles," and "white banners." The architecture is not a dominant element that drives Tolkien's story, partially because, as Emily Augur argues, he believed in describing fantasy more literally than visually.[195] Of course, using the fantasy genre permits freedom to borrow medieval architecture and invent a story set among a uniquely refabricated Middle Ages. By keeping his architecture references vague, it allows Tolkien's *LoTR* to exist as more of a fantasy and mythical narrative, rather than a story that is overwhelmingly medieval or even quasi-historical. On the contrary, both Lee and Jackson use medieval architecture more noticeably.[196] Jackson hired Lee to help conceptually design the movies, and Lee's illustrations were highly influential to the writers who worked on the script.[197] At times it even appears like Lee embedded his drawings into Jackson's landscapes.[198] Jackson clearly models some of his architecture off Lee's illustrations, especially Isengard and Moria. Lee's illustrations often stay faithful to Tolkien's descriptions and texts, but both he and Jackson derive their architectural designs more deliberately from several cultures and styles in medieval Europe.

One significance of *LoTR's* is its placement of medieval architecture within specific contexts. This demonstrates how the selection of medieval architecture for *LoTR* may encourage medieval stereotypes and continue to support a homogenized perception of the Middle Ages. Viewers see various medieval architecture from different western countries spanning several hundred years. But they do not likely comprehend what they see is a fourteenth-century French gothic church or an Icelandic turf house from the twelfth century. Instead, they may see *LoTR* as an alternate fantasy version of the Middle Ages – Tolkien's sub-creation – and not an accurate representation of medieval history.

The fantasy genre of fiction is commonly tied to the Middle Ages. And it is fair to ask here what is so harmful about how the architecture in *LoTR* is not always historically accurate? After all, Alan Lee and Peter Jackson's selection of architecture in the films is a way of honoring Tolkien and his sub-created world of *LoTR*. The medieval architecture in *LoTR* is an example of something that facilitates how we often continue to interpret the Middle Ages through our current perspectives. Yet, it can be prob-

---

194 Woodward and Kourelis, "Urban Legend," 190.

195Emily E. Auger, ""The Lord of the Rings"' Interlace: Tolkien's Narrative and Lee's Illustrations," *Journal of the Fantastic in the Arts* 19 (2008): 70.

196 Woodward and Kourelis, "Urban Legend," 190. Jackson's movies possess more "architectural associations."

197 Snyder, Making of Middle-Earth, 244.

198 Woodward and Kourelis, "Urban Legend," 193.

lematic to interpret the Middle Ages through our modern lenses, because certain perceptions of the Middle ages may influence cultures, societies, and politics. The novel and films have been accused of being "obviously obsessed with the difference between certain races."[199] Literary, media, and cultural scholar Niels Werber states that the story's realms, and the characters who populate them, "mirror these [racial] differences."[200] Additionally, the novel and films present subliminal messages of Nazi Third Reich modes of thinking.[201]

Perhaps an even more relevant and influential societal source that distorts medieval themes is the website *Return of Kings*, a pro-masculinity internet forum. Also thought of as misogynistic, the website has many supporters.

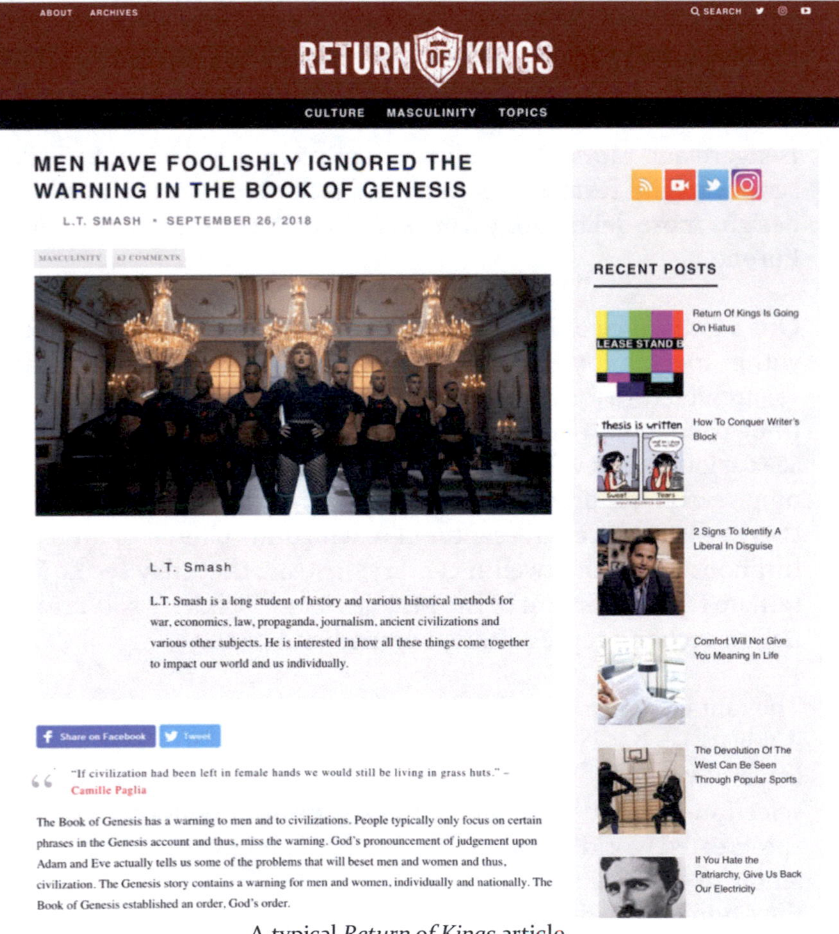

A typical *Return of Kings* article
(returnofkings.com/194915/men-have-foolishly-ignored-the-warning-in-the-book-of-genesis)

---

[199] Niels Werber, "Geo- and Biopolitics of Middle-Earth: A German Reading of Tolkien's 'The Lord of the Rings,'" *New Literary History* 36, no. 2 (2005): 227.
[200] Werber, "Geo- and Biopolitics of Middle-Earth," 227.
[201] Werber, "Geo- and Biopolitics of Middle-Earth," 227.

"Return of Kings" is dedicated to a space that allows men to fantasize about a time where men were men and women knew their place. Many of the website's writers love Tolkien and promote his literary supremacy.[202] Unfortunately, these writers also distort Tolkien's medieval themes, which attracts some who fall in that realm of pro- "white power." For instance, the article "4 Reasons Why Aragorn is a Great Man" states, "by virtue of [Aragorn's] bloodline, he had traits of the great Kings of Men that came before him," which contributes to "the nature of [Aragorn's] rise to power."[203] In Jackson's films, Aragorn is played by a blue-eyed Caucasian man. So essentially, one could easily confer that the author states that a white, blue-eyed man is what helps a great king rise to power. Aragorn is a born leader of purest blood. Some scholars are deriving from Tolkien's novel that purity of race is always a sure sign of superiority of knowledge, wisdom, and strength. In sum, themes in *LoTR* such as mysoginy and racism are found among pop culture websites and academic publications. These themes now exist is a world that Tolkien subcreated, and they have spread among popular culture because of the global popularity of *LoTR*.[204]

Lee and Jackson probably had good intentions when making Tolkien's story visually come to life in novel illustrations and on film. Illustrating how *LoTR* uses medieval architecture is one way that demonstrates how easily one can misunderstand or modify the Middle Ages. The comparative architectural analysis serves as a vehicle to illustrate that Tolkien subcreated a fantasy realm that has medieval characteristics that people are still currently taking out of context and appropriating. Some scholars and pop culture writers are misinterpreting Tolkien's subcreation. Themes are manipulated and the Middle Ages are revised. Because of its popularity, one hazard of *LoTR* is that it has a higher tendency to encourage medieval stereotypes, all of which might effortlessly spread throughout popular culture.

References

Auger, Emily E. ""The Lord of the Rings'" Interlace: Tolkien's Narrative and Lee's Illustrations." *Journal of the Fantastic in the Arts* 19, no. 1 (72) (2008): 70-93.

Chance, Jane Author. *Tolkien the Medievalist*. London; New York: Routledge, 2003.

Coldstream, Nicola. *Medieval Architecture*. Oxford: Oxford University Press, 2002.

Hammond, Wayne G. and Scull, Christina. *The Art of The Lord of the Rings by J.R.R. Tolkien*. New York: Houghton Mifflin Harcourt, 2015.

---

[202] For example, see Quintus Curtius, "The Fascinating Linguistic Journey of J.R.R. Tolkien," April 10, 2017, https://www.returnofkings.com/118902/the-fascinating-linguistic-journey-of-j-r-r-tolkien.

[203] John Smith, "4 Reasons Why Aragorn is a Great Man," December 1, 2014, https://www.returnofkings.com/49231/4-reasons-why-aragorn-is-a-great-man.

[204] Ernest Mathijs, *The Lord of the Rings: Popular Culture in Global Context* (London; New York: Wallflower Press, 2006): 1, 6.

Harty, Kevin J. *The Reel Middle Ages: Films About Medieval Europe.* Jefferson, NC: McFarland, 1999.

Haydock, Nickolas. *Movie Medievalism: The Imaginary Middle Ages.* Jefferson, N.C.: McFarland, 2008.

Jackson, Peter, dir. *The Lord of the Rings: The Fellowship of the Ring*. 2001. New Line Cinema, 2002. DVD.

Jackson, Peter, dir. *The Lord of the Rings: The Return of the King*. 2003. New Line Cinema, 2004. DVD.

Jackson, Peter, dir. *The Lord of the Rings: The Two Towers*. 2002. New Line Cinema, 2003. DVD.

Leibiger, Carol A. Review of *Picturing Tolkien: Essays on Peter Jackson's The Lord of the Rings Film Trilogy,* by Janice M. Bogstad and Philip E. Kaveny. *Journal of the Fantastic in the Arts* 23, no. 3 (86) (2012): 517-21.

Mathijs, Ernest. *The Lord of the Rings: Popular Culture in Global Context.* London; New York: Wallflower Press, 2006.

Mathijs, Ernest., and Pomerance, Murray, eds. *From Hobbits to Hollywood: Essays on Peter Jackson's Lord of the Rings*. Amsterdam: Editions Rodopi, 2006.

Synder, Christopher. *The Making of Middle-Earth: A New Look Inside the World of J.R.R. Tolkien.* New York: Sterling, 2013.

Tolkien, J.R.R. *The Fellowship of the Ring*. 2nd ed. Boston: Houghton Mifflin Company, 1993.

Tolkien, J.R.R. *The Monsters and the Critics: And Other Essays.* HarperCollins, 2007.

Tolkien, J.R.R. *The Return of the King*. 2nd ed. Boston: Houghton Mifflin Company, 1993.

Tolkien, J.R.R. *The Two Towers*. 2nd ed. Boston: Houghton Mifflin Company, 1993.

Walter, Hugo G. *Magnificent Houses in Twentieth Century European Literature.* New York: Lang, Peter, Publishing Inc., 2012.

Werber, Niels. "Geo- and Biopolitics of Middle-Earth: A German Reading of Tolkien's 'The Lord of the Rings.'" *New Literary History* 36, no. 2 (2005): 227-246.

SOCIETY

EMOTIONS AND EVERYDAY LIFE
SOME CULTURAL ISSUES FROM A SOCIOLOGICAL PERSPECTIVE

MASSIMO CERULO

1. WHAT DO WE TALK ABOUT WHEN WE TALK ABOUT EMOTION IN CULTURAL SOCIOLOGY?

In contemporary society, an explosion of the ambivalences of social acting and a multiplication of the logic of trends and countertrends typical of modernity seem to be emerging. In this context, the role of emotions in social relation is subjected to relevant changes. In particular, emotions are considered endowed with a visibility and a weight external to the inner reality of the subject: it can be modeled, controlled, and managed based on the cultural roles of the context that subjects act in, from time to time (Hochschild 1979; see Stets, Turner 2014, 2007; Turner, Stets 2005; Bolton 2005; Francis 1994).

Thus, we can identify a series of factors that reveal some features of emotions if studied from a cultural/sociological point of view (Sandstrom *et al.* 2013):

- emotions, like other aspects of human conduct (attitude, ideas, behaviors), are subject to the social effect;
- emotions are activated directly by social interaction;
- there always is a regulatory component based on what any society has as its own rules to judge the emotions as acceptable or not, about how they can be exhibited according to situations, public or private. Those laws regulate the exhibition and the control of emotions (coping) so that each social organization gets its own uniformity in expressing emotions;
- emotions and their own exhibitions change from time to time according to the specific context of daily life as well as to relational practices and as mental constructions that accompany them change;
- emotions have to always be distinguished by their own exhibitions;
- emotions have an important cognitive functions;
- emotions become a double hermeneutical tool: they are needed to study the collective and individual behavior in a social reality also crucial for the processes related to the discovery of ourselves in order to approach to our own awareness.

In short, a circular process between emotions and awareness is created—we can enter, act, and exit through emotional expressions from different social situations typical of our daily life. By controlling these states according to the context we live in, it is possible to interact with other subjects and, at the same time, immerse ourselves in a self-reflexive process to discover ourselves (see van Zomeren 2016; Fields *et al.* 2007). This process brings the researchers to consider emotions as a "social product":

they are defined and redefined within the interactions which take shape in society and according to the subjective meaning that these interactions acquire based on the different participants. In this way, emotions and their exhibition are considered as "cultural products" (Shott 1979, pp. 1319–1320; see Mead 1982), because they will be interpreted, named, and communicated based on the rules of that culture, belonging to both the subject involved and the specific situation.

Emotional experiences then are made up of two separate moments: 1) the physiological activation of a feeling; and 2) the resulting cognitive definition. For instance, if we feel excitement it can be read as joy, anger, or anxiety based on the social situation in which we live. The social structure in which we live—cultural level, social role, rules, etc.—will help us to interpret it.

The subject, even if involved in physiological sensations from one side and structural features of the society on the other, keeps his/her own space of interpretation and definition of the feeling and consequent expressive modality. Emotions are then configured as a physiological element but are also affected by the interpretation of the subject and the social influences that the subject can suffer or exercise (Shott 1979, p. 1323). In this process, there is the application of a social way of acting which implies a judgment: towards the self and towards the others.

The guidelines listed above make clear how emotions are made by a sort of "double presence"—on the one hand, they are connected to physiological processes and reactions that take place in the body, so the emotions we feel as a consequence of our actions and our interactions are experienced, fundamentally, in the body. On the other hand, in order to become emotions, these processes and reactions must be interpreted in terms of symbols and social categories: according to the theory of symbolic interactionism, they must be *named*, and we must make meaning about the physical sensation, as well as our surrounding environments (Sandstrom *et al.* 2013, p. 200; see Denzin 1983).

The process of naming allows us to organize particular sensations and give a meaning to them. It also allows us to see ourselves as "emotional man" (angry, joyful, happy, sad, etc.) and to act upon ourselves in light of that definition, reflecting on and deciding how or whether to express that feeling in a given situation.

In terms of sociology of culture and symbolic interactionism, it is possible to propose the best definition of emotion as a research and study tool. We will use the "four-factor model of emotion" of Peggy Thoits to explain that, although dated, it turns out to be the best model for an interactionist analysis of the emotional experience (see Redmond 2015; Charon 2007; Lively, Heise 2004). According to Thoits, emotions have to be considered like a subjective experience made by four interconnected

components: a) situational stimuli, b) physiological modifications, c) expressive gestures, and d) a definition of the emotion that serves to identify this specific configuration of components. According to the following scheme (Thoits, 1990: 141-142):

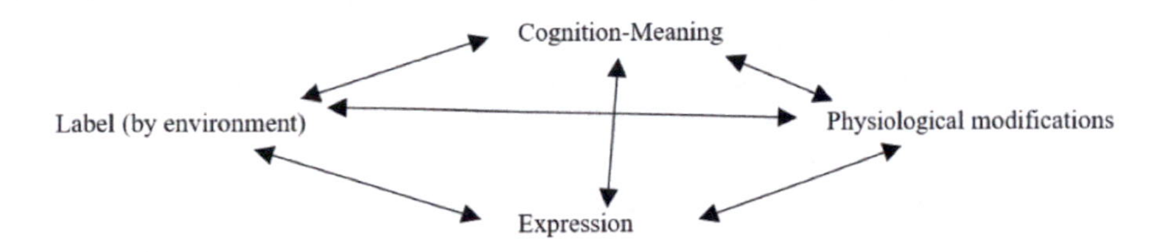

A model of the subjective emotional experience (Thoits 1984, 1989).

In fact, emotion is always culturally defined and categorized: from being identified as positive or negative according to the culture of belonging or the social context in which it is found, up to calling it different names depending on the perspective used by the subject that lives it and the consequent physical expressions he/she uses to manifest it (see Hassin *et al.* 2013; Von Scheve 2012; McCarthy 1992). Thus, what I call sadness can be called anger, melancholy, or even happiness by a friend of mine who interprets or lives the same situation from a different perspective.

It all depends on what is called *emotional culture* in sociology (Gordon 1990, 1989, 1981; see Mills, Kleinman 1988). Emotional culture is directly dependent on the type (or types) of emotional socialization that the subject has experienced in the course of his/her life. The names given to the emotions and their expressions are culturally codified: they depend on the socialization and interaction habits existing within a given historical-social context. The specific culture of an environment (group, city, nation, etc.) influences both the beliefs that people have for emotions and the laws that regulate their own expression. This emotional culture is not, therefore, innate, but it is learned by the subjects in the course of their experiences of socialization, from infancy to the existential paths that are experienced in adulthood. You then become emotionally competent on what emotions can be exhibited based on your cultural origin and the historical-social context in which you act (in such a process, emotional culture can also generate a series of prejudices, for example on the basis of gender construction: women are more sentimental than men, or man is the strong sex and should not cry in public, etc.).

In short, to analyze the concept of emotion in sociological terms, it is therefore necessary to perform a work of multiple interpretation:

1) symptoms and sensations tested according to the event that occurs;
2) social context and historical epoch in which we find ourselves (beliefs, habits, expectations, etc.);

3) emotional expressions implemented according to the biographical trajectory of the subject interacting and/or the feeling experience, etc.;
4) emotional culture of reference (both in the context in which we act, both the subject who acts or with whom we have an interaction).

2. THE (EMOTIONAL) AMBIVALENCE OF EVERYDAY LIFE

According to Georg Simmel's analysis of social acting forms in modernity (Simmel 1908, 1903, 1900), the two poles, between which everyday experience is developed, seem now to be stronger. Between the inescapable need of solitude and individualism and the existential necessity of acknowledgment, of encounters, and of relations with the others, it seems, indeed, that a sort of relevant stiffening is emerging. In other words, the necessary balance for the individual seems to wobble, also because of media and technological process in which the subject seems to become a "murmuring cane": exposed to the late-modern wind, carrying flashes of techno-science and echoes of traditions, the contemporary individual tries to whisper his/her discomfort, using several forms of experience in order to spread his/her emotionality—on the one hand, forms of what I would call "social eremitism;" on the other hand, forms of forced collective identification. This ambivalence represents, indeed, a kind of key entering the folds of the contemporary world (Hillcoat-Nallétamby, Phillips 2011), and it plays a central role in contemporary interactions and in the social building of reality—attempting, in this context, to reflect as well on the role played by some of the media in the transmission and construction of the social emotions.

As Simmel taught us, we would like to be free, but at the same time, we feel the need to have someone next to us. We would like to be alone, but when we really are, we feel lost in the dark (see Haroche 2008). Someone, then, tries an extreme solution in order to compromise with his/her own emotional status. This is the case, for instance, of over a million of young Japanese: they, refusing to adapt to the rules of society concerning their appearances and refusing to consume emotions that they do not recognize as their own, decide to lock themselves at home and to relate to the exterior world just through mass media (Allison 2013; Zielenziger 2006; see Furlong 2008). The enquiry in question sheds some light on the phenomenon called "hikikomori," a Japanese term meaning "the confined" or "the pulling away," identifying those teenagers who have decided to "not live under the sunlight anymore." Withdrawing from the rules of mass society, of the social window dressing, of the consumption of "commercial" emotions and, most of all, of the constant adequacy of social situations imposed by the context of belonging, they have decided to go back by one step and look at the world only through the windows opened for them by the media. This process is defined as a sort of "social eremitism," and I believe that it has to be related to the necessity of putting our own emotional capital at work, that is, to the relation established by anyone with his/her emotions. No communitarian obligations, no labelling fears. Paradoxically, then, these young Japanese choose to

BLACK
またたわく
ながった。
絶望
マリオ

relate to the exterior reality through the virtual reality: personal websites, blogs *in primis*, thanks to the potential social freedom they warrant, are the field within which their emotional capital is shown and, continuously, newly shaped. The virtual reality, then, becomes the "place" for the manifestation of emotional status, for the sharing of feelings, and for the solidarization with the other, free from the fear of receiving an unwanted social label (Benski, Fisher 2014; Döveling *et al.* 2011; Castells 2003; see Balbi, Magaudda 2018). In such cases, "looking inside yourself," that is, to deal with your own consciousness and share your emotions with others, seems much easier than in face-to face interactions. It becomes possible, in other words, to open the Pandora's box of our own emotionality, since the boundaries and the social conventions of other situations are lacking. We can speak, then, of an arrival, where we can attribute a therapeutic value to life on the web, since it is a place within which life is mimed, or really lived, or, to put it another way, a site where life is verified, tested, in order to accept it and in order to dilute the continuous and material brutality and violence (Turkle 2011).

Narrations applied to the web represent one of the main places for the manifestation of emotions—a place for confessing ourselves, listening, sharing, communicating. In any event, then, a place of narration: a social practice in which two or more people put a story in common. What the internet adds to such a definition is a free and unbound character for the narrative act. We could then affirm that, in our late-modern reality, that web narration is a social practice in which two or more people put a story in common and manifest their own emotional status, not giving in to moral or social bounds, if not to the one (purely hypothetical) foreseen by the virtual space. Here is an ambivalence: on the one hand, a sort of emotional drying up caused by the rhythms and by the late-modern contingencies; on the other hand, emotions and feelings find new places in which to manifest themselves. Nevertheless, we should not forget that the internet, in fact, certainly allows a continuous spatiotemporal interaction and a fully free expression, but at the same time it risks to alienate the subject from the "immediate" reality to create a process of googlization (Vaidhyanathan 2011).

3. EMOTIONAL CAPITALISM

The link between emotions and capitalism's development produces a highly specialized "emotional culture" that is dependent on the commercial rules of capitalism itself. From the second half of the twentieth century onwards, capitalist culture seems to be characterized by a double tendency, apparently contradictory: on the one hand, it gives to rationality the main role in human action; on the other hand, it involves an unprecedented intensification of emotional life. Rationality and emotion should therefore go hand in hand in intensifying their presence in contemporary society. So, even the intimate-personal sphere of the individual, by definition quite emotional, follows the rules and models interactions and relationships imposed by a rational, efficient, and productive market, inspired by Weber's theories: a rational faculty used by individuals with the sole aim of accumulating social approval from daily activities, sacrificing even the most intimate and personal moments of existence (Hochschild 2013, 2012). It is the so-called "emotional capitalism" that produces a culture in which emotional and economic discourses and practices mutually shape each other, thus instigating a vast movement in which feelings are placed as essential components of economic behavior and in which emotional life follows the logic of economic relations and exchanges (Illouz 2006; Hochschild 2003). In this context, as it appears in recent enquiries (Illouz 2017; Zelizer 2005), the consumption plays a central role, since it allows the marketing of emotions previously sketched (in the public sphere as well as in the private one).

Emotional capitalism harnesses the irrational part of emotions, placing them at the service of instrumental action: it rationalizes them, taking away what is poetic or unforeseen causing commodification, giving a price, a value, and an exchange value to each individual emotional state depending on the social context in which one finds oneself acting. Although the subjects believe they are free in their expressions and emotional expressions, actually they often reproduce only the emotional cultural models imposed by the capitalist market. This causes, in intimate and private interactions, a strong rationalization of emotions that leads to a separation between felt emotion and the subject that provides it: a reification of emotions—artificial emotions depending on the contingency; "shock emotions" with no time for sedimentation (Lacroix 2001). The individual, then, is no more in contact with his/her own inwardness—always more resembling a no man's land—but he is in contact with an artificial self, shaped by an emotional theatricalization show in order to gain profit. Thus, intimate relationships become "fungible," meaning they are transformed into commodifiable and interchangeable objects on the market of emotional relationships. As if they were struggling with a sort of self-service of emotions, contemporary subjects of the Western world are forced, through the logic of the capitalist market, to get excited on command, using the emotional mask best suited to the socio professional situation in which one acts and the goal we set ourselves.

Modernity—Simmel shows—is ambivalent. Such ambivalence seems to be marking late modernity: the individual, being the protagonist of the increasing rationalization of the public sphere, starts to be recognized as emotional subject but is compelled to the control of its own emotionality, if not to a real repression of emotions. Yet not everybody is able to repress himself. This effort made looking for the balance allowing to go on day after day requires a partial removal from the "authentic" emotional-sentimental side marking the social action, and this, following

Simmel's terminology, makes the subject always more intellectualized (Simmel 1903).

Every day we live and have experiences looking for an interior balance towards a not-too-difficult existence, towards an acceptable everyday life with not too many anxieties. In this process, the individual is inclined to inhibit positive emotions, since they hypothetically lead to "social disorder"—meanwhile, he or she has to adapt to the negative ones, and maybe these are the most influential for human contemporary behaviors (Turner 2011). Not incidentally, already twenty years ago Stearns and Stearns analyzed American society through a historical perspective and talked about "emotionology"—an enquiry and a practice of everyday emotions aiming to keep under control more profound and "authentic" ones, so as to manifest the more suited according to the circumstances (Stearns, Stearns 1985). This behavior appears today amplified and strongly marked by the logic of capitalism and global market.

References

Allison, A. (2013). *Precarious Japan*. Durham: Duke University Press.

Balbi, G., and P. Magaudda (2018), *A History of Digital Media: An Intermedia and Global Perspective*. London and New York: Routledge.

Benski, T., and E. Fisher (2014) (eds.). *Internet and Emotions*. London and New York: Routledge.

Bolton, S.C. (2005). *Emotions Management in the Workplace*. Basingstoke: Palgrave.

Castells, M., (2003). *The Internet Galaxy: Reflections on the Internet, Business, and Society*. Oxford: Oxford University Press.

Charon, J.M. (2007). *Symbolic Interactionism*. Upper Saddle River: Pearson Prentice Hall.

Denzin, N.K. (1983). "A Note on Emotionality, Self, and Interaction." *American Journal of Sociology* 89 (2), pp. 402–409.

Döveling, K., C. von Scheve, and E.A. Konijn (2011) (eds.). *The Routledge Handbook of Emotions and Mass Media*. London and New York: Routledge.

Fields, J., M. Copp, and S. Kleinman (2007), "Symbolic Interactionism, Inequality, and Emotions," in J.E. Stets and J.H. Turner, *Handbook of the Sociology of Emotions*, pp. 155–178.

Francis, L.E. (1994). "Laughter, the Best Mediator: Humor as Emotion Management in Interaction." *Symbolic Interaction* 17, pp. 147–163.

Furlong, A. (2008). "The Japanese Hikikomori Phenomenon." *The Sociological Review* 56 (2), pp. 309–325.

Gordon, S.L. (1981). "The Sociology of Sentiments and Emotion." In M. Rosenberg and R.H. Turner (eds.), *Social Psychology: Sociological Perspectives*, pp. 562–592. New York: Basic Books.

——— (1989). "The Socialization of Children's Emotions: Emotional Culture, Competence, and Exposure." In C. Saarni and P.L. Harris (eds.), *Children's Understanding of Emotion*, pp. 319–349. Cambridge: Cambridge University Press.

——— (1990). "Social Structural Effects on Emotions." In T.D. Kemper (ed.), *Research Agendas in the Sociology of Emotions*, pp. 145–179. Albany: State University of New York Press.

Haroche, C. (2008). *L'avenir du sensible: le sens et les sentiments en question*. Paris : PUF.

Hassin, R.R., H. Aviezer, and S. Bentin (2013). "Inherently Ambiguous: Facial Expressions of Emotions in Context." In *Emotion Review* 5 (1), pp. 60–65.

Hillcoat-Nallétamby, S., and J.E. Phillips (2011). "Sociological Ambivalence Revisited." *Sociology* 45 (2), pp. 202–217.

Hochschild, A.R. (1979). "Emotion Work, Feeling Rules and Social Structure." *American Journal of Sociology* 85 (3), pp. 551–575.

——— (1990). "Ideology and Emotion Management: A Perspective and Path for Future Research." In T.D. Kemper, (ed.), *Research Agendas in the Sociology of Emotions*, pp. 117–142. New York, State University of New York Press.

——— (2003), *The Commercialization of Intimate Life: Notes From Home and Work*. Berkeley: University of California Press.

——— (2012). *The Outsourced Self: Intimate Life in Market Times*. New York: Metropolitan Books.

——— (2013). *So How's the Family?: And Other Essays*. Berkeley: University of California Press.

Illouz, E. (2006). *Gefhüle in Zeiten des Kapitalismus*. Frankfurt am Main: Suhrkamp Verlag.

——— (2017) (eds.), *Emotions as Commodities. Capitalism, Consumption and Authenticity*. London and New York: Routledge.

Lacroix, M. (2001). *Le culte de l'émotion*. Paris : Flammarion.

Lively, K.J., and D.R. Heise (2004). "Sociological Realms of Emotional Experience." *American Journal of Sociology* 109 (5), pp. 1109–1136.

McCarthy, E.D. (1992). "Emotions Are Social Things: An Essay in the Sociology of Emotions." In D.D. Franks and E.D. McCarthy (eds.), *The Sociology of Emotions: Original Essays and Research Papers*, pp. 51–72. Greenwich: JAI Press.

——— (2017). *Emotional Lives: Dramas of Identity in an Age of Mass Media*. Cambridge: Cambridge University Press.

Mead, G.H. (1982). *The Individual and the Social Self: Unpublished Work of George Herbert Mead*. In D.L. Miller (ed.), Chicago, University of Chicago Press.

Mills, T., and S. Kleinman (1988). "Emotions, Reflexivity, and Action: An Interactionist Analysis." *Social Forces*, pp. 1009–1027.

Redmond, M.V. (2015). "Symbolic Interactionism." *English Technical Reports and White Papers*, p. 4.

Sandstrom, K.L., K.J. Lively, D.D. Martin, and G.A. Fine (2013). *Symbols, Selves, and Social Reality: A Symbolic Interactionist Approach to Social Psychology and Sociology*. Oxford: Oxford University Press.

Shott, S. (1979). "Emotion and Social Life: A Symbolic Interactionist Approach." *American Journal of Sociology* 84 (6), pp. 1317–1334.

Simmel, G. (1900). *Philosophie des Geldes*. Leipzig: Duncker & Humblot.

——— (1903). *Die Großstädte und das Geistesleben*. Dresden: Petermann.

——— (1908). *Soziologie. Untersuchungen über die Formen der Vergesellschaftung*. Leipzig: Duncker & Humblot.

Stearns, P.N., and C.Z. Stearns (1985). "Emotionology: Clarifying the History of Emotions and Emotional Standards." *The American Historical Review* 90 (4), pp. 813–836.

Stets, J.E., and J.H. Turner (2007) (eds.). *Handbook of the Sociology of Emotions*. New York: Springer.

——— (2014) (eds.). *Handbook of the Sociology of Emotions: Volume II*. New York: Springer.

Thoits, P. (1984). "Coping, Social Support, and Psychological Outcomes: The Central Role of Emotion." *Review of Personality and Social Psychology* 5, pp. 219–238.

——— (1989). "The Sociology of Emotions." *Annual Review of Sociology* 15, pp. 317–342.

——— (1990). "Emotional Deviance: Research Agendas." In T.D. Kemper (ed.), *Research Agendas in the Sociology of Emotions*. New York: State University of New York Press, pp. 180–205.

——— (1996). "Managing the Emotions of Others." *Symbolic Interaction* 19, pp. 85–109.

Turkle, S. (2011). *Alone Together: Why We Expect More from Technology and Less from Each Other*. New York: Basic Books.

Turner, J.H. (2011). *The Problem of Emotions in Societies*. London and New York: Routledge.

Turner, J.H., and J.E. Stets (2005). *The Sociology of Emotions*. Cambridge: Cambridge University Press.

Vaidhyanathan, S. (2011). *The Googlization of Everything: And Why We Should Worry*. Berkeley: University of California Press.

van Zomeren, M. (2016). *From Self to Social Relationships*. Cambridge: Cambridge University Press.

von Scheve, C. (2012). "The Social Calibration of Emotion Expression: An Affective Basis of Micro-Social Order." *Sociological Theory* 30, pp. 1–14.

Zelizer, V. (2005). *The Purchase of Intimacy*. Princeton: Princeton University Press.

EMOTIONAL SOCIAL ACTS
A SOCIOLOGICAL INTERPRETATION OF THE PRECOGNITIVE CHARACTER OF EMOTIONS

LORENZO BRUNI[205]

ABSTRACT

The aim of this paper is theoretical and regards the social theory of G.H. Mead as it relates to the sociological understanding of emotional phenomena. This is the main hypothesis of the paper: in order to develop the unexpressed sociological potentialities of Mead's theory for the comprehension of emotional phenomena, it is necessary to relocate both the cognitively intentional and sensitive or perceptual-subjective moments of emotion within a broader intersubjective and practical dimension, in some way antecedent and constitutive of both. In the first part of the paper, we will see how Meadian theory has been transposed by two important contributions to the sociology of emotions, those of Susan Shott and Thomas Scheff. In the last part, we will try to clarify in what terms these positions may be considered partially unsatisfactory from the standpoint of a Meadian interpretation of the relationship between the precognitive and cognitive dimensions of emotions.

Keywords: Mead, Emotions, Shott, Scheff, Intersubjectivity

1. INTRODUCTION

The aim of this paper is purely theoretical and regards the social theory of George Herbert Mead as it relates to the sociological understanding of emotional phenomena. Furthermore, this paper is part of a larger effort to construct a Meadian theoretical interpretation of the intersubjective nature of emotions, whose first steps have been taken in some sporadic and brilliant contributions to contemporary social theory (Lupton 1998; Engdahl 2005).

Emotions are composed of a cognitively intentional moment: the emotions are intentionally directed at some object, such as, for example, the fear of losing one's home or the love for Eastern European literature—and a sensitive moment—the corporeal perceptive feeling of them. The hypothesis that this text will try to support is the following: in order to develop the sociological potential of Mead's theory for the understanding of emotional phenomena, it is necessary to place both the cognitive intentional moment and the sensitive or subjective-perceptual moment within a broader practical and intersubjective dimension, within, that is,

[205] Lorenzo Bruni (PhD) is a Research Fellow in Sociology at the Department of Political Sciences, University of Perugia.

a social infrastructure that is in some way antecedent and constitutive of both moments. By moving beyond the simple summation of the subjective-sensitive moment and the sociocognitive moment, the paper proposes a sociological understanding of emotional experience as a fundamentally cooperative experience, never separated, even in its precognitive sensitive component from a social dimension of practical intersubjectivity.

The hypothesis that I wish to demonstrate by reference to Mead's thought is that it is not the definition of the cognitive character of emotions that defines their social connotation (contents of belief, norms of expression, etc.) but that the irreducibly social character of emotions constitutes the basis for their potentially more highly determined cognitive development. In this sense, Meadian theory seems to be a decisive help in pursuing a course of theoretical research that contributes to reconfiguring the direction of the relationship that connects the cognitive character and the social character of emotions. The objective is to demonstrate that it is not necessary to move from the definition of the cognitive character of emotions in order to establish their social character, as has been proposed, among many others within the sphere of the theory of emotions, by Marth Nussbaum (Nussbaum 2001). The necessary move, on the contrary, is to start from the intrinsically social dimension of emotions to then investigate their eventual further cognitive development. Meadian theory is a classic of sociological thought, which still has a great deal to tell us in this regard.

2. THE USE OF THE THEORY OF GEORGE HERBERT MEAD IN THE SOCIOLOGY OF EMOTIONS: SUSAN SHOTT AND THOMAS SCHEFF

Within the realm of the classics of sociology, it can be said that the theory of George Herbert Mead, by concentrating on the centrality of intersubjective relations for the purposes of the emergence of self-consciousness and subjectivity, is a sort of exception (Blumer 1969; Habermas 1987, chap. 8; 1992; Couch 1989; Joas 1997; Crespi 1999; Carreira de Silva 2006, 2008), marking a discontinuity with, among others, Durkheim. Mead submits, in fact (Mead 2015), that the dynamics of socialization and those of individuation are processually coextensive phenomena. Processes of individuation, that is, do not necessarily constitute a residual and anomic moment. To account for this dynamic phenomenon, as is well known, Mead speaks of two components of the Self, which he defines as Me and I. Mead proposes to define socialized individuality—directed, on the one hand, towards social meanings and, on the other, towards itself as an object for itself—by way of the concept of Me. Within the same process of socialization, there also emerges what Mead defines as I, the exquisitely subjective component of the socialized individual (Mead 2015). Subjectivity is not resolved in a mere product of socialization—it does not coincide with the social—but there is a component of subjectivity itself that potentially escapes what has already

been socially established, without, however, being reduced to a solipsistic component. The I defines the subjective component of subjectivity, beyond the social, but is never completely separated from it. Meadian theory, therefore, is both a sociologically typical and peculiar reference in that it is centered on the role of intersubjectivity understood as concrete practical cooperation between codetermined Selves.

Mead's thought has not been canonized only within the sociological discipline understood in the broadest sense (Blumer 1969; Habermas 1992), but it has also found a place within the narrower field of sociological reflection on emotions. We will now examine two positions that constitute two particularly representative moments of the effort to place Meadian theory within the parameters of the sociology of emotions. These two positions are not presented with any pretense to exhaustiveness or completeness but with the primary purpose of pointing out some interpretive limitations in the reception of Mead's thought, so as to then advance a hypothesis with regard to the further sociological potential that Mead's theory offers for the understanding of emotional phenomena.

Let's begin by focusing on the central role played by Mead in Susan Shott's elaboration of the sociology of emotions. According to Shott, emotions are constituted by the combination of two coextensive dimensions: physiological-sensorial and cognitive. Social structures are the necessary dimensions that allow the individual to connote emotions as cognitive phenomena. Put differently, emotions become the object of sociological study to the extent that the physiological arousal is consolidated in a cognitive definition. This passage occurs primarily through the mediation of the social role played by current and effective social norms (Shott 1979, p. 1323). The aspects of Meadian theory that contribute to Shott's sociological interpretation of emotions seem to be concentrated above all on this interpretation: emotions play a fundamental role in social integration/control. This function can be performed by emotions to the extent in which the subject, in the course of interaction, takes on the role of partner. The reference to the Meadian concept of the generalized other is obvious: for Shott, role-taking emotions are those whose expression and manifestation depend on the individual's capacity to make his own the attitude of others towards himself. Emotions—Shott mentions especially guilt and shame—are functional in ensuring a form of self-control aimed at the social integration of the subject. Role-taking emotions may also include, in any case, positive emotions such as, for example, pride. Here too, however, it would seem that the role of alterity is primarily functional to the internalization of normative rules of emotional expressiveness, which, in the end, ensure the social integration of the subject.

Although perhaps less evident than what we have seen in relation to Shott, Mead's influence is also traceable in the reflections of Thomas Scheff, who has devoted a significant part of his sociological study of crying Blocked

emotions to shame. The theme of shame involved nearly the totality of Scheff's scientific production (1995, 1996, 2000, 2003, 2004), but the most significant of his contributions which illustrate his fundamental theses are probably "Shame and the Social Bond" (2000) and "Shame in Self and Society" (2003). In this case too, rendered in extreme synthesis, Mead's influence reverberates primarily in the idea of understanding shame as a cognitive objectification of others' attitudes towards the self. Scheff's sociological theory of shame turns around two large hypotheses. The first is that shame comprises an extended family of emotions. The second is that shame represents a feeling of threat to the strength of the social bond (2000, 2003, 2004). In Scheff, shame becomes an extended family of emotions that includes embarrassment, shyness, feelings of inadequacy or failure, and the phenomenon of humiliation in the broad sense. The common denominator of all of these different emanations of the same root emotion consists in the recognition that shame is to be considered, at bottom, as a manifestation of a threat to the social bond.

3. INTERPRETIVE WEAKNESSES IN THE REPORTED POSITIONS AND FURTHER SOCIOLOGICAL POTENTIAL OF MEADIAN THEORY FOR THE UNDERSTANDING OF EMOTIONAL PHENOMENA

3.1. CRITICAL WEAKNESSES

While it is certainly undeniable that Susan Shott's reflections deserve credit for connecting the contribution of Meadian theory with a program of empirical sociological research, her sociology of emotions seems to be focused selectively on some specific aspects of Mead's theory at the expense of others. Although Shott refers explicitly to the Meadian derived concept of empathetically understanding the emotions of others and how that empathetic dimension gives rise to a double process of individuation and social integration (Cerulo 2018), it would seem, at first glance, that the role of intersubjectivity and reciprocal recognition is defined primarily as a cognitive relationship (1) between constituted subjects (2), and, consequently, it is mainly reduced, even if not exclusively, to a dimension oriented towards social integration and possible definitions of role expectations (3). The decisive technical point is that Shott seems to interpret the extended Meadian concept of social act in cognitivizing terms, underlining its prevalently discursive and symbolic-linguistic connotation. The key point in terms of interpretation is that the Meadian concept of social act is glossed as symbolic interaction, albeit emotionally connoted (1), while, as I will argue in more detail in the following section, the concept of social act, as Mead himself defines it, goes well beyond symbolic interaction. The consequence of this first interpretive operation is that Shott is led to consider intersubjectivity primarily as a sort of containment of possible individuating and disintegrating tendencies, rather than as a resource for open-ended reconfigurations of individual identity (2–3).

As regards Thomas Scheff, whose reliance on Mead, as we have already observed, does not take on the systematic centrality that it has for Shott, it is also possible to underline some hypothetical weak points of his reception of Meadian thought for the purposes of his sociological study of shame. While it is certainly true that shame can constitute a threat to the endurance of the social bond, by undoing the ties that hold society together, it must also be seen at the same time as a sign of a social order that has become crystallized, thus diminishing the chances of open-ended experiences of individuation and shrinking the possibilities for its own transcendence. Compared to Scheff's thesis, a Meadian hypothesis on shame might instead give equal weight to a dual risk, to a threat that is not univocal but ambivalent. A Meadian reading of shame might, that is, identify a double and coextensive risk that comes with the emergence of this emotion—on the one hand, the risk to guaranteeing a stable social bond, as it is contemplated by Scheff himself; on the other hand, however, it would be mistaken to overlook the risk related to the conditions of self-realization. Shame is not only evidence of a threat to the endurance of the social bond, but it can also indicate a weakening of the social possibilities for self-realization, thus compressing subjective expectations of social recognition (Bruni 2016).

3.2 FURTHER POTENTIAL

In Mead, intersubjectivity is not defined only in cognitive and "symbolic interactionist" terms, but above all in practical and contingent terms. The practical and contingent nature of the definition of intersubjectivity within the Meadian theoretical system traces an anti-dualistic overcoming of the subject-object dichotomy, in epistemological terms; of the sensitive-cognitive dichotomy in pragmatic-phenomenological terms; of the society-individual dichotomy in sociological terms. Intersubjectivity shows itself, that is, to be a constant and unceasing source of innovation for social life and not merely a limitation and adaptation to already codified and currently valid social roles and determinants of meaning.
The sociological potentialities not fully explicated by the previously considered approaches can be related schematically to the emphasis of the following aspects of Meadian thought: 1) the pragmatic and social solution of the subject-object relationship; 2) the declination of intersubjectivity understood as practical and cooperative act, processually constitutive of subjectivity and social norms; 3) interpretation of emotional experience as a precognitive interactive and, at the same time, potentially cognitive dimension. The treatment of the relationship between the individual and social dimensions that is so peculiarly characteristic of Meadian theory stands out against the background of its broader reflections devoted to the pragmatic solution of the relationship between subject-object, between sense perception and objective world. There is not enough space here to go into these aspects in depth. That which according to Mead leads to the formation of symbolic meanings and the elaboration of our fully cognitive social practices is already an original and constitutive perceptive dimension of a social and interactive nature, operating in the dynamic relationship between organism and physical world. Cognitively reflective and linguistic human communication "is

rooted in the social nature of primitive impulses whose affective content is represented by emotions" (Mead 2010, p. 7). Physical sense perception is already social (Mead 2010).

Sense perception is produced in acts, in social conduct, Mead says, and emotions are a central component of social conduct. Social conduct is practical and precognitive, but at the same time, it is already fully social. The interactive material of emotions constitutes an unreflective social conduct, which precedes the specialization of this same material in symbols and cognitions. It is, therefore, already social even if not fully cognitive (Mead 2010, 2015). There is not a dualistic dilation between sense perception and social reflectivity. Thus, emotions have a processually intersubjective unfolding; they are the outcome of a continuous decentralization and of an unceasing disposition to take on the point of view of one's partner in the interaction, even if just by anticipating it through imagination. Emotions emerge and find their open expression in the pragmatic development of human activity, as the anticipation or imagination of continuous reciprocal responses (Mead 2010a).

Mead's theoretical reflection is radical: every act, even a nonsocial act, is characterized, therefore, by sociality. Primitive consciousness, even of the physical world, is social, Mead affirms in "The Social Character of Instinct" (Mead 2010). Consciousness, insofar as it is continuously social, finds its first base in socially emotive intersubjective communication. Emotions thus play the role of anticipating the response that the organism is about to deploy towards the stimulus, in such a way that the organism to which that response is directed can in turn respond. Emotions assume socially meaningful character as the social basis from which a more abstract meaning can potentially emerge, since only an intersubjective reference to others' reactions has allowed them to emerge. As has been seen, moreover, emotions are socially meaningful in that they emerge intersubjectively in a reciprocal fine-tuning of gestures and since they are bound to a component that is not only physiological nor wholly intentionally cognitive.

4. CONCLUSIONS

Emotive conduct is a constitutive part of social conduct. Social conduct is definable as a practical intersubjective dimension of both a sensitive and cognitive nature, and it cannot be understood only in reductionist terms as the social cognitivization of presocial sensitive dimensions. The expression taking the role of the other, as it has been taken up in some contemporary sociology, brings with it, then, a sort of misunderstanding of what Mead appears to mean. The social process of taking the role of the other is not the mere mechanical adhesion to what has already been socially integrated, but rather is always bound to the practical and social sensibility for what the other may do—to us, with us, or against us. The identities of the subject and his interlocutors are never completely determined before the social act takes place and unfolds, since it is precisely

by way of the intersubjective affective-social relationship that those identities emerge and develop.

With this contribution, I have tried to illustrate the kind of approaches to interpretive digging that the fundamental points of this classic theory still allow us to glimpse in terms of the sociology of the emotions. Emotions must not necessarily be associated with either an individual / sensitive axis or a social / cognitive axis. The social dimension of emotions can thus continue to be interpreted, with the assistance of Mead, as an intersubjective infrastructure of social life. The sensitive and the potentially cognitive coexist in a nondualistic way by virtue of an intersubjective infrastructure of social life. As recently reiterated by the philosopher Michael Tomasello, referring explicitly to Mead's theoretical legacy in his much-debated studies in evolutionary anthropology, it is not possible to reach cultural and linguistic cognitive processes without reconstructing their preexistent cooperative social infrastructure. Emotions constitute in turn the infrastructure of that social infrastructure. They are its integral component, which lays the groundwork for more fully cultural and cognitive social processes (Tomasello 2014). Insofar as the present contribution is partly inspired by philosophical anthropology and social theory, it aims to indicate some possible interpretive approaches for a pragmatic sociology of the senses that dialogues with the sociology of emotions.

References

Blumer, H. (1969). *Symbolic Interactionism: Perspective and Method.* Berkeley: University of California Press.

Bruni, L. (2016). *Vergogna. Un'emozione sociale dialettica.* Napoli-Salerno: Orthotes.

Carreira De Silva, F. (2006). "G. H. Mead in the History of Sociological Ideas." In *Journal of the History of the Behavioral Sciences* 42 (1), pp. 19–39.

——— (2008). *Mead and Modernity: Science, Selfhood, and Democratic Politics.* Lexington: Lanham.

Cerulo, M.(2018). *Sociologia delle emozioni.* Bologna: Il Mulino.

Couch, C.J. (1989). "Comments on a Neo-Meadian Sociology of the Mind." *Symbolic Interaction* 12, pp. 59–62.

Crepsi, F. (1999). *Teoria dell'agire sociale.* Bologna: Il Mulino.

Eengdahl, E. (2005). *A Theory of the Emotional Self.* Örebro: Örebro University Press.

Habermas, J. (1987). *Theory of Communicative Action, Volume Two: Lifeworld and System: A Critique of Functionalist Reason.* Boston: Beacon Press.

——— (1992). "Individuation through Socialization: On George Herbert Mead's Theory of Subjectivity." In *Postmetaphysical Thinking: Philosophical Essays*, pp. 179–204. Cambridge: MIT Press.

Joas, H. (1997). *G. H. Mead: A Contemporary Re-examination of His Thought.* Cambridge: MIT Press.

Lupton, D. (1998). *The Emotional Self: A Sociocultural Exploration.* London: Sage.

Mead, G. H. (2015 [1934]). *Mind, Self and Society: The definitive Edition.* Annotated by D.R. Huebner and H. Joas. Chicago: Chicago University Press.

——— (2010), "The Social Character of the Instinct." In Id., *Essays in Social Psychology*, edited by M.J. Deegan. New Brunswick and London: Transaction Publisher.

——— (2010a) *Essays in Social Psychology.* Edited by M.J. Deegan. New Brunswick and London: Transaction Publisher.

Nussbaum, M. C. (2001). Upheavals of Thought: The Intelligence of Emotions. Cambridge: Cambridge University Press.
Scheff, T. (1995). "Shame and Related Emotions: Overview." In *American Behavioral Scientist* 38 (8), pp. 1053–1059.
——— (1996) "Self-Esteem and Shame: Unlocking the Puzzle." In R. Kwan, *Individuality and Social Control: Essays in Honour of Tiamotsu Shibutani*. Greenwich: JAI Press.
——— (2000). "Shame and the Social Bond." *Sociological Theory* 18, pp. 84–99.
——— (2003). "Shame in Self and Society." *Symbolic Interaction* 2, pp. 239–262.
——— (2004). "Elias, Freud and Goffman: Shame as the Master Emotion." In S. Loyal and S. Quilley (eds.), *The Sociology of Norbert Elias*. Cambridge: Cambridge University Press.
Shott, S. (1979). "Emotions and Social Life: A Symbolic Interactionist Analysis." *American Journal of Sociology* 84 (6), pp. 1317–1334.
Tomasello, M. (2014). *A Natural History of Human Thinking*. Cambridge and London: Harvard University Press.

SOME OF THE THINGS WE KNOW

ALBERTO ABRUZZESE

1. THE BIT ROUTE

With human existence historically still within the linear and historical time of human existence, the bit road is its latest and most adventurous "spice road." As with then, it is not only the precious, sophisticated nature of highly innovative products for everyday life that is at work, but above all their circulation in the world and the progressive richness their application provides.

Nevertheless, by means and because of digital languages, it is linear time itself that unravels: before the squares and theaters and their great communal ceremonies, and before the highways of mainstream media the streams, swarms, and nodes of the networks come into play, dragging with them the forms of historical time and the accumulation of their systems.

This inconsistent and ambiguous transition is inevitable when managed and experienced by social decision-makers and players in competition and conflict with one another, often with unrelatable experiences. In the face of this, is it still reasonable to try and express judgement on the first thirty years of the internet? To reason in terms of places and years over such a vertiginous, boundless crushing of time and space? Can this portion of time still make any sense steeped as it is in calendars and periodization appropriate to wars, states, and parliaments? Thirty years of internet means having something to say about those who have been using it and how, but much less about what has really happened to it, mainly without our knowledge. Is it about speaking "within" a time frame superimposable with those usually practiced in the lexicon inherited by the culture of the modern? According to this time-obsessed lexicon, the noun *era* indicates a period of years, centuries, or even millennia marked by an event that in virtue of its catastrophic nature has turned upside down the preexisting perception of the world. The reverse side of catastrophe is apocalypse, *revelation,* so do these thirty years of the continuous advance of the internet already mark a new beginning or are they still at the end of an era? Might they even mark a future with no other era than an absolute present, without a possible end to an era? A time when everything has been revealed?

The internet has been preceded by the technoscientific hope to free the world from the recent horrors of the first half of the twentieth century: to liberate the world by no longer calling upon the "what can we do" of ideology and politics but by turning to the virtuous resources of information technology, nonhuman resources and therefore without defects and emotions—resources so precise, efficient, and clarifying compared to the redundant noise of the real world and the interferences wreaked upon it by human passions. The internet has never realized such a project, nor could it have done: it was too ambitious to suggest that its advent would mean the end of philosophy itself. That utopia of a peaceful and fertile coexistence of peoples, modern and postmodern, both technological and hyperhumanist, has if anything encouraged the ordinary routine of war and peace (between nations, capitals, markets, culture, peoples): different regimes of power with the same need of resources based on the possession and management of the media. In accord with the colonial spirit of American civilization, the boundless space of the internet was immediately called the "new frontier"—a frontier made in the image of that "first" one, the Far West: virgin lands, ownerless and lawless, violently snatched from their bison and indigenous peoples. A few centuries before, these territories were called "new world" by European conquerors. Magic lanterns, the imaginative technology from which cinema would be born, had also been defined as "new world."

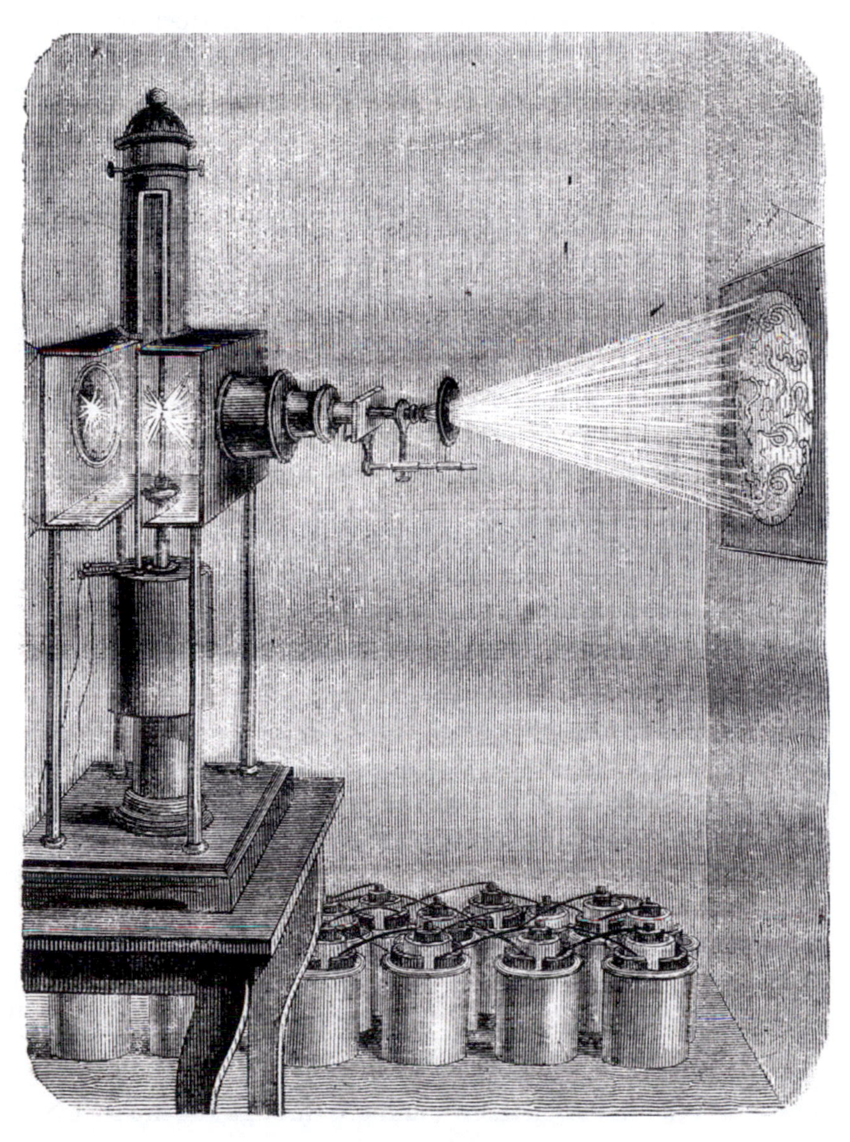

The internet therefore grew thanks to infinite social and personal reasons finding their most fertile and reactive outlet in information technology. But it also developed in unison with the proliferation of new local wars, now of global interest, waged with every kind of material and nonmaterial weapon—a strange kind of permanent world war, permanent but breaking out in temporally and territorially periodical flare-ups, always an umpteenth war compared to the peace of the previous day or the day itself or even the day after.

It is here that the very nature of the West unfolds from being the world of a conflict with no possible end if not by its own extinction—a West that does not die but always sets upon itself, over its own ruins: a spiral movement rotating more and more fully, completely but pointlessly, around its own axis of values. Even the recent brazen formula "peace wars" has been desperate, an expression of it as if it was as semantically "happy" because it implies a recognition of peace as a form of dominion, a weapon. Do the digital networks therefore have the same amphibian nature, the same double vocation, as the "pacifier" and therefore for this reason the "imperialist" of the world-system we call the West, and are they its latest manifestation?

If by "technique" we mean the way humans in their social systems increase their own power and satisfy their need to assert themselves, then I cannot be entirely convinced by the idea that the spirit of innovation, any social innovation, more or less powerful and territorially "invasive," is not in itself violent. In other words, I think innovation can never be completely outside the spirit of weapons. If, as political scientists used to say, war is the highest form of politics, then without forcing the issue, the opposite theory can be ventured. The globalization of politics, facilitated and introduced by innovative networks and their languages, is the highest form of war. I think a proof of this thesis lies in the fact that today the planet shares a war where the firepower of democratic regimes and that of antidemocratic regimes face each other but contaminate each other. Their different but equal desire for power drives them to use the same weapons.

2. SHOULD THE INTERNET BE CELEBRATED?

Is it celebration when we pay tribute to this innovation as we greet the arrival of an honorable guest visiting our house for the first time? Or when, according to ancient tradition, a crowd of citizens and a host of dignitaries were called to open the gates of their city to a sovereign from faraway unknown lands, showering them with symbolic gifts? The most reliable etymology of the adjective "celebrated" derives it from the roots which mean "to frequent, possess, live in, be of importance" and "to exalt" or "to glorify" and finally "to be on everyone's lips." Thus, in answer to the question whether the internet should be celebrated or not, I reply with a paradox: it is the world expressed by human beings' desire for power to celebrate itself more and more through the internet—to have found in the internet the most advanced form of self-celebration, more efficient and penetrating but also more ambiguous than ever before. It is a double-edged sword. Yet today we still puzzle over it as if our systems had been thrown into a sudden state of emergency, as if an event had fallen on us, that "us" we can also call public opinion: *hic et nunc*, here and now.

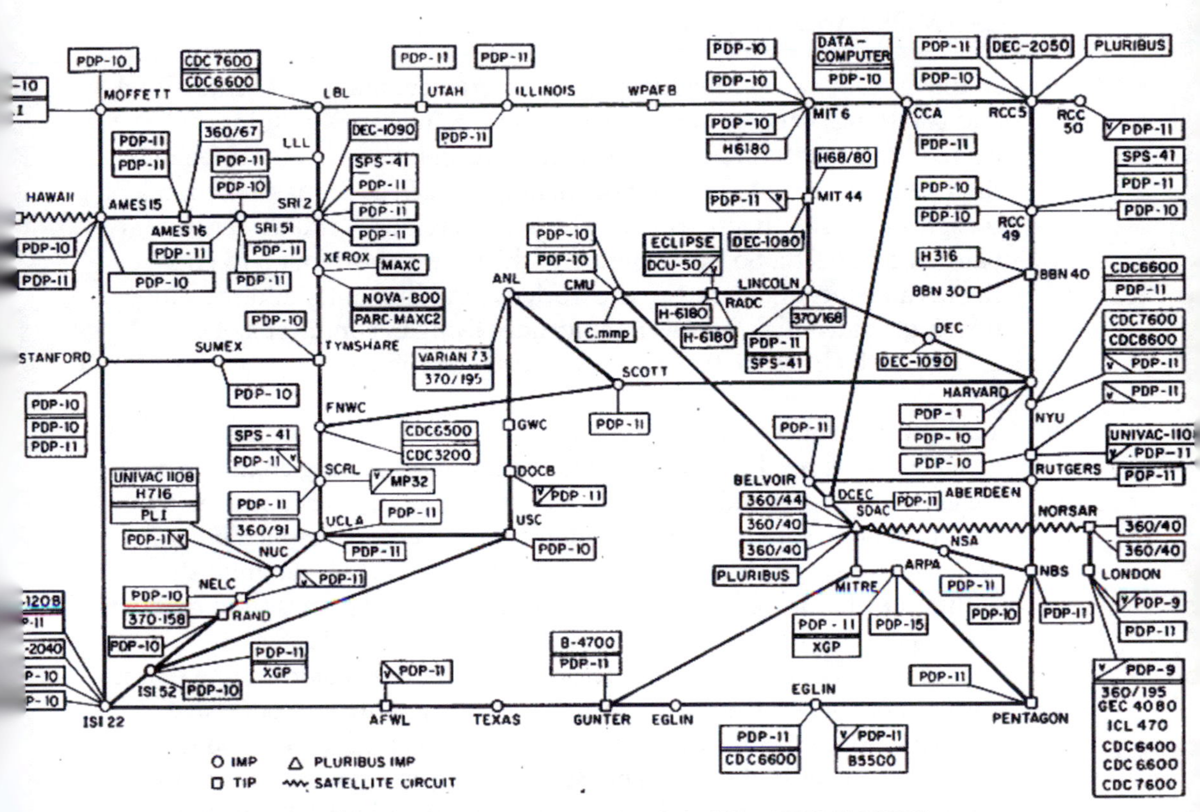

We have been speaking about the information technology revolution for decades even if it has been happening at great speed. Still today, most traditional media continues to speak about it as if it happened yesterday, going over the same questions again and again and—which is worse—over the same answers or at least the same doubts. Yet in the face of the communication explosion in a land no longer flat, plain, and plannable, but spherical, no longer tridimensional, deep but also of the same size and intensity, someone still straddling the society of the book and that of the new digital languages manages to be fascinated by the opposition of real and virtual as extreme relativization and then dissolution of the values of modern reality. A "way out of the world," from its unbearable weight. Yet equally sensitive authors—I am thinking here of Pierre Lévy—soon found the way to come back into the world, to read it again and to regenerate it thanks to the power of technology. It would be helpful to go back to a literature on the new media where a complexity of views, hard to find today, could be gleaned. Perhaps the invitation to celebrate thirty years of internet could be understood through this interpretation, by going back to the classics that preceded it and discussing the relationship between culture and technology: Walter Benjamin was one of these, an intellectual escaping Nazism and who committed suicide.

3. THE "US" SPEAKING INSIDE OURSELVES: "NO MORE AND NOT YET"

In thirty years, each individual person experiences many incidents. Good and bad things happen, sometimes traumatic and overwhelming, expected or unexpected. But this person's face and manners, their feelings, their general way of being, do not change as much as to make them unrecognizable. That happens only after a very serious accident, serious impairment, or death. So, can a period of thirty years be enough to change the ins and outs of the part of the world we call humanity and therefore not single individuals but even multitudes living in conditions of space and time which are so disparate, so *unjustly* distributed?

To answer the question—how and for how long the internet has marked an advent—we need to keep in mind the meaning of a big change, one we greet as epoch-making, -shocking, and -upsetting, perhaps even before it becomes so. To answer this question and accept responsibility for doing so, we have to ask ourselves what process it is that distinguishes it within the plurality of ourselves and of the things we possess and that possess us. This begs the question of what it alludes and refers to. Is it a simple curvature or contortion, retortion, of the world along a line that is always more or less the same, always consequent, always progressive compared to itself? Or, when its direction is perceived to veer too much, snap, stop, hover in the void, and then, turning around, it seems to move towards an opposite and unknown direction, should we ask ourselves if it is not actually a *refounding* instead? And then, recognizing it or believing it to be a refounding, we can finally say "no more": the world is no longer as it was; we are no longer as we were.

But can we really say "no more," or do we immediately tend to confess "not yet"? It is an age-old dilemma and has its late-modern tradition. It was a question posed by the most authoritative humanist culture, witness to the unlimited violence of the modernizing processes of the twentieth century. It was when the formula of "no more and not yet" circulated in Europe, marking a phase of human civilization tragically suspended on the edges of an indecisive and ever more undecidable present, a present produced by the Holocaust and the atom bomb. It was a present poised between the certainty that the past was, or at least should be, forever extinct and yet the fear that the future had not yet come about, had not yet turned into the present. Still today the modern world lives with the uncertainty and suspicion of the "not yet": meanwhile, much if not everything continues to be as before.

To name and assess our time, we have to decide on the indicators to use: whether to take the usual road of frontal oppositions, dialectic mediations, or come out of it altogether, to attempt a way of thinking that remains outside it. It is precisely at this point that while trying to find some representation to help me express this way out, I came upon a rather special trail. It was the adjective *chiral* and its etymology, which goes back to the Greek word χείρ, or *hand* (*manus* in Latin). With the verb "smanettare" (*to use our hands to mess around with*) we mean using the computer: a "what can we do," the application of politics as the art of living, the ability to handle things of the world in a different way from their usual *maintenance*. But we can only begin discussing this after stating certain premises.

4. THE MOEBIUS BAND: ORDINARY MAINTENANCE OF THE WORLD

By "ordinary maintenance," I mean the collection of processes for conserving, continuing to give time and space to the same dominant content however protean: their belonging to the same desire for power and violence by human beings. By "maintenance," I therefore mean the incessant work carried out by the institutions of thought and their social practices in order to guarantee long-lasting life to Western tradition. Thus through the strategic, targeted use of its own partisan conflicts, its greater or smaller epochal crises, its regime changes—an affirmative use (of survival and development) of human civilization thanks to the dialectic and implacable synthesis of affirmative theories (visions and positive movements) and negative theories of society. Affirmation and negation, between divergent religions, different politics of sovereignty, human work, capitalism, and technology, coagulate together in one substance, one quality.

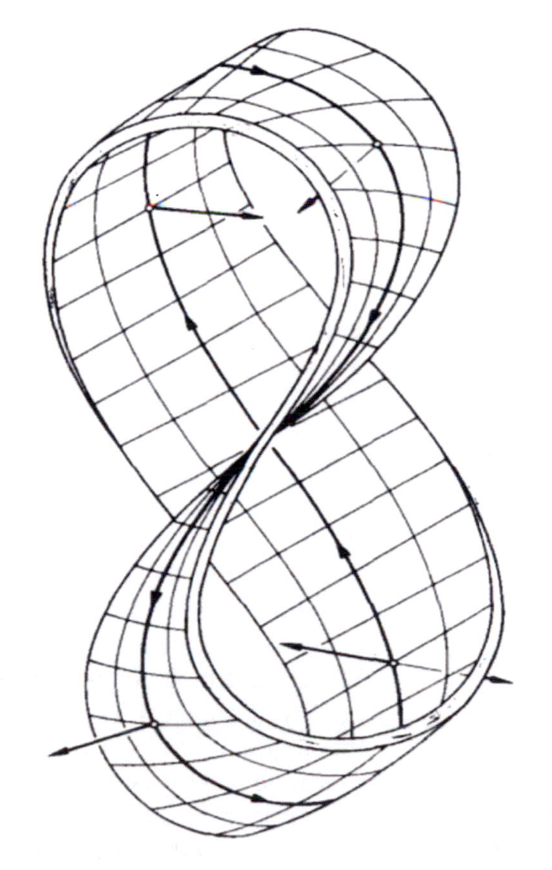

The formula "no more and not yet" mirrors all the social systems engaged in promising human happiness in the name of a basically unfounded hope, a promise, a "principle" which cannot become reality. This is the truth "par excellence," never *actually* denied through history, though always fed anew by it and considered its real inspiration. All sides in conflict comply with this "false conscience" that undermines the foundations of any model of human existence offering to be a solution to violence due to its state of necessity and survival. This is a condition of life in which every ordinary material need succumbs to the excess of desire—outsized excess, immense, "off-loaded" by modern politics onto capitalism. This is the variously recurrent theme in the juxtaposition between goods that are necessary and those that are superfluous, between authentic needs and induced ones, as well as between the diversification of fashions and the standardization of people.

To tackle the dilemma between "no more" and "not yet" that grips and blocks the identity of the modern individual, it might be useful to accept the meaning of a paradoxical object such as the Moebius band. It is a band with only one border, only one side; its right side always leads to its apparently wrong side and then to its apparently right side again: an infinite loop. Unsurprisingly, it is a figure much loved by art because its "real beauty" lies not in promising something but in declaring its "impossibility." Can we therefore distance ourselves from the "hope principle" and from the "false conscience" to which the humanist tradition of religion and nations has accustomed us—nations arisen precisely from that excess of desire that has taken the name of capitalism and the West? Can we finally confess to ourselves that at every refoundation of human civilization, any substantial leap in quality, any promised land, does not and never will happen?

Every refoundation, entailing the intention to force the world to deny its own nature and ours, not to see and not to accept its infinite loop, is an act of violence on itself and by itself on others, an act that reveals its same nature, the nature it has in common. Every refoundation has always meant violent crimes, suffering, and death: the burgeoning humanity of Cain and Abel, the monotheism of Moses, the imperial Rome of Romulus and Remus, the Jewish and Christian God, the reason of the Enlightenment, the fury for identity of Romanticism, the subject of modern revolutions, Hitler's pure race, Hiroshima's peace. The desire to refound the world beyond the world *as it is* harbors the same violence it claims to free itself from. All this shows how wondering about the meaning of the internet forces us not to limit ourselves to simply considering the latest novelties of communication media. It involves the act of feeling responsible for the human condition as such. The internet really is a great leap in quality: if the term "revolution" did not smack so much of modernity and ideologies of progress, and the term "beauty" did not have such a whiff of aesthetic sanctity, we could say that the internet has been, or better still promises to be, a beautiful revolution. In the most sophisticated vocabulary of twentieth century philosophy, this traumatic leap from one regime of meaning to another is known as *permutation of values*, or more scientifically, *paradigm shift*.

5. MESSING AROUND WITH OUR HANDS

Captured by the paradoxical reality of the Moebius band and the particularity of the hand, where the left is just like the right limb, symmetrical but not superimposable, I would like to get to the crux of the matter and give my view on these thirty years of digital life by saying something about the word "mano," *hand*, etymologically present in many compound words. Starting from the jargon version of "to manipulate," "smanettare," or *to mess around with your hands* (a familiar term to users of media devices such as personal computers, mobiles, and tablets, with all their numerous applications), to arrive at the exact opposition between maintenance and manipulation.

The connection to the function of the human hand as a fundamental technological stage is very useful in understanding what happens to us when we work on the computer and what we produce, consume, and reproduce with this work. As we know, reference to the hand is a classic node in the theoretical reflection on the relationship between living beings and technology. It is at the root of this relationship, well before social relationships and even before the so-called human being. The "history" of the hand allows us to understand the metamorphosis of the organic life of a body, the organized life that a system can undergo within so as to better equip itself to face the positive and negative stimuli received from the outside world. The genesis of the hand shows us that it introduces us to metamorphoses born thanks to strictly relational processes. In the concatenation of psychosomatic transformations of an animal equipping itself better and better to shelter from the dangers of an inhospitable world, the advancing of the hand has made an extraordinary contribution to the shaping of the progressive coming into being of Homo sapiens. In virtue of its specific executive qualities, it has continued to modify its own functions with the modifications of the functions carried out by the human body as a social actor. This is exactly what happens with the development of means of communication: a single medium continually redefines its functions to fit the development of the media system it belongs to.

Interacting with the surrounding environment, such a relationship is conceived much more deeply compared to the mechanistic idea according to which the relationship between the human body and technology is between two dominions, two separate distinct parts. Conversely, it is presented as one single process of transformation of work necessary to the reproduction of the environment through itself—an environment that will only become truly "surrounding" when, thanks to the conflicting nature of their specific needs, human beings will be able to see it, name it, and have the power, even if within certain limits, to dominate it. It will be an environment that has acted as a placenta to the human body, given birth to it, and fed it thanks to the same series of innumerable technical intermediations with which a fetus is born from the full reciprocity between the birthing body and the body birthed. The separation of the latter from the body that conceived it, hosted it, and fed it, is purely apparent.

The hand is one of these "births." Formed to put pressure on the points of resistance and hostility arising from the body dealing with itself and with the things it must make habitable, the hand is the most suitable tool for satisfying needs of survival and desires for supremacy. The hand demonstrates the work the body of a four-legged animal has done to transform the technology of its own foot into a more highly evolved technology. Achieving a prehensile function, it allows the mouth to pursue other aims and to fulfill other needs. We know very well, and there is a vast body of literature to prove it, how much human beings owe their access to language skills to this technological leap. Freeing them from the limits and confines of the condition of an animal, they enabled humans to access that progressive dematerialization of living realized in the technology of communication: from cave drawings to the word and writing, from photography to radio, right up to the information revolution. The word *territory* is the definition of a place which is actually always interconnected between the potentiality of the human body and that of the outside world, one that lies in opposition to it, resists and collaborates with it. Individual body and collective body: we talk about the body of a god, a king, an individual, and an army. Community and corporate dynamics have been the consequence of a metamorphosis of the world due to its progressive subjection through a continual flow of vast processes. From the hand to the mouth to the thousands and "a thousand plans" of network communication, again and again it has been a matter of powering a territorial expansion of the subject, becoming and gaining territory.

This is and has been the grand plan, design, project, of a Westernization of the world rendered feasible by the hypertechnological nature of human wish for power.

Moving away from this interpretation, the recurrent ethical question of technique collapses and even crumbles away (where this, as separate and autonomous means with respect to human beings, would only be legitimate if it expressed a socially legitimate end). The distinction between what good or evil is done by the Western individual disappears and has to be entirely reformulated. It is precisely these categories of good and evil that individuals have built for themselves, the ruptures and synthesis of their positive and negative dialectics, that have built them as a technique, to build their vast striking power over the existence of the world—to form their capacity for dragging every single person into their own destiny, whatever their creed and their passion. The same form of fundamentalism lives, covertly or overtly, in every one of their religions of Western life.

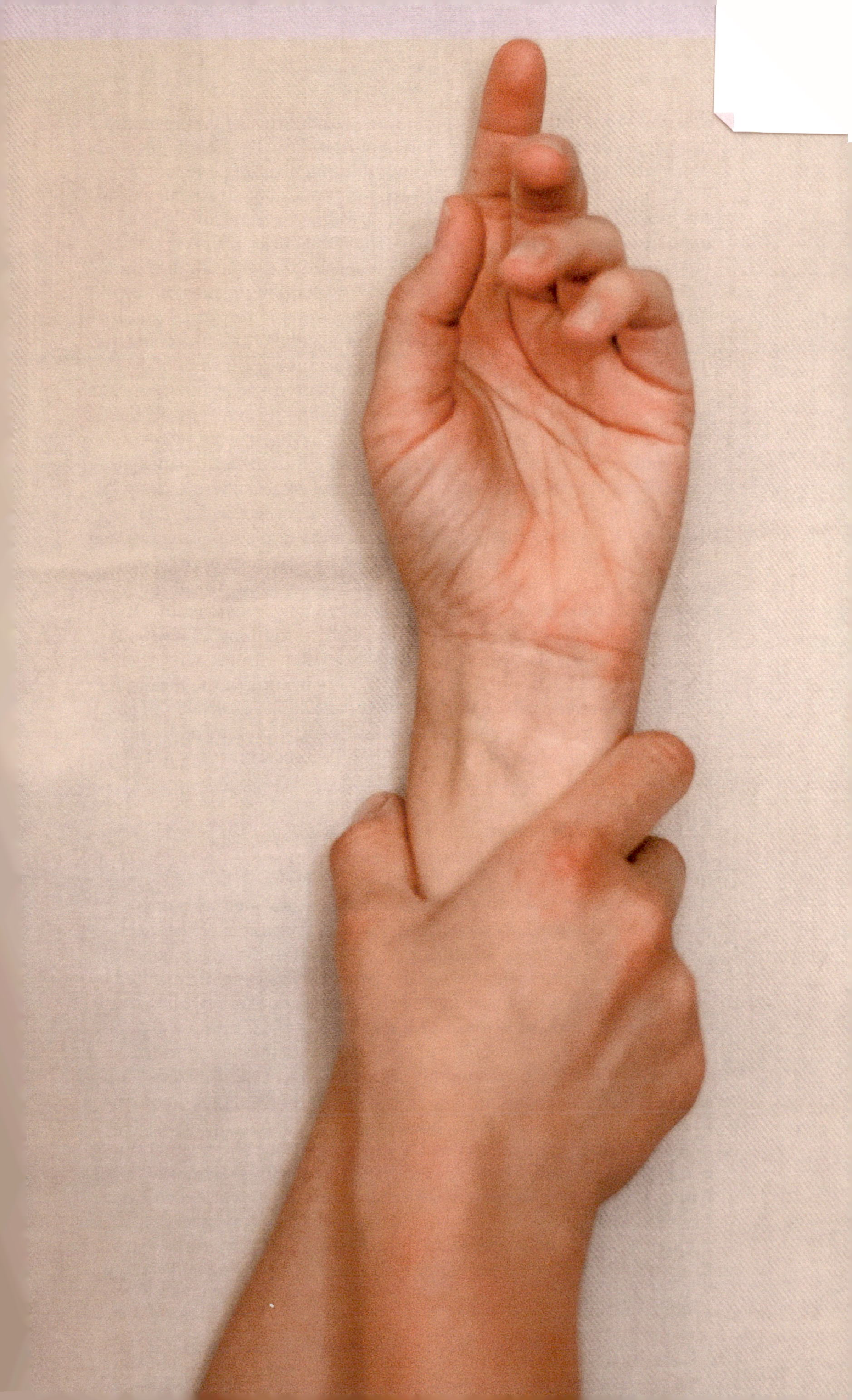

Messing about with our hands, therefore. Jargon words that break into ordinary language are a real revelation of how difficult it would be, if not impossible, to use other terms without diluting meaning. Messing about with our hands is one of these expressions: it does not stress the means, the technology, but the hand using it, its affinity with technology. It highlights not the unknown nature of the "black box" of the computer but the wholly habitual nature, habitudinal and domestic, of the hand. "Messing about with our hands" refers not to something alien but intimate, "animal." At work in these neologisms of rapid territorial success is the sphere of the unconscious. There is the aggregation in the collective imagination of instinctive sensibility, and for this reason viral, of every single person in front of something unprecedented and at the same time emotionally recognizable as something desirable. It is recognizable as an emotive condition that perceives it as always having existed before it in time and space. There is a famous song from 1961 that says "mani senza fine," *hands without end*. Romantic stuff in the erotic sphere: exactly the cultural pool Facebook is born from. And it is a set phrase like this, "hands without end," that helps us understand our personal engagement online. Glorified as it is on the sentimental and amorous horizon on which it floats, this little "refrain" is collective work, reciprocal, and now information technology, algorithmic, between body and machine. It is the regime of a sense that we need to go back to thinking of when not only the hand but also the eye and the voice will no longer serve their specific purposes. The aim will have as it were overtaken the means, its bodily prosthetic: it will be the territory of neural networks. It will be something more than the expanded "flesh" which, announced by an ever more marked attention to the intensive and extensive tactility of the skin, today we call "the internet of things."

Perhaps we should remember that one of the greatest ideas from Western philosophical thought goes back to Eros. The Judeo-Christian and humanist axis, thanks to which modern individuals learned and educated themselves, is founded on a creative force, a system of desire, that has managed to develop and perfect, through its own collective ideologies, the most stable and stabilizing principles of traditional modern society, that of human solidarity—from democracy as a regime dictated by love for the other, to translating and then betraying and revealing those same values in ever more dynamic, more laissez-faire and sovereign principles of globalization. That is a more and more dehumanized dimension of the eroticism of goods. A world the most moralist or more opportunistic say is full of temptations without wanting or realizing that temptation itself is another way of evoking hands as tools of communication of primordial, archaic origin and of undisputed dominion. The temptation comes from searching by touch, exploring by touching—games with the hands, in fact—as with sport and gambling ever more widely broadcast from "television networks," expanded by every means and everywhere on interactive digital networks. Here the use of pornography, more and more widespread on the networks, is the continual staging and interior revelation of a thin interface dividing the society of eroticism from the unsocial nature of the real pleasures of the flesh. I think we have to insist on temptation as the life force of the hand and with it of bodies and their flesh. Together with manipulation, temptation is the object of severe criticism from those who understand them as forms that adulterate and betray a historically and socially given reality, this strategically built and recognized truth that the powers, institutions, decision-makers, and mediators of the same world they belong to attribute to reality. But all these handiworks can find a radically different collocation if instead of being referred to by the conservative and restorative idea of maintenance, they are adopted and adapted to the subversive idea of manipulation: the heroic era of the arrival of the computer revealed rows of people messing about with their hands who wanted to break its "black box," get inside it, to access it freely from "the outside." The boundaries between digital experience and the experience of drugs has become blurred; Western and Eastern values contaminated each other; hackers build ethics of transgression from socially imposed ethics.

6. *PIAZZA MANO PIAZZA, CI PASSÒ UNA LEPRE PAZZA*

(An Italian nursery rhyme similar to the rhyme in English "This little piggy," where each finger has a role.)

Within the framework of one of the first great cultural promotional events of the bursting social impulse of the networks (by no coincidence *I Cibernauti*, 1994, in Bologna), I decided to talk about the net with reference to an old nursery rhyme for children: "Piazza mano piazza, ci passò una lepre pazza. ... Square, hand, square, there went across it a crazy hare." The square is the device of the urban civilization of the Renaissance, center of sacred and profane power: crossroads of the escape and access route between the inside and the outside of the city walls, of its ordered regime—harmony of anthropomorphic perspectives that was a map for daily life until the latter was invaded by the territorial dimension of the metropolis, the intrusion of the foreigner, the crazed crowds. The hare then tears around and all over the place, driven by fear and survival instinct, incarnating the living.

Then going from the public sphere of the city, with the square as its symbol, to the private sphere of the family, the reproductive cell of human existence, the nursery rhyme tells us the working of each finger of the hand, each with its sensory function, concluding "and the little one eats from the plate." Certainly then I could not have imagined that gastronomy would reach both its terminal feature and the beginning of its current symbolic dimension, educational but dissipating, but the nursery rhyme is really delightful for those who share the teachings of McLuhan. Anyway, in times such as those, there was the direct juxtaposition between hand and *square,* which is what I wanted to point out. There where the square was changing more and more into television life and into a virtual square, and the hand was slowly becoming the ever more perfect tool of the remote control, the history of digital networks began.

Perhaps instead of giving my opinion on thirty years of internet, instead of passing judgment, I should give evidence of how messing about on Facebook and other devilries of the time-space present is increasingly transforming from within itself every social relationship and the very everyday life of people. I say "evidence" because the best approach to hitting the mark of the range of digital languages is that of starting from how and how much I have myself changed when using the expressive platforms in which I have decided to live and, I could say, have lived in me. Research and monitoring of quantity and quality, results and motivation going on in social networks, are growing all the time these days, but I think it makes more sense, and perhaps has more strength, to confess the "minimal" experiences that I have progressively realized I feel by getting used to living on Facebook.

For example, a more and more acute sense of intolerance for capital letters and punctuation: unexpected for me, and even unpredictable and contradictory, because at the same time I try not to give up, or I cannot give up, on my habits of good style and complexity of content. But I have come to understand that I am troubled, tested, by those who have immediately become familiar with writing incorrectly and in a simplified, elementary, and emotive way, because I have always suffered with every rule of lexicon and syntax of academic writing. In other words, I have realized that I was starting to speak with my hands. Moreover, I realized I was starting to begin to think in a conversational, choral way the more I was writing in a collaborative way (the interval between I who say and the other who answers can be so small, and at times before I have even spoken). The future of such mental changes is incalculable.

At school, books and teachers are unlikely to let you go your own way. The point is that the net on one hand gives me access to a really unlimited quantity of information and text, but on the other it pushes me to communicate like an idiot, helping me to forget the education I received, stripping me of the rules the role given to me by society imposes. Traditional culture must step in—the culture of the alphabet and the image, also linked to writing and social narrative. This emerging ignorance is called returning ignorance, or the fruit of progressive regression from the levels of literacy gained at school and from its tools of classroom and book. It seems to me that this ignorance could instead be opening before an objective full of promises.

It could be better defined by returning to the idea of culture: a culture long submerged by civilization, based on elementary questions, instinctive needs—a culture able to face the mind-blowing growth of complexity of modern knowledge; a self-reflective and self-critical culture, not one imposing and prescriptive. The only culture that could tackle the most frightening effects of knowledge embodied in civilization.

There is a tradition of aristocratic thought, made to become *vox populi*, that has become the current voice in the social production of public opinion, and it passes itself off for proven truth. This voice of slander insists on the damaging effects of progress while enjoying its benefits and failing to reject its promises. The hierarchical relationship between means and ends or content and technology is here in question, a power conflict between diverging visions. It is absolutely not certain that it is technology that represses, atrophies, or deviates the individual from assuming their responsibility towards the present world. Its impoverishment cannot be attributed to the fact that individual or collective society messing about online distances the individual from the values of civil society. We need to remember that these are values produced from expressive platforms, ways of being and of living all there long before the advent of networks and already largely obsolete in late-television society.

The opposite could be true: it is not TV that damages children destined to become citizens, but children educated by and for citizenship that damage TV. Similarly, we can assume that it is those very human traditions most eradicated in civil society that harm technology and that *make it turn it evil*. The increasingly impenetrable hard crust, the armor of a dominant social culture, might now be obstructing and deviating the use of networks. This in its improper attempt to make them correspond with greater efficiency and speed to the old expectations of the past and not to the very different expectations that its unrealized, failed future has produced and continues to produce. Significantly, among those first "bewitched by the net," many have become its most bitter enemies, proof that the forms of conflict in modern society continue to be the two faces of the same ancient regime.

7. NETWORK LANGUAGES AND MENTALITY

Some further basic considerations: by the very nature of its language becoming more and more fluid and participatory than the traditional face-to-face regime of modern communication such as the press and entertainment, the advent of the internet contributes to breaking down barriers built by the spirit of nations, but less so the differences arising from the spirit of places where roots are much deeper. The digitalization of operative languages favors reforms and revolutions of geopolitical and structural impact more than rapid changes within the people who have to produce and live them. This means the internet does not have the power to redefine the qualities of the individual in the same times and forms with which it can crumble social systems and organizations. It can intervene on the working of their systems but not on the anthropological-cultural background of their social actors. In this shift, the varied quality of devices that power has is measured: large government devices and an extreme variety of personal and local devices.

For a while now, the journey of the internet towards global social regimes destined to subvert the national ones has been quite clear and is constantly active: it only had to pick up the *torch* of the previous political and economic processes of globalization. But, given the anthropological time it takes, the journey the web will have to undergo to transform the individual is much longer. The individual gathers in their own flesh more than their own body, the uniqueness of their naked sensory and affective life more than the communitarian plurality to which they belong, their remote conscience rather than their memory and social education. Rather than rule, it is instinct, animal life, generative, more than and before civilizing eroticism.

The individual and their traditional existence have been subjected to the processes of modernization started up by capitalism's need to develop through a network of national and international sovereignty. People have supplied the energy required by the strategies of socialization put to work by the modern subject, and as catalyst and cultural glue they have been involved with the construction of national states.

They have functioned and been made to function by massive psychosomatic power accumulated while they took root physically and mentally, within the limits posed by their own land of foundation or adoption. The land where one is born and learns to believe in something other than oneself in order to survive, works, sets up family, gains habits and suffers pains, produces, purchases, owns, defends, and passes on goods, dies hoping for the continuity of one's own chosen place and *heritage*. Becoming the content of the apparent uniformity and linearity of the processes of globalization, a dominant content if divided up into infinite localisms, has been the very rooting of the individual in lived traditions that came long before the values of the first industrial modernization, long before the first instances of globalization. The territories of national imprint and destiny are born from the transformation of places in territories just as people have transformed themselves and have been transformed into places. Globalization favored by the internet has in itself the qualities of both movements: localization of people in places and of places in territories. The breaking apart of the territories of modernity before the digital ones needs stimulation and exploitation of the individuals as a basic device (consisting of conscious and unconscious memories immersed in the human bustle of daily life), a device that has guaranteed the construction of places that in turn would build nations.

The networks are still a Babel of national languages in contrast or parallel; some maintain that their transmutation into one single imperial language or jargon is not certain. They even maintain that such contradiction can reach very particular Babel-like character: national languages that "regress" with respect to their historical hegemony till they become dialects and dialects "freeing themselves" till they become transnational languages—centrifugal phenomena compensated by the tendency of images to become able to transform themselves into universal grammars and syntaxes. These

Prospectus Turris Babylonicæ ex Præscripto R. Adm: Patris Athanasy Kircheri Soc: Jesu.
TURRIS BABEL

transitions happen at different times for each of the actual linguistic areas of the planet. They are verbal spheres that not even the internet ones can include within themselves; rather, they depend on them without being able to trace them completely.

Every past forecast on the present and future is turned upside down. More than ever, now the possibility of being able to share the image of a homogenous future is at stake, both along the long ridges of Western civilization and within its most varied flare-up points. The value and political system of every traditional or innovative culture ends up confused and produces desires and passions in contrast with their present—people, groups, and systems divided within themselves between safety and insecurity of their own regime of affirmation and survival. Thus the use of networks superimposes on its own nature of instrument of innovation a totally opposite vocation: the erosion of its every dialectic imagination, therefore the negation, the disarmament, of each of its instruments and historical form of power.

One could put forward the hypothesis that surfing the web alienates surfers from the expressive platform they have at their disposal. The reputation of digital languages has grown where it has quickly shown to work as replacement and compensation for the limits shown by technology produced for and by mass society.

We can then say that in accessing the complexity of postindustrial society, the instrumental rationalism of the modern individual has continued to find comfort and, at least on an ideological level, a relative, if difficult, continuity and compatibility. It is different in the areas of knowledge, where humanist values have been a historical counterpart to the modern development of technosciences and their social applications. They have been therefore the counterpart of areas such as the production of material goods, the organization of work and territory, market economies, service structures, training in functional skills for such services, and so on.

Thinking of society as one great machine, we can say that with the advance of digitalization and the web, it has begun to suffer from a fracture between its hardware and its software. The hardware, with its great bureaucratic, institutional but also social and political resistance, has progressed thanks to the discontinuous but progressive assimilation of the resources of information technology, of their operative efficiency, their global and local dimension. On the contrary, software has stayed linked to its past historical and social roots.

The current situation of training in schools and universities and all the other agencies, basically tied to the same system and function, suffers from the mental habits they acquired during the long time they belonged to the more traditional regimes of their training and profession. Founded on these old vocational regimes, the reputation itself, gained through time by their mentality in the public sphere of civil society, of its institutional and political hierarchies, its means of information, so of the visibility that these agencies have had and still can confer, still obtains consensus and credibility even today.

Visibility, however, that not only comes from the top but also from the roots popular culture and culture of a patronage nature with the same old mentality, if only because this is exactly what taught them and has had ideological exclusivity. To the difficult affirmation of network cultures, we therefore need to add the fact that a mentality lingering over itself and on its reproduction is often believed to be and confused with a completely different type of position that is not forms of misunderstanding and resistance to the quality of digital languages as such, but expression of alternatives expressed within them by those that anyway share their meaning.

MEDIA

LRG
LRG

FROM BEATS AND RHYMES TO POETRY SLAMS
HIP HOP AND POETICS

JENNIFER CARTER

In the highly personal and politicized world of hip-hop, lyrics serve as much of a window into culture as they do in helping transform the culture itself. The origin of hip-hop grew out of the function of live performance entirely dependent upon face-to-face social interaction and contact. As Greg Dimitriadis has it, "[i]ndeed, the event itself, as an amalgam of dance, dress, art and music, was intrinsic to hip-hop culture" (Dimitriadis 1996, p. 179). The very foundation of hip-hop began as communal production, flexible wordplay, and community dance. It was born out of a flexible art form, relating to the exchange of floating verses and rhymed couplets. The spontaneity present at these types of live performances allows for open-ended and engaging social experiences.

Early hip-hop comprised "a number of interlocking and integrated practices, including graffiti writing and ... breakdancing, arts which reflected hip hop's rough and abrupt 'cut and mix' aesthetic in visual art and physical movement" (Dimitriadis 1996, p. 182). The cross-fertilization of styles and aesthetics is what gave hip-hop its initial large draw. Anybody could participate—it became a joint practice between several different parties, each with their own contributions, coming together for a singular purpose, sometimes merely entertainment and other times in order to facilitate and distribute broader, politically engaged messages.

According to Heather E. Bruce and Bryan Dexter Davis (2000), "[Poetry] slams ... provide Hip-hop poets with performance alternatives to rap." Spoken word poetry is a genre of performance poetry in which stage performance is just as important as the words which make up the poem. Unconventional angles, slang, and surprise twists add to the unique elements of the slam poetry style. Borrowing from traditional hip-hop elements, most slam poetry is made up of rhythmic flows, rhymes, exaggeration, and wordplay, lines cascading into each other with pause and effect, creating powerful emotions. Slam poetry borrows from hip-hop's evolution around concepts of flow, layering, ruptures in line, and repetition, using their own poetic tropes and devices. Much of this has been drawn from dub poetry, "a rhythmic and politicized genre belonging to black and particularly West Indian culture" (Wikipedia). As an expressive means to challenging authority, many slam poets also find themselves to be activists. The movement has been called "the democratization of verse" in regard to slam poetry's potential for political weight and influence of activism. The personae or masks that black poets regularly choose to adopt allow them to provide commentary on African

American history and society. Bob Holman, notorious poet and activist, puts it this way:

> The spoken word revolution is led a lot by women and by poets of color. It gives a depth to the nation's dialogue that you don't hear on the floor of Congress. I want a floor of Congress to look more like a National Poetry Slam. That would make me happy (Wikipedia).

Slam poetry can also be attributed to several movements throughout the arts and musical scenes and their varying cultural influences. The Harlem Renaissance and, in particular, the works of Langston Hughes have been seen as having "laid a rebellious aesthetic foundation that would be emulated by generations of poets to follow" (Kientz, 3). Jack Kerouac, during the Beat Generation, "mirrored the improvisation of black American folk music in his spontaneous writings" (Kientz). Even the Black Arts Movement in the '60s and '70s liberally employed performance art and spoken word as a tool for demanding black liberation and self-determination. Culminating in the popular rise of rap and hip-hop, with epic lyric "battles" of skill and talent, slam poetry arose within the hip-hop community—hip-hop being a kind of precursor with its back-and-forth rhyming competitions accompanied by music. The idea of creating a "lyrical boxing contest" quickly grew to immense proportions, rooting itself in urban areas where hip-hop exploded, mainly Chicago, New York, and San Francisco. The rhythmic and contentious performances make profound statements on both the audience and the performers.

Given that the cultural phenomenon of slam poetry has grown exponentially over the years, especially among hip-hop fans, it's interesting to learn that it grew up in Chicago around twenty years ago, spawned by a white man named Marc Smith. Ironically, while Smith birthed this format as a response to the poetic elite who ignored the message of everyday people, he himself scorned the spin-off of the more lyrical slam poetry associated with hip-hop and, specifically, Def Poetry, an HBO series produced by hip-hop music entrepreneur Russell Simmons. Adopting the style of performance artist without background music, slam poets have adapted versions of hip-hop and molded it into their own voice of poetry, casting themselves as orchestrator and using lyricism to the sound of an inner beat. Hearing most poetry slam performances, one would think the pieces were actually derived from actual hip-hop songs. Certainly, hip-hop music could benefit from the extension of performance slam poetry fused into the music industry. The two are not so different—when stripped of their outer coats, they emerge as the same organism with similar messages. Devices such as homophonic word-play, repetition, singing, call and response, and rhyme are frequently used on the slam stage.

There is also much to be said in regard to the listener's felt experience in both hip-hop and performance poetry. The success of a piece is largely attributed to it having "some kind of 'realness'-authenticity ... that effects a 'felt change of consciousness' on the part of the listener" (Somers-Willett 2005, p. 53). These pieces serve as "confession[s] of lived experience. ... If authenticity is ... the criterion for slam [and hip-hop] success, then convincing audience members of the authenticity of one's identity is a major component of [an artist's] success" (Somers-Willett 2005, p. 53). Together, as well as separately, the two genres of hip-hop and performance poetry slams can be viewed as representational practices which authenticate certain voices and identities. They have the capacity to generate the very identities which poets and audiences expect to hear.

The visionary aspect of slam poetry has extended itself into various philanthropic endeavors, especially in creating a voice and safe space for children and teens to express themselves and encourage youth literacy. Urban Word NYC is "founded on the belief that teenagers can and must speak for themselves ... and has been at the forefront of the youth spoken word, poetry and hip-hop movements in New York City since 1999" (Urban Word NYC). Their programs include literary arts education and youth development in creative writing, spoken word, college prep, literature, and hip-hop. These uncensored workshops are free to youth in the community and are a shining example of the fusion between hip-hop and other forms of creative expression.

More specifically, women have been able to find a niche within the context of poetry slams, a place where they're able to stand up and scrutinize their personal as well as political lives and not be judged by it, but rather applauded. There have been some highly successful female poetry slam artists to emerge over recent years, including Zora Howard, New York City's inaugural youth poet laureate, and Sonya Renee, internationally renowned slam poet. What began as a scored contest quickly became more about these women telling their stories and developing their craft. Performance poet Lauren Zuniga says,

> Women still have such a hard time giving themselves permission to create art, be vulnerable or be, heaven forbid, ugly on stage. ... Slam poetry offers us a platform to be celebrated for our grit, our intelligence, our strength. It will always be a reflection of our society, absurd and beautiful, but when women poets come together in a safe space, we give each other permission to do whatever our creative hearts need to do. Even if we just need to explode (Zuniga 2010).

Slam poetry has given women a shared space in direct juxtaposition with mainstream hip-hop aesthetics and culture. While the two share very similar words, passions, and lyrical inflections, poetry slams offer a safe haven for women of hip-hop to explore their own boundaries and selves, away from the prying eyes and conflagration of sexually biased images and stereotypes which haunt women in hip-hop, both as performers and as accessories.

URBAN
WORD
NYC

POETRY IS LIT

FREE WORKSHOPS

NYC YOUTH POET LAUREATE

FIRST DRAFT

SLAM

WHAT WE DO

Urban Word NYC
presents free & uncensored writing &
performance opportunities to teenagers,
in creative writing, spoken word, college
preparation, literature & hip-hop. UWNYC
also provides professional development
services for educators, artists
& community organizers.
–Est. 1999–

AGES 13-19

OPEN MIC

COLLEGE PREP

GET DOWN AT

WWW.URBANWORDNYC.ORG

As a successful education tool, slam poetry has also been utilized in inner-city classrooms and youth organizations as a curriculum for fostering creativity and diversity. Urban Word, WritersCorps, and Youth Speaks are some of the more prominent organizations which have emerged in an effort to "help spread this urban pedagogy through slam poetry, bringing in-class and after-school poetry programs to inner-city schools across the country" (Kientz, p. 5). Poetry as music is being used here to bridge the gap between verse and song lyric, spoken word, and hip-hop, linking slam poetry to the broader realm of urban culture. By viewing their daily lives as inspiration and material for their poetry work, students are able to make stronger connections to not only the real world but also the larger world of literacy, education, and community. Youth utilizing slam poetry as a means of expression and control over their own lives gain a sense of personal power through the spoken word, rather than feeling powerless and consumed within their environments.

Yet another common thread seen in both hip-hop and performance poetry is the manifestation of the performance space itself as a site for the reclaiming of identity. In mainstream hip-hop, we see this manifest mostly in terms of ideas, images, and symbols, specifically within a process of the artist visualizing him/herself. Within both contexts, or media, the artist's identity emerges from the ability to think of him/herself in terms of the community into which s/he has been socialized.

This conscious and creative performance ethnography is a presentation of self in everyday life. Lyrics and beats are specifically designed to produce particular images, a conscious orchestration of sights and sounds in order to elicit certain reactions. There are also specific "rhetorical techniques slammers use to gain legitimacy and authenticity on stage—including call and response, repetition, sampling, rapping, beat boxing, and effusive rhyme—all of which one can recognize from black popular music" (Somers-Willett 2005, pp. 68–69). Whether it's music or poetry, performance is the vehicle in which messages are being sent to the audience, the medium in which personal and political sacrifices are being made. Performance artist Guillermo Gómez-Peña makes the claim that

> performance art is a conceptual 'territory' with fluctuating weather and borders; a place where contradiction, ambiguity, and paradox are not only tolerated, but also encouraged. ... We converge in this overlapping terrain precisely because it grants us special freedoms often denied to us in other realms where we are mere temporary insiders (Gómez-Peña 2001).

Interestingly, he goes on to say that within performance art is the willingness to defy authoritarian models and dogmas and to keep pushing the outer limits of culture and identity. If hip-hop serves at least one

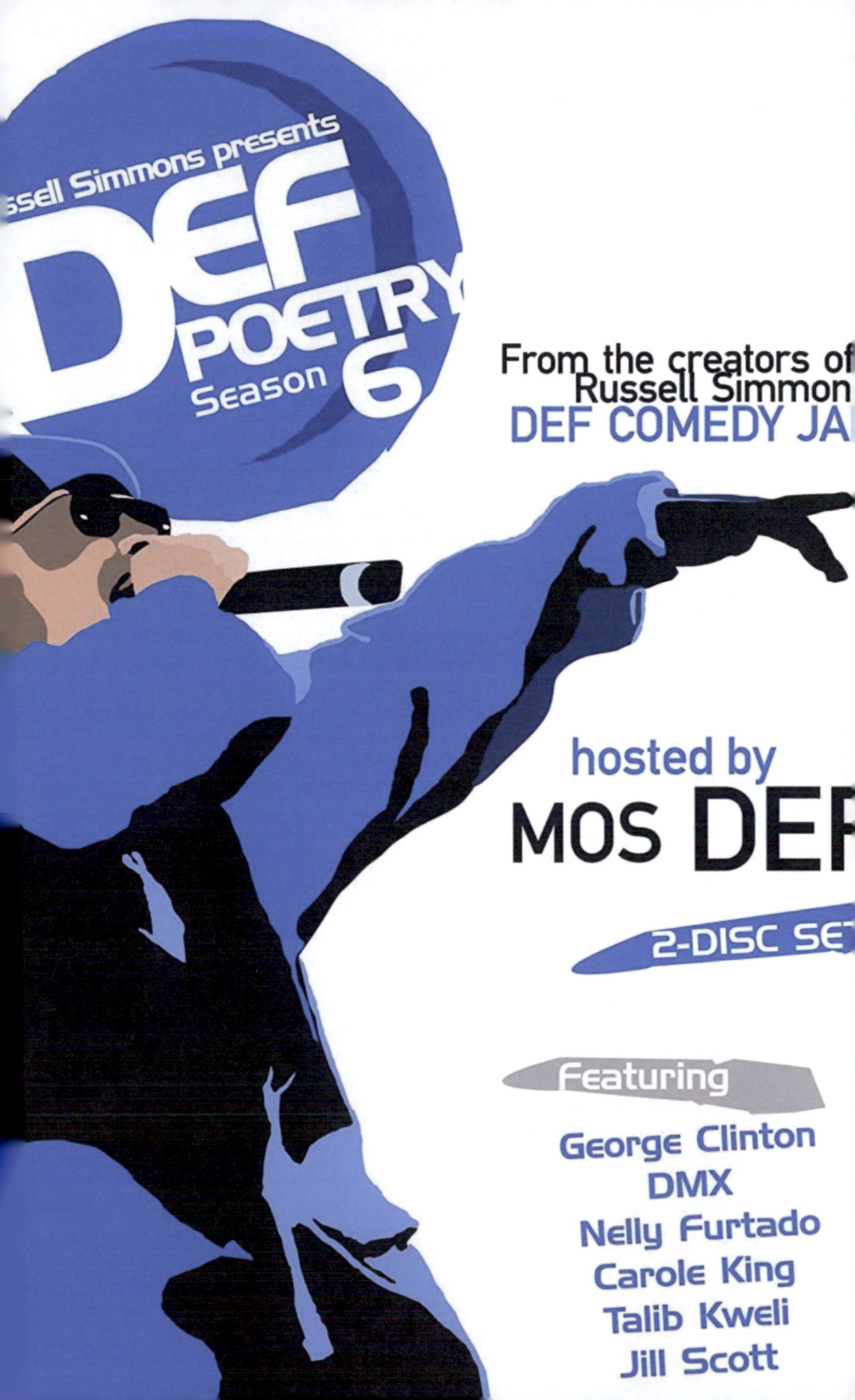

ssell Simmons presents
DEF
POETR
Season 6
From the creators of
Russell Simmon
DEF COMEDY JA
hosted by
MOS DE
2-DISC SE
Featuring
George Clinton
DMX
Nelly Furtado
Carole King
Talib Kweli
Jill Scott

thing, it can be said that it serves as an intersection of culture and identity within the framework of music and art. So, too, can poetry slams be given the same weight.

Black identity has become the main focus and most widely performed identity within poetry slam, often being compared to hip-hop in this sense. "Media projects, such as the feature-length film *Slam* and the ... HBO series *Russell Simmons Presents Def Poetry*, have presented slam poets to mainstream audiences alongside hip-hop artists and against the backdrop of black urban culture" (Somers-Willett 2005, p. 58). Interestingly enough, the slam poetry audience has been shown to be predominately white and middle-class, in a parallel we also see often among hip-hop consumers. Both mediums can be billed as counterculture performances for a primarily younger audience. Focus in both arenas has been commonly narrated on issues of racial diversity, race relations, and racial inequities. Both hip-hop and poetry slams open up the conversation for interracial dialogue in a creative and powerful manner. Through the power of lyricism, artists of both kinds are able to harness their emotions and use words to rebel against the injustices found within urban life. Poverty, oppression, and political injustice are large issues touched upon in both hip-hop and slam poetry. Both are regarded as a "counter-cultural force," meaning they are "characterized as the art form of the literary and social underdog" (Somers-Willett 2005, p. 41). As far as authenticity, in order for the countercultural messages to be heard and accepted by audiences, the artist/performer must have experienced some level of hardship, that they have something important to say and that their messages are authentic. Hip-hop and slam poetry have become a fusion of inner-city culture as well as a vehicle for social change.

A fusion of two art forms come together in new workshops and performances being planned for future hip-hop and poetry connoisseurs. What began for slam poetry as an extension of hip-hop culture has now evolved into a new amalgamation consisting of all of the above: poets, beatboxers, dancers, and DJs. Well-known DJ Reborn, who has spun for popular artists such as Common and The Roots, has been known to host slam poetry events, as has international emcee and beatboxer Baba Israel. The attention of poetry slam has not gone unnoticed, even making its debut at the White House in 2009, a gathering that Michelle Obama had been wanting to facilitate since the beginning of the Obama presidential term. The president and first lady looked on as "poets and playwrights, actors and musicians packed the ornate East Room, delivering cool jazz and glorious spoken-word poetry, sprinkling a bit of hip-hop and a bit of the heroic couplet" (Brown 2009). As a way to invite diversity and inspiration into the White House, the spoken-word evening was a way of celebrating the power of words with the help of James Earl Jones, jazz musician Esperanza Spalding, novelist Michael Chabon, among others. Spoken-word poets Jamaica Heolimeleikalani Osorio and Joshua Brandon Bennett delivered emotional performances, with lyrical references to hip-hop. Even poetry slam champion Mayda del Valle "delivered a personal narrative to the cadence of a hip-hop beat" (Brown 2009).

To close, then, it appears that both hip-hop and slam poetry explore, in the words of Susan B.A. Somers-Willett, "the dynamics of power between [artist] and audience in the real world. ... The complexity of the [artist]-audience relationship precludes the dynamics between [artist] and audience from being any one thing—fetishistic, revolutionary, essentialist, liberating, entertaining. Instead, the performance of [each genre] is usually a host of these forces working together (70)." Slam poetry can be regarded as the sister to hip-hop, both art forms in their own right and both pieces of inner-city culture. Both of these art forms challenge the traditional notions of poetry and music, respectively, by allowing and encouraging participation from all types of people, as well as involvement from oppressed and/or impoverished artists. The proliferation of slam poetry "has expanded the popular notion of what poetry is and has brought a wider, younger public to a form long associated with intimidating erudition" (Kientz, p. 20). Poetry is extended from a verbal composition designed to convey experiences, ideas, or emotions in a vivid and imaginative way, characterized by the use of language chosen for its sound and suggestive power. Hip-hop, as the popular subculture of urbanites, also uses the same design. Together, they are a celebration of the art of oral interpretation with lyrics about rebellion and experimentation. They have become the living, breathing pulse of the street.

References

Brown, DeNeen L. "Obamas Host Speakers, Musicians for White House Poetry Jam." *The Washington Post*, May 13, 2009. Web.

Bruce, Heather E., and Bryan Dexter Davis. "Slam: Hip-Hop Meets Poetry—A Strategy for Violence Intervention." In *The English Journal* 89 (5), *A Curriculum of Peace* (May 2000), pp. 119–127. Print.

Dimitriadis, Greg. "Hip Hop: From Live Performance to Mediated Narrative." In *Popular Music* 15 (2) (May, 1996), pp. 179–194. Print.

Gómez-Peña, Guillermo. "In Defense of Performance Art." Literary Archives, Pocha Nostra. www.pochanostra.com/antes/jazz_pocha2/mainpages/in_defense.htm. Web.

Kientz, Sarah C. "Poverty Slam! How Slam Poetry Transforms The Lives of Impoverished Youth." In *Poverty Capstone*. Print.

Somers-Willett, Susan B. A. "Slam Poetry and the Cultural Politics of Performing Identity." In *The Journal of the Midwest Modern Language Association* 38 (1) (Spring, 2005), pp. 51–73. Print.

Urban Word NYC. "About Us." www.urbanwordnyc.org/new-folder. Web.

Wikipedia. S.v. "Poetry slam." http://en.wikipedia.org/wiki/Poetry_slam. Web.

Zuniga, Lauren. "Girls—and Women—Are Slamming the Poetry Scene." *On the Issues*. www.ontheissuesmagazine.com/cafe2/article/114. Web.

CREDIT WHERE DEBIT IS DUE
FINANCE AND CONSUMER CREDIT IN POPULAR CULTURE

RALPH CLARE

In June of 2017, NBC News reported that Americans currently held the largest amount of credit card debt in history, totaling over $1 trillion.[206] Carrying on average over $15,000 a family, Americans hold more credit cards than they ever have before and can be sure to find offers for new cards arriving in their mail or e-mail on a daily basis. And all of this not even ten years after the 2008 financial crisis in which major banks and investment firms failed or had to be bailed out by the government while millions of Americans lost their savings and millions more their jobs and homes. Estimates of the total losses carry into the trillions.[207] In the end, the government's bailout of the banks on the basis that they were "too big to fail" meant establishing a kind of corporate welfare at the expense of taxpayers. All the financial industry's risk and loss was effectively "socialized"—that is, society paid for it—while a majority of the wealth and gains were "privatized"—that is, it went primarily to those who worked in the industry and to wealthy investors. What happened, we might wonder, to the sense of responsibility on the part of the banks and financial industry as a whole?

This question was all but left unanswered as experts simply threw their hands in the air and complained about the whims of the market or deadbeat customers. We can nevertheless detect the tacit "answer" to this query in the ideological currents of popular culture. For popular culture has lately represented and responded to the banking industry, finance capital, and consumer credit in interesting and telling ways. If we look at several films that appeared in the wake of the 2008 financial crisis, including *Wall Street: Money Never Sleeps* (2010), *Margin Call* (2011), *The Wolf of Wall Street* (2013), and *The Big Short* (2015), we often find attempts to address the larger systemic causes of the crash that are unable to extend their critique of the financial system further than the individual industry insiders and professionals. Finance and credit are problematic, these films tell us, but there isn't much that can be done. Perhaps more intriguing are a range of credit card ads that promote a particular type of fiscally responsible subject in contrast to the depictions of bankers, traders, and financial experts in the post-crash films. Taken together, these films and ads suggest that the financial system sees the answer to

---

[206] Lamagna, Maria. "Americans Now Have the Highest Credit Card Debt in U.S. History." *Market Watch*, August 8, 2017. www.marketwatch.com/story/us-households-will-soon-have-as-much-debt-as-they-had-in-2008-2017-04-03.

[207] Porter, Eduardo. "Recession's True Cost Still Being Tallied." *The New York Times*, January 21, 2014. www.nytimes.com/2014/01/22/business/economy/the-cost-of-the-financial-crisis-is-still-being-tallied.html.

MICHAEL
DOUGLAS

SHIA
LaBEOUF

★★★★★

"Douglas is superb"

– *The Mail on Sunday*

its problems as more of the same. Yet this course of action demands a further disciplining of consumer-subjects who must continue to rely on credit to make constant purchases that keep the whole capitalist system churning. To this end, these ads promote a sham rhetoric of "individualism," "freedom," and "individual fiscal responsibility" that is at odds with what is practiced by those experts who work to game the system. For every credit is, in truth, a debit. As David Graeber writes, "A debt [. . .] is just an exchange that has not been brought to completion."[208] What is really being sold here is the same old financial system and its projected nonfuture of endless purchases in which the bill never arrives, a future that could nevertheless be much different than is currently promised.

HIGH FINANCE IN LOW TIMES

Popular culture's films about the 2008 financial crisis tended to produce movies that were at pains to explain what went wrong in the crash by dramatizing the actions of major firms and players. While these post-crash films don't appear to be about consumer credit, indirectly they are (it is often mentioned by a character once a film in passing). Indeed, the crash was also referred to as a "credit crunch" since ailing banks refused to lend money in such a risky market and finance froze up. Films such as *Margin Call, The Big Short, Wall Street: Money Never Sleeps*, and (though not about the 2008 crisis) *The Wolf of Wall Street* seek both to instruct their audiences about the deep problems with the financial system and offer (with the exception of *The Wolf*) some degree of moral condemnation of the system and its main players.

Oliver Stone's *Wall Street: Money Never Sleeps* is surprisingly more interested in the family fortunes of the Gekkos than it is in the financial crisis, and it does so in a realistic-dramatic fashion set mostly in around the time of the crash itself. The film follows Jake Moore (Shia LeBeouf), a trader for a firm that goes under in the crash and whose fiancé is the daughter of Gordon Gekko (Michael Douglas), a financial wizard jailed for insider trading and the subject of the first *Wall Street* (1987). Gekko is released from jail just in time to write a book called *Is Greed Good?* The book is a condemnation of the financial industry's current practices, a foretelling of the coming crash, and a seeming turnaround from Gekko's claim that "greed is good" in the first film.

Gekko, however, is merely bitter that the "real" crooks of the housing bubble are getting away with more than he ever did and is secretly focused on rebuilding his empire (he eventually steals money from his daughter from whom he is estranged to set up his business again). Though the film registers the ills of the financial industry through Gekko's speeches—we are apparently all equally to blame, according to Gekko/the film—it never lets us see anything outside the drama of its biggest players.

---

[208] David Graeber, *Debt: The First Five Thousand Years* (Brooklyn, NY: Melville House, 2011), 121. See Graeber's book for a fascinating analysis of different kinds of debt and bonds throughout human history.

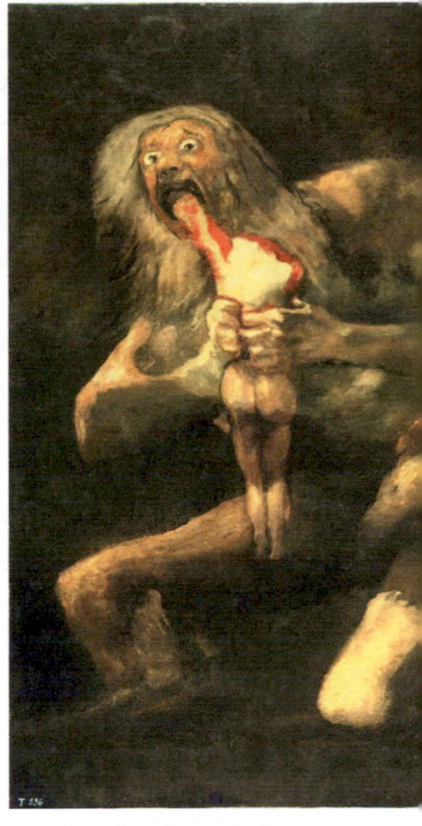

Because the relationship between fathers and sons (and one daughter, Winnie Gekko [Carey Mulligan]), or to be more precise, symbolic fathers/mentors and protégés, is the true focus of the film (Francisco Goya's painting *Saturn Devouring His Son* is even referenced twice), its critique of finance remains ultimately incestuous and cynical. Not only is much of the financial world reduced to personal issues and gripes—it's "all about the game between people," Gekko says[209]—but Stone brings "blood" and "money" together in a way that normalizes the ups and downs of capitalism itself—as if it were one big family drama. By the film's end, for instance, each of Jake's three symbolic fathers represents different ways of approaching money and finance. When, in a ridiculous twist ending, Gekko returns to mend the split between Jake and Winnie and return the stolen money, we are soon treated to scenes of their marriage during the credits. Gekko is there, and the new union between good money (Jake's renewable energy project) and philanthropy (Winnie's nonprofit mentality and truth-telling website), and Gekko (greed now put to good) is solidified. The suggestion is that the financial industry can be, or has already, been fixed, though we have no true cause for such optimism. The soap bubbles that the camera follows (which also appear near the start of the film) are thus related not only to the various financial bubbles that Gordon traces back to the seventeenth-century Dutch Tulip crisis, but also to what Jake, in his voiceover, attributes to "the mother of all bubbles," which is the "Cambrian Explosion."[210] In other words, bubbles are really a part of evolution, as Jake and the film conflate nature with capitalism (a man-made economic system) and falsely suggest that capitalist economic cycles are as natural as the weather and is simply how things will always be. Apparently this particular bubble of belief in capitalism as a natural force has yet to burst, and the film certainly isn't interested in poking holes in it.

Adam McKay's *The Big Short* is another film that represents the crisis through a realistic-dramatic lens, though at times with a cinéma-vérité or documentary feel (the shaky camera work and real photographs of people suffering from the crash, etc.), and once again plays out during the crisis. The film follows the fictionalized pursuits of several real-life insiders and investors who foresaw the impending crash and sought to profit from it by shorting the market (betting against the housing market

---

[209] Stone, Oliver, dir. *Wall Street: Money Never Sleeps.* 2010; Century City, CA: Twentieth Century Fox, 2010. DVD.
[210] Ibid.

going up). We are thus introduced to "good guys," one of whom (Mark Baum [Steve Carrell]) is angry at the market and wants to punish the banks, and "bad guys" who simply want to profit from the financial chicanery. In short, while we find ourselves rooting for the "good guys," in the end, as one trader suggests, the amorality of the system makes one group fairly indistinguishable from the other. "Punishing" the banks for their unethical practices means ultimately profiting from the losses of everyday people. In this way, *The Big Short* is far from cathartic.

Even more so than *Wall Street: Money Never Sleeps* and *Margin Call, The Big Short* is intent upon explaining the ways in which the system developed and broke (the movie is based on true events and real people); it is even narrated by investment banker Jared Vennett (Ryan Gosling). Indeed, at times the film feels at times like a carefully arranged set of interviews in which Baum asks people supposedly in-the-know (who are often not) questions as he tries to discover whether the housing bubble really is ready to burst. Ironically, most of the people he speaks to happily and candidly tell him about how things work and confirm his suspicion of the incipient crash. These interviews make it appear as if an amoral finance is speaking purely through the mouths of its foolish proponents, of speaking for its proponents themselves, instead of the other way around. Here, even those benefiting from the system sometimes appear to be dupes.

The Big Short also departs occasionally from the realistic-dramatic structure in scenes that function as meta-asides in which celebrities satirically explain what complicated financial terms and arrangements actually mean. In one such scene, Margot Robbie, enjoying champagne in a bubble bath, tells us about how mortgage-backed securities began to take on subprime loans.

Money has always been sexy, but mortgage-backed securities? Margot Robbie talks "assets."[211]

---

[211] McKay, Adam. *The Big Short*. 2015; Hollywood, CA: Paramount, 2016. DVD.

In another scene, Anthony Bourdain explains Collaterized Debt Obligation (CDO):

Set within the realistic narrative, these scenes comically function as absurd moments in which the language of the financial industry and their elaborate schemes (from CDOs to credit default swaps) are shown to be rhetorical obfuscations—essentially lies. But for all the metajokes—including a false, happy ending in which bankers go to jail and government regulation comes to the rescue—the film ends with the Pyrrhic victory of a few investors getting filthy rich after the crash, rather than during or before it, at the expense of the overextended banks. The film remains cynical, rightly so, as its localized look at the financial system and mortgage-backed securities offers no answers as to the future, and its metajokes begin to look like desperate attempts to laugh instead of cry about it.

Even more sobering is J. C. Chandor's *Margin Call* (2011), another realistic drama that takes place during (mostly just before) the crisis. The film's atmosphere and tone is gloomy, even noirish at times, and the music is often appropriately haunting. The film opens with an investment firm's rash of brutal firings and trading floor head Sam Rogers (Kevin Spacey) feeling upset about his dying dog—whom he buries at the film's end. Much of the action occurs at night and follows several characters that have just discovered their firm is overleveraged in mortgage-backed securities in a market about to crash. The moral dilemma of the firm is whether to dump what they know are all but valueless securities on their customers and competitors in order to limit their own losses—thus spreading the losses to others and throughout the market and world. But the "conscience" of the traders is really limited to Sam, whose only "reservation" is that unloading assets in this fashion will only spread the problem and ultimately hurt the firm's reputation and future. Nevertheless, he helps to do so for a massive payout.

This grave look at financial capitalism at it meanest reveals that those at the top have little idea how or why trades are made and underscores the problem of attributing of agency or blame in systems. When CEO John Tuld (Jeremy Irons) arrives by helicopter to solve the crisis, he asks for the situation to be explained to him as it would be to "a young child" or a "golden retriever."[212] He is, he later tells Rogers in justification of his plan to offload the bad assets, simply a salesman. A market, in his reading, is simply an unregulated trading floor with no guarantees of anything. Yet as Rogers points out, Tuld does have a choice at this point. Tuld's later claim that capitalist crises are the norm (he lists off just about every year of crisis in capitalism's history) and that one has no actual choice but to "react" to the market's forces is thus shown to be not entirely true. The players could be playing the game differently.

True to the film's dark tone, however, even the traders and managers are fairly unhappy in taking part in an economic system that is more premised on finance, branding, and consuming than it is producing, manufacturing, or building things. A recently fired risk manager waxes nostalgic of a commuter bridge a company that he once worked for built. The bridge is physical thing that tangibly saved commuters driving time and represents the very opposite of what his work in the financial industry has turned out to be. One young trader who is eventually fired asks throughout the movie about other peoples' salaries, the only way they he can gauge value in this world. Yet when one successful trader, Will Emerson (Paul Bettany) lists what he spends/wastes the money on, the pursuit of wealth seems fairly empty. Moreover, the veteran, world-weary Emerson is a ground-down cynic, eventually blaming "real people" for the mess—taking out credit beyond their means—and even sees himself as a kind of God who makes the entire capitalist world spin, a quasi-Randian *Atlas Shrugged* position. If these are indeed the nihilistic Gods of finance, looking down at the rest of us from their mirrored towers, we ought to declare them dead.

At their worst, post-2008 crash films simply reproduce the ecstasy of their subjects—reproducing, perhaps even romanticizing, the pathological thrills of chasing and gaining wealth. Martin Scorsese's dumbfounding *The Wolf of Wall Street* is such a film. Though it covers a 1990s Wall Street scandal perpetrated by trader Jordan Belfort, the film cannot help but be a comment on the 2008 financial crisis. However, the film treats

---

[212] Chandor, J.C. *Margin Call*. 2011; Los Angeles, CA: Benaroya Pictures, 2011. DVD.

Belfort's shady dealings and meteoric rise in Wall Street as a drug-fueled, sophomoric buddy comedy. Scorsese's depiction of Belfort (the script is drawn from Belfort's memoir) is comparable to his portrayal of the somewhat sympathetic protagonist of *Goodfellas* (1990). Both films employ the protagonist's first-person narrative to create a personalized, insider look at how a particular "business" works, yet *The Wolf of Wall Street* lacks any ethical context or the operatic weaving of family, history, and questions about American identity that form the substance of the earlier domestic mob film. Instead, the film indulges in Belfort's self-triumphant story and feels as much a rip-off as Belfort's financial dealings.

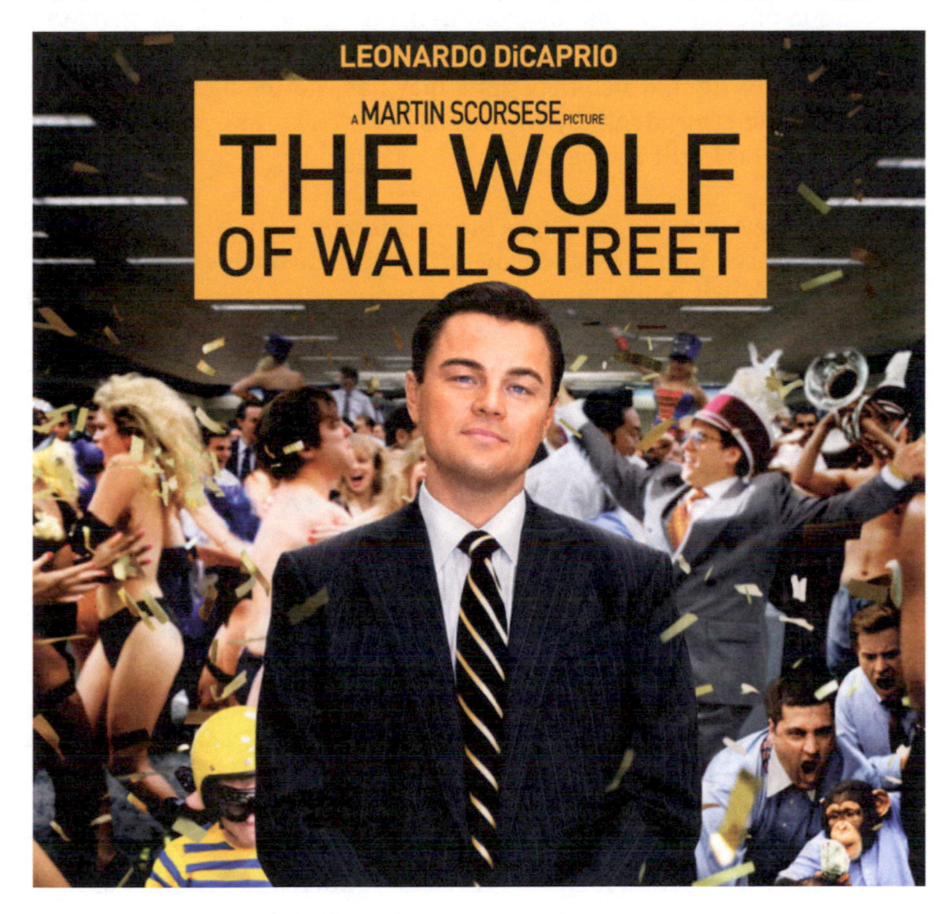

As a film that comments upon the 2008 crisis, then, *The Wolf of Wall Street* is troubling at best. Its amoral, "greed is good" endorsement (even if we take much of the film as satirical, yet fairly true to events) is the antithesis of the handwringing, pessimistic moralizing of *Margin Call* or the laugh-or-go-crazy cynicism of *The Big Short*, nor is its satire contextualized in a way that undercuts or challenges Belfort's narrative, as *The Big Short* does. Ironically, for this reason *The Wolf of Wall Street* is perhaps an even "truer" film about the amorality of the system than those films that lodge a protest of sorts. From this perspective, the real blackness of the comedy that comprises *The Wolf of Wall Street* is that it's actually no joke at all—or the joke's on us.

In the end, these films, whether conscientious or frivolous, are limited in their critical scope and attempts to address systemic causes of the crash. The focus on the major players involved in the crash—real or invented—leaves out the deeper consideration of those regular people (though some characters deride them) who were direly affected by their actions and who lost savings, pensions, houses, and jobs. This "insider" point of view thus results in a kind of "systemic" view that is actually limited both to explaining the rules and glitches of the financial system by those who think they understand it best and/or to claiming that the system is all but unstoppable. This means that a longer and more reasonable consideration of capitalism and its discontents is off the table. The insiders depicted in these films are believers, to one degree or another, in the capitalist system, which they think may have gotten a little out of control, though they don't, by any means, have any idea or much interest in how to fix it in the future.

We also never get to imagine a future different from what has been since, because each film's realist-dramatic structure demands that we stay mostly in crisis mode. In general, the action unfolds just before, during, and after the crisis and heightens the sense of drama—there is a "realistic" effect of re-creating the panicky feeling of the crisis itself, the information moving in real time, etc. The viewers are pulled along often at breakneck speed, watching characters make crucial "real-time" decisions in an effort to avoid or profit from the crisis. The crisis begs for immediate action, not solutions, and a more sustained reflection never arrives.

Thus, despite their look into the proximate systemic causes of the crash (by describing the failures of regulation, loopholes, CDOs, leveraging, greedy traders, etc.), postcrash films actually lack any kind of larger historical context from which to consider the deeper systemic issues of financial capitalism. Without a more properly historical context of capitalism—which in its six-hundred-year or so existence has periodically experienced crashes and financial boom and bust cycles[213]—these films ensure the continuing disconnect between the world of high finance and everyday people that the films, as we have seen, already perpetuate (thus factors such as race, class, gender, and sexuality as it pertains to finance and consumer credit are never considered).

In a certain way, then, the insider characters—already overwhelmed by the system they've created and gamed—are let off the hook. In a crisis, decisions had to be made, the argument goes, and there's no point in hindsight. But what, we might wonder, would a postcrisis film (not a documentary) about the crash look like? Perhaps it wouldn't be a film

---

[213] See Giovanni Arrighi's *The Long Twentieth Century: Money, Power, and the Origin of Our Times* (New York: Verso, 1994) for a history of the expansion of nation-states' credit cycles and their demise. Charles P. Kindleberger's *Manias, Panic, and Crashes: A History of Financial Crashes*, 5th ed. (New Jersey: Wiley, 2005 [1978]) provides an excellent history of financial crises throughout capitalist history.

running in a cinema, but instead the spectacle of images on TV, in print ads, and on the internet that compose the seeming reality of our everyday lives.

MANAGING YOUR FINANCIAL "FREEDOM"

As a matter of fact, this film is currently running and has been for decades. It is composed of the ideological phantasmagoria of TV commercials and internet ads for banking, credit cards, and financial services that fixate on the individual as wholly responsible for his or her financial livelihood. While the films addressing the 2008 financial crisis suggest that systemic problems are essentially out of any one person's control, the ads for financial services and credit cards take the opposite position that an empowered consumer can make informed choices that may even be more astute than those of professionals.

This apparent raising of consumers to the status of the subject-in-the-know is surprising and fairly hypocritical considering the way in which the banks' losses after the crash were paid for with taxpayer money. Moreover, the tacit blaming of everyday people for their outrageous debt and for the existence of the financial system (as does Will Emerson in *Margin Call*) is also strange when considering that it was the dictates of finance capital that urged deregulation in key areas to allow for the growth of consumer credit. In the *Financialization of Daily Life*, Randy Martin provides an in-depth analysis of the expansion and marketing of consumer credit at the end of the twentieth century, the deregulation of financial services and banks, and the financial industry's increasing speculation and socialization of risk in a prescient book published several years before the 2008 financial crisis. Martin notes that, "Between the late 1970s and late 1990s, the average of monthly income charged increased from 3.4 percent to 20 percent."[214] A major consequence of this credit expansion and easy money was to move business risk to everyday life. Thus, "*financialization* aims to make life like an approach to business, and thereby return the protocols of work to daily life with a vengeance."[215] Once accomplished, the result is, "to concentrate wealth and the authority to dispose of it in a corporate stratum that lives by finance. At the same time, however, financial self-management becomes a general feature of life for millions linked indirectly through their investments, in turn enlarging the scale of capital."[216] To be sure, capitalism—and particularly our current hyperfinancialized neoliberal form—demands the expansion of finance-credit in order to keep itself going.[217] This means, as Martin has shown, that the oft-derided "regular people" must be properly "educated," buy into the idea of credit, and quite literally be sold a bill of goods—and as 2008 showed, this bill will harbor more risk than reward.

---

[214] Randy Martin, *Financialization of Daily Life* (Philadelphia: Temple UP, 2002), p. 30.
[215] Ibid., p. 35.
[216] Ibid., p. 192.
[217] See David Harvey's *The Limits of Capital* (Brooklyn, NY: Verso, 2006 [1982]), particularly Chapter 10, " Finance Capital and Its Contradictions," pp. 283–384.

Instead of embracing truly effective banking regulations, the banks seem more concerned with instituting new forms of customer self-regulation.

This free education comes through an array of ads for banks, credit cards, and financial services. Essentially what they all are selling is the idea of freedom—whether to purchase and do things or retire comfortably. This is often the obvious message of credit card ads, such as Chase's Freedom card, which has been topped without a hint of irony by Chase's Freedom Unlimited card.

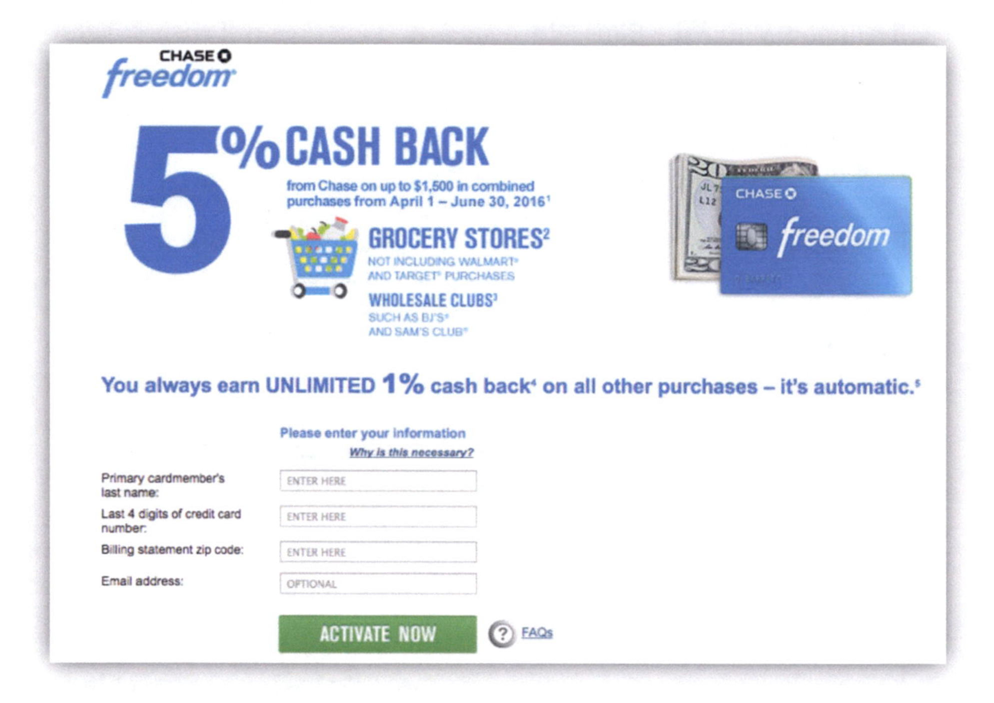

The "freedom" to pursue one's needs or desires—whether they be groceries, a new computer, a car, or a much-needed vacation—is equated solely with credit and, therefore, with going into debt. In a recent ad for the Unlimited Freedom card, the ever-smiling celebrity Ellie Kemper strolls through New York's Times Square, lauding the benefits of the card's rewards feature. "Free" and fluid movement is linked with the power of the card to access spaces and places. All the while, Kemper pokes fun at product placement ads and the ads in social media feeds. How dare ads try and be subtle; we're too smart for that, right? Yet at the commercial's end, Kemper catches herself in a TV ad and jokes that celebrity endorsement ads are terrible.

There is a dizzying mise-en-abyme quality to an ad that is trying to sell us something (a card) so that we can purchase the somethings that these other "hidden" ads that are being made fun of are selling. The self-aware, metaquality of this commercial, which ends with the spunky Kemper's ironic recognition of her paid involvement in ads, doesn't quite erase the claustrophobic and solipsistic dimension of Kemper meeting Kemper.

Chase and Kemper would like to "reward" you by acknowledging you're too smart to fall for the ads.

Unlimited freedom, in this case, feels like unlimited narcissism. Such freedom cannot purchase ad-free peace of mind either.

In short, these ads pitch going into debt as attaining "freedom." But as the record-high American credit card debt shows, this debt is a serious drag on many an individual and family. According to the ads, Americans are "free" to go on vacation, buy luxury items, and pick fresh fruits and vegetables from high-end grocery stores. Closer to the truth is that many cannot get out from under credit card debt and/or are often using credit to buy necessities or pay off other bills—you certainly never see someone in the ads gritting her teeth because she has no choice but to pay for medical bills or fix her car on credit! Of course, those in the financial industry leveraged debt to make more money, but the same is not true of consumer debt. Furthermore, gaining this "freedom" requires you to get credit, to use it frequently, and to "build" your credit. It becomes a project and paying for a large purchase by cash or check is often difficult to do. The "freedom" of credit—the freedom to debt-fund your life—is thus a very curious freedom indeed.

This freedom might instead be thought of as a particular form of control. In his provocative "Postscript on the Societies of Control," Deleuze argues that what Michel Foucault once dubbed disciplinary societies have reached an apotheosis at the end of the millennium. A new kind of control has arisen instead, a more fluid and undetectable type that comes from the technological developments of the end of the twentieth century. The rise of the internet is exemplary of such a shift, in which "the numerical language of control is made of codes that mark access to information or reject it."[218]

[218] Gilles Deleuze, "Postscript on the Societies of Control," *October* 59 (Winter, 1992), p. 5.

Whereas disciplinary societies counted the individual and its ability to form a resistant mass, in our more corporatized society, "individuals have become '*dividuals*,' and masses, samples, data, markets, and '*banks*.'"[219] Think, for instance, of Facebook and other social media platforms that see "individuals" as simply users whose data is tracked, mined, and marketed algorithmically. Indeed, Facebook builds nothing physical; it "builds" social networks and makes money by selling information to advertisers and keeping its stock value high to please investors and the market (speculation). As Deleuze states (in a statement that resonates with the films mentioned above), "This is no longer a capitalism for production but for the product, which is to say, for being sold or marketed."[220] Nowhere does Deleuze's notions of the "dividual" and the ways in which control works in today's society appear more clearly than in a somewhat terrifying recent ad campaign for Discover cards. The commercials each feature a customer calling Discover with a problem or question. The customer service representative at the other end of the line instantly takes their call and listens earnestly to the customer's concern. Strangely enough, it is soon revealed that the Discover representative is none other than the customer herself. The customer is talking to her double.

In one such commercial, a man calls Discover in order to beg forgiveness for missing a payment. The customer worries over having to pay a late fee and that his interest rate will be raised. Luckily, his twin has nothing but empathy and good news for the customer, giddily announcing that Discover won't assess a charge for this late payment because he is (or they are) such a valued customer and usually pays on time. The customer is thrilled, and the two begin an inconsequential chat about how great everything is. "Discover," the voiceover announces, delivering the tag line. "We treat you like you'd treat you."

[219] Ibid., p. 5.
[220] Ibid., p. 6.

The *dividual* treating himself to his second job at Discover.

Yet what appears in this Discover commercial (and others) to be a fluid customer experience—free of waiting on-hold on the phone, free from annoying penalties and hidden fees, free of the individual facing an uncaring corporation or the finance industry itself—is in reality a perfect example of the carefully orchestrated forms of control that the finance industry wields over everyday life. Here you are constantly checking up on or monitoring yourself or, more precisely, your algorithmic credit score, which is actually "you" divided from yourself (Deleuze's "dividual"). The strange schizosplit of the subject here suggests a bizarre self-monitoring that is experienced (or at least pitched) as giving one more freedom through empowerment or transparency.

The financialization of regular people, or dividuals, who must become the CEOs of their corporatized lives also means that they must be "educated" in order to "manage" their debt properly. Thus another form of subtle control is evidenced in recent ads highlighting special card features that are meant to "educate" (read discipline) consumers. This includes the bizarre claim that "cash back" features allow you to "earn" more rewards or money by spending more—of course, you just go into more debt. Just as baffling are the ads in which companies brag about their free service of not charging you if false charges appear on your statement. Perhaps most telling of all is the way in which companies like Discover have recently trumpeted the fact that they will let you know your FICO score for free once a month. This score is essentially an indication of one's credit rating or worthiness. The FICO score, however, is clearly a subtle form of financial control in which the customer regulates herself—checking her score monthly to make sure her payments and financial decisions are blessed by the market or credit rating agencies. Indeed, what appears to be an instance of more choice and agency for the subject/customer is actually the very opposite. Firstly, since using cash will not improve your credit rating, though it will keep down your debt, you are compelled to use credit again and again. Thus being in manageable debt is the entire point. Secondly, not only has self-regulation taken effect here, but the customer still has no clue how her credit rating is set or who sets it (a computer program, of course). Matters seem even worse when we consider that credit rating agencies have long engaged in corrupt practices when it comes to rating banks and bonds that directly contributed to the 2008 crisis. Nobody regulates the regulators, but they still regulate everyday people carefully when they wish to. Hence the FICO score functions as a kind of control mechanism that appears as its opposite, as personal freedom.

Moreover, it is as if Discover—or any credit card company, bank, and the financial system itself—doesn't exist in the usual sense. It appears as merely a neutral platform delivering free services and credit to you. Nowadays we somehow work for the companies—making sure we keep ourselves properly creditworthy, inputting data for them, and make our payments on time. Consider that many a bank's website has also become a platform full of ads posing as news stories (essentially "fake news")—stories about investing, retirement savings, etc. all geared toward getting the customer to purchase more of the bank's services. We must manage ourselves the way a bank or corporation manages itself.

BREAKING THE FINANCIAL CYCLE

Financialization's invisible form of control signifies, as Deleuze states, that "Man is no longer a man enclosed, but man in debt"[221] (181). Building upon Deleuze's declaration, Maurizio Lazzarato theorizes that our debt-driven, neoliberal world has given rise to "the subjective figure of the 'indebted man.'"[222] For in a world in which freedom is conflated with finance, you must finance your freedom to gain financial freedom to finance your freedom, etc. in a sort of vicious circle. But this is merely freedom (credit) masking control (debit). Instead, we are tied, through credit/debit's time loop, to deliver on a promise in a future that has already been bought and sold. So too do capital's cycles' purchase and bet upon a future that must deliver more products and payments. Trapped by such cycles, it's no wonder the postcrash films are so cynical about the possibility of change. Indeed, as Lazzarato argues, "debt simply neutralizes time, time as the creation of new possibilities, that is to say, the raw material for all political, social, or esthetic change."[223] A system that wishes to blame the system itself is therefore one that insists that change or a different future of any kind is impossible—time as *potential* and *change* is neutralized. This is because, Lazzarato writes in a point similar to Martin's about financialization, that, "finance's goal of reducing what will be to what is" entails "reducing the future and its possibilities to current power relations."[224]
Debt neutralizes time in the way a cycle can. In the Discover commercials, one talks to oneself, watches oneself in a strange, solipsistic feedback loop in which time is only the time of credit/debt cycles or

[221] Ibid., p. 6.
[222] Maurizio Lazzarato, *The Making of the Indebted Man: An Essay on the Neoliberal Condition*, trans. Joshua David Jordan (Los Angeles: Semiotext(e), 2012), p. 38.
[223] Ibid., p. 49.
[224] Ibid., p. 46.

capitalist boom and bust cycles (as in the films). In the Chase Freedom Unlimited commercial, the purchasing of products to purchase more products suggests an endless loop in which the check never arrives (it always does). Even celebrities are ultimately faced with their own empty, branded images. In these ads, the future is just products churned out by the same old system.

Now consider the bleak futures portrayed in many dystopian novels, films, or TV series that have flooded popular culture since 2008. All fear one disaster or another—often a conflation of economic, environmental, or political causes. One figure in a subgenre of these disaster films, the zombie, could well be a prime example of Lazzarato's "indebted man." Reduced to a bare, brainless existence and stuck in an endless, non-time, the zombie is the remainder or unpaid debt of a used-up human resource in a system gone bankrupt—the subprime being who has nothing left to do but to bring her worthless assets back to the debtors, to show that their promise is a lie.[225] In the end, the bankrupt futures depicted in such texts address our collective lack of a *positive* or *different* version of the future than capital allows. As Mark Fisher puts it in *Capitalist Realism*, there is a "widespread sense [. . .] that it is now impossible even to *imagine* an alternative to [capitalism]."[226]

But why wait for the grim, mass zombie retribution that only marks a different kind of apocalypse? And why be so cynical about the future? After all, capitalism is a man-made system, not the weather, so it can be changed. What if the belief in and cynicism about capital's cyclical ups and downs actually serves capital's purposes—creates a bitter excuse to keep the cycle going, a commercial's nod-and-wink that we all get how this really works, so just grin and bear it? In that case, we ought to recognize that the bonds of debt we have to one another are very different than the financial kind that reduces relations and time to money and finance, effectively neutralizing the creation of possible futures. "If freedom (real freedom)," writes Graeber as he speculates about wiping away all debts, "is the ability to make friends, then it is also, necessarily, the ability to make real promises. What sorts of promises might genuinely free men and women make to one another?"[227] Time to realize freedom isn't what it seems, to wake the zombies up, to think outside of the time of debt, to imagine something beyond this pessimistic, deadening cycle. It's time to give credit where debit is supposedly due.

---

[225] For more political readings of the zombie see, Daniel W. Drezner's *Theories of International Politics and Zombies* (Princeton, New Jersey: Princeton UP, 2014) and Henry A. Giroux's *Zombie Politics and Culture in the Age of Casino Capitalism* (New York: Peter Lang, 2011). Also see Colson Whitehead's postcrash zombie novel, *Zone One* (New York: Anchor Books, 2011).
[226] Mark Fisher, *Capitalist Realism: Is There No Alternative?* (Zero Books, 2009), p. 2.
[227] Graeber, *Debt*, p. 391.

ONTOBRANDING AS A DESTINY FOR FASHION

SOCIAL POLARIZATION, GRASSROOTS CREATIVITY, AND THE AUTOMATION OF EVERYTHING

NELLO BARILE

This essay aims to demonstrate how contemporary innovations in artificial intelligence applied to fashion is implementing and exasperating ideas that had been already developed during the evolution of the fashion system: here, creative inspiration comes not from the celebrated figure of the fashion designer, but mostly derives from an external context—especially from consumers recently reimagined as data sets.

The transformation of the whole system can be considered a direct consequence of the modification of the social structure that has evolved from the straight contraposition between higher and lower classes during the so-called *Golden Age of the Haute Couture* to the expansion of middle classes across the second half of the twentieth century, to a new polarization between the classes determined by globalization and by the latest financial crisis. Both are involved in what has been called the new war declared by the rich against the poor (Reich 2008, Gallino 2012) and in the creation of a highly entropic system of consumption.

So-called trickle-down theories help to describe a parallel process—the institutionalization of fashion as a productive system—which, since 1856, (when Charles Frederick Worth revolutionized the notion of fashion), has been breaking away from how it was examined previously. Worth's invention of the fashion show (in French called *defilè*) led to the overturning of the relationship between client and tailor (Lipovetsky 1990), the evolution from the tailor to the designer, and the placement of the figure of the designer above upper classes and common people. In other words, he becomes the center and the pillar of the whole system.

This is how fashion becomes a mutable phenomenon and autonomous, hegemonized by the designer that turned himself into a creative genius able to implement his peculiar taste (or bad taste as according to Veblen), completely emancipated from social conventions. In this period, fashion is also a sort of a "minor" art that, nearby her older sister, claims a total autonomy from the definition of taste in modern society. As a direct consequence, *haute couture* becomes a sort of pushing system, thrusting trends upon the whole world, according to the logic of "monocentric" transmission (Lipovetsky 1990, Davis 1992).

This augurs a phase of naked opposition between the upper and the lower classes, with the modest presence of a middle class that will grow remarkably after WWII.

On the opposite side of haute couture, there is a rudimentary form of industrial fashion, designed for the lower classes, that produces low-quality garments with quite common imperfections (Lipovetsky 1990). Although the notion of "low cost" fashion is quite contemporary, we can find its origin in the availability of an enormous quantity of cotton imported from the New World to the old one, a development that played a fundamental role in the new pedagogy of consumption. This new hyperproduction of garments drastically reduced prices so that even the poor could wear these products. As Baines writes back in 1835: "[i]t is impossible to estimate the advantage to the bulk of the people from the wonderful cheapness of cotton goods. ... The humble classes now have the means of a great neatness, and even gaitey of dress as the middle and upper classes of the last age (358)."

Apart from the rising material levels of clothes production, the parallel development of the fashion system contributed to an evolution or metamorphosis of what we might call a *logic of distinction* and conspicuous consumption. The social innovations proposed by Coco Chanel and company revealed a fashion machine that was able to interact with everyday needs. More than Poiret, who started an emancipation of the female fashion industry and created a new style in opposition to Victorian restrictions on the female image, Chanel aimed to overwhelm the differences between social classes, proposing new materials and new styles inspired by and targeted at the lower social classes.

Futurism tried to change the sense of fashion as submissive—considered as subservient to art—and augmented a dandyism aesthetic that wanted to aestheticize everyday routines. And here we come to the famous case of Balla's "*le vetement masculine futuriste*," lately translated in the Italian version as "*vestito antineutrale*": a radical change in the conceptions, functions, and forms of clothing that fracture the relationship between clothing and the external environment. As with the Simmelian vision, here fashion change animating the luxury market emerges as a vital and deadly mechanism—note here as well the etymological derivation of the Italian word "*lusso*" from the Latin "*lug-lutto*" (Calefato 2014).

The last example of the classic relation between creativity and fashion is revealed by the work of the designer that, according to Polhemus (1994), represents the last example of a fashion devoted to the celebration of the "new": Christian Dior. In fact, "time is a social-cultural concept which reflects and expresses a society's or a person's real or ideal social situation" (Polhemus and Procter 1978, p. 13). Indeed, "traditional anti-fashion adornment is a model of time as continuity (the maintenance of a status quo) and fashion is a model of time as a change" (Id.).

Christian Dior's New Look impacted powerfully on the fashion imagery of the Fifties, since, as was said at the time, every taxi driver in New York wanted to know about the latest New Look coming from Paris (Steele

1997), but it also augured the end of the "Golden Age" of fashion, built around the core value of the "new."

As Steele underlines, the powerful image of French housewives ripping a young woman's "Corolla" dress, designed by Dior, presents to us the clear image of a contrast between a traditional vision stressed by the austerity imposed by the Nazi occupation and a future characterized by wastage, with the consumption of large amounts of fabric, necessary to create the Corolla effect.

The cultural reaction against this new style (but also a new social ideal of femininity) were able to trigger huge waves of moral outrage, with new forms of youth subcultures and riots. When the New Look arrived in the United States, several newspapers and civic organizations expressed a violent counterreaction. In 1947, the English Labour Party Member of Parliament Mabel Reidalgh attacked the New Look on the pages of the *Daily Herald*: "Ridiculous stupidly exaggerated waste of material and manpower. ... The New Look is reminiscent of a caged bird's attitude. I hope our fashion dictators will realize the new outlook of women and give the death blow to any attempt to curtail women's freedom" (David 2012, p. 190).

The Sixties are the period in which the democratic ideal of luxury was born, promoting a new wide variety of services and goods available to the new middle class, expanding themselves since at least 1956 when the amount of so-called white collar workers overwhelmed the number of blue collar workers in the West (Bell 1973). The paradox of this period is that, while Western societies were entering the postindustrial age, the fashion system was developing through the application of industrial standardized methods to the production of garments. Shifting from the general logic of new consumption to the dynamics of the fashion system, the democratization of fashion rides the invention of a new business model.

In this context, *prêt-à-porter* [ready-to-wear garments] represents not just an increased offering of high-quality, industrially-produced outfits, but, in fact, a new ideology determined by the impressive explosion of the middle classes during the Sixties. The fashion system in this period amounts to a series of solicitations that come from the streets of the diffused that change the basic rules of the system. This new productive model is triggered, on the one hand, by the development of a new mass society and, on the other hand, is naturally opened to the cultural innovations coming from the subcultures and street styles—so it is that Mary Quants, during this age, had to admit that the new duty of the designer was to follow the street styles, especially the mods generation (Barile 2005, 2013).

If the creativity during the Golden Age of Fashion was ruled by the Simmelian and Veblenian trickle-down schemas, this new democratic luxury can be analyzed according to Blumer's collective selection *as a transversal model that combines forces from different directions of the social system*:

> The fashion mechanism appears not in response to a need of class differentiation and class emulation, but in response to a wish to be in fashion, to be abreast of what has good standing, to express new tastes which are emerging in a changing world (Blumer 1969, p. 268).

The same democratic logic of the new *prêt-à-porter*, an ideal combination of creativity and bureaucratization (Lipovetsky 1990, p. 84), is the sign of a transformation under the mark of a new middle-class "dictatorship." Here recall Gundle's suggestion that the concept of glamour is consolidated in the social sectors starting from this age (2009, p. 253). At the same time, the idea of grassroots creativity coming from the street completely changes the top-down approach to fashion into a more open modality.

New values coming from an ethic of leisure (Morin 1962) and also from counterculture movements are added to the traditional conception based on power, ostentation, and respectability. When in 1968 Yves Saint Laurent declares "*non au Ritz, vive la route*," luxury absorbs values of transgression and sexual provocation coming from the youth culture and street styles. After a the long decade of the Seventies—split into a critical first half where the ideals of the countercultural critic to consumption were absorbed even by mainstream fashion, and a second half when the nihilism of the Punk era was just able to set the field for a new hedonistic age—the Eighties were a controversial decade in which creativity, luxury, and deconstruction were mixed in several powerful solutions.

During the '80s, a pact of blood between fashion and art is signed. Artists used designers as a subject of their portraits, turning them and other protagonists of the system (models, photographers, advertisers, etc.) into celebrities. While fashion firms made a huge investment in the artistic field, firms invested in patronage in order to consolidate their relationship with the artistic universe and to regenerate their brand image even later when, in the second half of the Nineties, both the fashion system and the global brands system suffered a huge crisis of credibility. The figure of Versace is probably is infused with what can be seen as a democratic need, especially in the U.S., of being understood by common people. This is why he created a controversial style—fueled by a creative contradiction between cultural inclusion and social exclusion. Gundle's view here is to the point:

He was a modern Merlin with the power to make dreams come true. ... The lavish window displays of Versace stores on London's Bond Street, Rodeo Drive in Los Angeles, and Via Montenapoleone in Milan confirmed that the brashest and sexiest garments in the world could be bought by anyone. ... The price tags on Versace's ready-to-wear clothes were eye-wateringly steep and the models stunningly attractive. For most, therefore, his creations remained tantalizingly out of reach. But this too was part of the glamour (Gundle 2008, p. 5).

Versace's clothes aroused an immediate aesthetic understanding, but they were too expensive. Luxury democratization (and also his relationship with art) led to a decomposition due to an excess visibility. The Eighties marked the point of a maximum exaltation of the relation between displayed luxury and mass consumption. Luxury needed to touch new peaks to accumulate a symbolic capital that could be spent through strategies of brand extension. At the same time, the "Aristo" trend (Steele 2000, p. 109) signaled the relaunching of *Haute Couture* which coincides with the success of prêt-à-porter all over the world. This is clear when in the second half of the decade, Christian Lacroix launched his new luxury collections based on an extreme and sophisticated style that declared the end of the "democratic" phase. "In the 80's, luxury goods that used to belong to the upper class became visible, recognizable, and accessible to the public. Hence, the market for luxury goods went through an enormous demand growth spurt, and developed into a significant economic sector in the 90's" (Stegemann 2006, p. 60).

One of the protagonists of the Eighties cultural scene was the social type called "the yuppie" or young, upwardly mobile, (preppy) professionals. The yuppie ideal epitomizes an extremely pragmatic ideal of fashion summarized in the famous formula "Dress for Success." In this new hedonistic view, the main purpose of life is pleasure, which compels people to invest in their bodies, fitness, travels, etc.—but at the top of these prerogatives, there is power, obtained via frenetic and obsessive intense activities. In this sense, the luxury claimed by yuppies is diametrically antithetical to the one of the upper middle class of the nineteenth century, still based on bourgeoisie traditional respectability and the Protestant work ethic. Or better put, more than just wasting or speculating with time we could say that this new social type was also *consumed by time*. Yuppies in the Eighties are the social type that embodied the values of the Reagan neoliberalism: young, up to snuff, ruthless, and, above all, social climbers. Yuppie's ethics (urban, young, powerful) was a non-ethic. Yuppies, as the arrival point of the capitalistic ideal and as a point of no return, represent the epilogue of a lifestyle devoted to excess, where an attention to self-image is the precondition for a rapid and exciting career.

The Yuppie's bible movie was the first version of *Wall Street* (1987) directed by Oliver Stone. His main character, Gekko, is a ruthless broker who is obsessed with power and used every means, his women included, in order to manipulate the life of his colleagues and competitors. More than the praise of a definitively Italian elegance (with dresses in the film designed by Nino Cerruti), the fellowship between finance and contemporary art depicted in the movie calls out for scrutiny. In the sequel of the movie directed by Oliver Stone (2010), the topic of *wild capitalism* is combined with a newer trend of a *responsible capitalism* as embodied by Gekko's son-in-law, who aims to invest in a company of renewable energy resources. In contrast to the positive view presented by the media of that time, this new social type is seen in a negative way by the cultural world. In *American Psycho*, Ellis (1991) described his main character as the quintessence of the narcissism and of violence. A landscape created by the most glamorous brands of that time helps the main character to build his identity. The list of the brands include Valentino, Brook Brothers, and Canali, etc. and express a simple and powerful idea: fashion, art, and design are the only contents able to fill the empty space of an individual's identity. Another emblematic and imagery figure of the decade is the pop icon of the Joker. This villain, narrated in the first Batman movie, is committed to ruining the most important pieces of modern art, ripping them while he's dancing to the music of Prince. He's a sort of a "post-modern dandy whose philosophy of art for art has been transformed in consumption for consumption"(Abruzzese 2006, p. 14).

In the second half of the '90s, H. M. Enzensberger examined this contemporary fetish for luxury and elaborated a conservative vision that, according to other analysts of the same period, emphasized the importance of the intangible assets of luxury. From his point of view, there are six key concepts of the new for defining this contemporary sense of luxury: time, attention, space, serenity, environment, and security (Enzensberger 1996, pp. 50–51). In this way, the notion itself shifts from a utilitarian/monetary dimension to an existential one. Anticipating also some trends of digital culture, contemporary luxury moves from a public perspective to a private and daily one. The paradox of his reflection is that while he is criticizing the state of contemporary material culture, he simultaneously, according to the six new values, points out the new trénds of the evolution of this concept. In other words, he defines the trajectory of the so-called new luxury, representing a step forward compared to the democratic luxury achievements that, thanks to mass production, have increased the opportunities of consumption for a large part of the population. When we talk about accessible luxury, we can include in this concept new consumption practices and overall a new kind of consumer, called "*bricoleur*," who can mix trading up with trading down (Fiske and Silverstein 2008), according to economic capacity and fancy. New luxury has been able to incorporate the critical cultural values of the end of the 90s, with fashion that was able to propose an additional differentiation: not just the brand coupled with a fancy style but also culture and ethics. The myth of accessibility is just a part of the story; in fact, if we consider the minimalism of the 90s, we immediately find an overturning of what was said about Versace's style of the 80s. In fact, the minimalism itself is

> The need to construct an identity that spoke of both fashion and cultural capital was acute at a time of anxiety and insecurity. Both the etiolated, androgynous chic of Lang's work, and the ironic subtly of Prada's inverted status symbols, enabled consumption that was gratifyingly fashionable, yet eluded the taint of obviousness (Arnold 2001, p. 20).

This radical overturning of the main values that inspired fashion during the previous decade was particularly clear in the notion of a new "snobbery" introduced by Rebecca Armstrong and analyzed by Arnold. This specific cultural attitude can be applied both to minimalism but also as well to several manifestations of the so-called new luxury.

> Eighties snobbery may have been simplistic, but ... it was democratic, early grasped by everyone. This new version, by contrast, has taken to its heart a completely different system of status symbols that, far from being recognized, from the other end of Bond Street, couldn't be identified from next door (Armstrong 1995, Arnold 2001, p. 21).

Seen in another light, this minimalism participates in a straight redefinition of the notion of luxury, as it has been developed by the so-called new luxury. Helmut Lang's disruption of clothes or the more extreme contamination with mildew and bacteria is a radical and symbolic example of a creativity aiming to disown its original mission. Clothes are not able to be consumed anymore: they are inhospitable like the objects managed by beggars and clochards at the end of the cycle of values. In this highly culturalized idea of fashion design, creativity is sublimated and adjoins more than before with the world of art.

The common denominator in the multiple manifestations of neoluxury is the authenticity that represents the key value not only of the high-range consumptions but also of all the brands and the strategies that are developed in this period under the sign of the "non conventional" marketing (Cova and Saucet 2010). The economy of experience becomes a fundamental approach for comprehending the evolution of consumption in the last decade. It indicates the surpassing of the purely ostentatious conception of consumption toward the idea that buying products or brands means entering into an experiential world. Therefore, the quality of experience, which depends on the variable price but is also irrespective of the same, is the engine of consumption. According to Pine and Gilmore (1999), market evolution naturally leads to a declassing of the economic role of products and services, while emphasizing the new role of the experience, whose orchestration in a new design thinking has created the Total Living trend. If the total look of the '80s was the demonstration that brand power over the consumer is completely controlled by marketing and based on brand extension or stretching strategies, the total living deals with an integration of creativity in multiple expressions. In other words, it is an environmental conception of luxury that enfolds the consumer at the crossroad of the multiple creative languages (fashion, design, art, music, etc.).

In spite of the optimistic and maybe anachronistic intentions of a whole decade, when the immaterial and symbolic power of new consumptions was exalted, the years after 2000 restore the oldest ideal of luxury, pushing it to its extreme borders. Regardless of the successions of financial crises, which magically can solve themselves into increasing exponential profits for those that survive to them, the contemporary luxury is becoming reinforced due to a structural reason. On one side, the mechanics of new capitalism multiplies social injustice, especially if we consider financial speculation. On the other side, the dynamics of social networks, instead of the democratic rhetoric of the so-called digital capitalism, are based on a simple principle, perfectly defined by the network analysts as the so-called Matthew Effect: the "rich get richer"(Barabási and Frangos 2002; Buchanan 2004). On the same path, there are also researchers dedicated to the dramatic effect of the internet and of the so-called sharing economy on the disruption of the middle classes (Lanier 2014).

Creativity, unfastened from a functional base and more and more arbitrary with regard to its permutations and recombinations, is going back to traditional whims of luxury. It is also an indicator of the direction that high-symbolic-value consumptions are assuming in the phase that evolves from cultural change and ethics of neoluxury. Hence a new global superelite dedicates itself to unachievable consumptions for the rest of the global population. It lives in enclaves that require higher levels of protection, control, and safety, especially in the countries such as South Africa and those in South America. The separation of the new superclass does not affect only status, but it manifests itself in an increasingly marked distancing space that passes as the enclosure of the new class from the circuits of daily life. Besides the indicator of financial capital, which is the abstract equivalent whereby it is possible to permute any good, the objects that can define the super rich have a strong symbolic power. For this reason, Kempf (2007, p. 70) recuperates wisely T. Veblen's model because it regenerates the myth of infinite growth and the production-consumption cycle.

What unites the old and the new oligarchy, the old and the new ideal of luxury, is the idea that respectability concerns the class of peers, while the rest of society, inasmuch as they are followers, is not worthy of attention and interest. The excessive inequalities between the social classes are also condemned to increase ecologic issues. As spearhead of the consumption society, which survives only because of the idea that there are patterns of consumption unattainable, this ultra-bourgeoisie shows a schizophrenic mentality: it is, in the main, responsible for the destruction of the environment in a sort of "trickle-down" and disruption; and on the other side, it incorporates the new political vision and values coming from radical ecology. This hyper-luxury is an entropic luxury that brings waste up to the tips of the most extreme generosity and environmental and social impact, but it is also able to produce a specific storytelling that works as a cognitive air bag to soften the social impact of its lifestyle. Even though for a long time it was not fashionable "to be rich," now the money is returning to be flaunted in the most striking manner that still tries to enrich the credibility of a cultural project.

This is the case of the inauguration of the Francois-Henri Pinault museum in Venice, when, for the occasion, there were 920 of his friends "that reached the city by private jets ... in total 160, so the Marco Polo Airport was clogged and then they had to divert more flights to other airports" (Kempf 2007, p. 138).

Moving on the other side of the so-called hyper-luxury, an immense waste machine results that is even more alarming. The value of the low-cost sector triggered by so-called *fast fashion* is impressive. The speed of production and the capability of penetration of all the global markets make this model of business the most dissipative one.

Between them, the economy of experience has accustomed high and low consumers to a new relationship with the brand, products and these sites for consumption. While with *haute couture* the experiential dimension is completely designed under the guidelines coming from the brand identity and is ruled by the value of coherence (with few exceptions such as Givenchy's adoption of Donatella Versace as a testimonial for their latest campaign), in the case of fashion brands, the aim is to reconstruct a sense of authenticity surrounding products, places, and brands that are completely dehumanized by the need of speed. Consider, for example, the capsule collection, produced in a limited range and diffused in the stores as a purely dissipative event. The customer's emotional investment, waiting in line for the opening of the store, fighting with others to catch the best piece, spending his knowhow to identify quickly the best elements fitting with his taste, is something very similar to the new forms of cognitive exploitation recreated by the social networks, with the only difference being that the social network is completely free, while fast fashion has a cost. But in both cases, we see how the user is eager to fill the needs of consumption with his or her own emotions, relations, and experiences.

The centrality of this new interaction between digital devices and experiences reorients the creativity of contemporary fashion in the direction of what I've called "*ontobranding*." The notion of ontobranding (Barile 2009a, 2009b, 2013) defines a dynamic interaction between the user experience, the processes of consumption, and the redefinition of the identity of a place through the user's emotional capital (Illouz 2007). This interaction can be mediated, or, better said, "remediated," by the use of digital technologies, especially in the mobile and ubiquitous form of the smartphones.

Since we moved from a clear opposition between atoms and bits (Negroponte 1995) to a straight integration between them (Jurgenson 2011), digital media became a fundamental tool in the exploration and modification of real objects and contexts. As McLuhan somehow predicted in the 60s, the map is completely overlapped with the territory.

Technologies such as artificial intelligence, augmented reality, geolocalization, and internet of things redefine the brand experience in general and the fashion experience in particular. If traditional branding was just a tool in the hands of companies to build their own image and positioning, self-branding (Barile 2012) demonstrates how marketing is able to manage also the existential positioning of people (as in the formula: "be your own brand").

Biobranding means two different things: the marketing and communication of the biotechnologies, and the colonization of the consumer's everyday life made by global brands. Metabranding is the extension of the branding structure in the external context such as the competitive identity systems (Anholt 2007) for nations, cities, and regions. The more general category of branding—what I am calling *ontobranding*—combines branding with ontology, which is a complicated concept coming from philosophy, used in psychology and sociology. Ontology refers originally to a general theory of being as the total reality in which our local experience is situated.

Today the "ontological difference" (Heidegger 1978) is getting complex, and the modern distinction between virtuality and reality is vanishing.

But ontology is not just related to the dimension of space. It has a lot to do with the notion of time and the way in which the management of experiences today can be considered as a proper "time design." Even if we are still not completely in a world of intelligent things, the "age of Dilution" (Barile 2009a) shows us how communication and branding are flexible enough to colonize new parts of reality, giving them the opportunity to say something, "dialogue," and define their positioning in the worldwide market of identities. If classic science fiction insisted on the evolution of garments as the effect of a technical and material innovation, the contemporary value of fashion has more to do with a smart digital environment made of places, emotions, relations, and experiences (Schmitt 1999).

Never more so than today, the fashion system is undergoing major changes that, with good probability, will radically change its structure over the next decade. Many of the considerations developed during the 1990s—such as those of Ted Polhemus on style surfing and sampling and mixing practices (1994)—appear today not only widespread on a large scale but also driven to their most extreme consequences by fashion designers. On the other hand, instead, a more general socioeconomic dynamic is literally restructuring the market following the new polarization between social classes, that is between a superelite dedicated to unrivaled consumption and a parallel low-cost constituency content with an accessible, quality low cost equivalent (which was not available until recently).

In this situation, to say the least, composed in which cultural dynamics and economic dynamics conflict, the key role of digital technologies is grafted, which, mind you, are not simply mass media added to traditional media such as TV, radio, and the cinema. Digital is more than anything else a new environment capable of incorporating all the previous media and of reconfiguring social and economic relations both in quantitative terms and, above all, in qualitative terms. The digital pervades completely and deeply every area of culture, economy, and contemporary creativity.

Let's take as an example a particularly debated figure like that of the social media influencer. This new species of entity simply would not exist without the structure of social networks that changed our relationship with the world. If the old media regime was populated by opinion leaders, gatekeepers, or at most by trendsetters, the influencer works transversally at the same time on a small scale and on a large scale, as in the case of microblogging preferred by several marketers such as Brown and Fiorella (2013) in their reflection on the influencer-based strategies.

In the near future, the use of chatbots will tend to replace the relationship between brand and consumer with that between artificial intelligence systems and digital assistants. Although more used to show that they are in step with the times (from Armani to Tommy Hilfiger), chatbots allow you to automate the relationship with an increasingly profiled customer; this way, they help to move the focus of fashion from the designer's style to the performativity of the consumer-user. The great debate on artificial intelligence is not just about the idea that much of today's work will disappear in the next decade, because of the fourth industrial revolution, but also and above all the way in which our daily life will be totally transformed by automation not only in physical activities but also in cognitive ones. In fashion, artificial intelligence can represent a "disruptive" technology (i.e., of substitution) not only with respect to communication processes, but also with creative and creative ones—as in some applications by IBM, Watson, and Google that aim to replace the role of the designer by proposing models designed on the characteristics of consumers transformed into data flows.

The paradox of our age is given by the fact that the incredible development of technique does not sacrifice the emotional sphere, rather it urges it, amplifies it, makes it omnipresent. Today, all this high emotional density exudes, from design to retail spaces to communication campaigns, as in Spike Jonze's film *Her*, in which an extremely vintage environment (from costumes to interior design and technology) hosts the love story between the protagonist Theodor and an artificial intelligence system so advanced that it can manage the sentimental sphere. If we tried to replace Samantha's identity with fashion brands, we would understand perfectly where we are going. Jean Baudrillard argued that the problem of artificial intelligence is that it is without artifice, therefore without intelligence. Today, perhaps the picture is profoundly changing toward an AI that is increasingly empathetic and a fashion that will be more and more interested in the potential offered by this new technology. For a few years, I have been working on the concept of onto-branding that concerns the way in which branding on the one hand exploits the most advanced frontiers of automation (artificial intelligence), and on the other tends to exalt and nourish more and more of what is more alive, deep, and authentic in the human being: emotions, which in the meantime have become the currency of exchange for the whole system of communication and consumption.

In the end, we could define at least three scenarios describing the possible interaction between the new fashion system, artificial intelligence, and the consumer's situated experience.

- The first one is the new homology structured around the domestic nucleus. Delivery systems, e-commerce, Netflix, and a pseudodomestic design will build together a new extremely "isolated" lifestyle disrupting traditional consumption practices.[228]

- The second one is the integration between immaterial and material, digital and physical, virtual and real. New devices from the wearable technologies to the augmented reality will redefine the functions and the values of stores and physical places. The consumer profile will be the core of an experiential universe driven by AI.

- The third scenario describes a collective use of AI integrated with social innovation, codesign, and peer production to enhance the genius loci and turn the production process into a political action.

In the very near future, even social innovation processes will be integrated, if not driven, by AI, collecting and managing data extracted from the personal profile of users, during their collective interactions and geolocalized experience of products and brands. Despite the almost obsolete theorizations on the postmodern consumer, emphasizing the processes of fragmentation and hybridization, today we are witness to a new restoration in which classes and the associated lifestyles are polarized. Where the economic developments still divide, culture is still a belt or lasso, trying to keep together the new elites and the low-cost society. The role of the so-called Fourth Industrial Revolution in this case is still ambiguous. On the one hand, it exploits automation technologies that

[228] No doubt the impact of the Covid-19 pandemic will accelerate this sociological metamorphosis.

are able to replace the role of producers, especially the creative one. On the other hand, it creates a culture totally driven by consumers turned into sets of data.

References

Abruzzese, A. (2006). *L'occhio di Joker: cinema e modernità*. 1st ed. Roma: Carocci.

Aldana-Gonzalez, M. (2003). "Nexus: Small Worlds and the Groundbreaking Science of Networks; Linked: The New Science of Networks." In *Physics Today* 56 (n. 3), pp. 71–72.

Anholt, S. (2006). *Competitive Identity: The New Brand Management for Nations, Cities and Regions*. United Kingdom: Palgrave Macmillan.

Armstrong, L. (1995). "The New Snobbery." In *Vogue*.

Arnold, R. (2001). *Fashion, Desire and Anxiety: Image and Morality in the Twentieth Century*. United States: Rutgers University Press.

Baines. Cited in P. Rivoli, *The Travels Of A T-Shirt In The Global Economy*, below.

Barile, N. (2009a). *Brand new world. Il consumo delle marche come forma di rappresentazione del mondo*. Milano: Lupetti.

——— (2009b, October). "From post-human consumer to the ontobranding dimension: Mobile phones and other ubiquitous devices as a new way in which reality can promote itself." Paper presented at mobile communication and social policy conference, New Brunswick, NJ.

——— (2012). "The age of personal web TVs: A cultural analysis of the convergence between web 2.0, branding and everyday life." In A. Abruzzese, N. Barile, J. Gebhardt, J. Vincent, and L. Fortunati (eds.), *The new television ecosystem* (pp. 41–60). Berlin: Peter Lang.

Bell, D. (1973). *The Coming of Post-Industrial Society. A Venture in Social Forecasting*. New York: Basic Books.

Barabási, A.-L., and J. Frangos (2002). *Linked: The New Science of Networks*. 1st ed. Cambridge, MA: Perseus Books Group.

Bauman, Z. (2000). "Tourists and Vagabonds: Or, Living in Postmodern Times." In Joseph E. Davis (ed.), *Identity and Social Change*, pp. 13–26. New Brunswick, NJ: Transaction.

Baumgold, J. (1987). "Dancing on the Lip of the Volcano." In *New York Magazine*.

Blumer, H. (1969). "Fashion: From Class Differentiation to Collective Selection." In *The Sociological Quarterly* 10 (no. 3), pp. 275–291.

- D. Brown, S. Fiorella, *Influence Marketing, How to Create, Manage, and Measure Brand Influencers in Social Media Marketing*, 2013.

Calefato, P. (2014). *Luxury: Fashion, Lifestyle and Excess*. London: Bloomsbury Academic.

Cova, B. and M. Saucet (2010). *Unconventional Marketing: The Routledge Companion to the Future of Marketing*.

Crane D. (2000). *Fashion and Its Social Agendas: Class, Gender, and Identity in Clothing*. Chicago: University of Chicago Press.

David, Deirdre (2012). *Olivia Manning: A Woman at War*. Oxford University Press.

Davis, F. (1992). *Fashion, Culture and Identity*. Chicago: University of Chicago Press.

Ellis, B. E. (1991). *American Psycho*. Picador.

Hans Magnus Enzensberger, *Zig Zag: The Politics of Culture and Vice Versa*. New Press,1997.
Ewen, S. (1988). *All Consuming Images*. Basic Books.
Fiske, N., and M.J. Silverstein (2008) *Trading Up: Why Consumers Want New Luxury Goods—and How Companies Create Them*. Portfolio Trade.
Fuchs, C. (2013). *Digital Labour and Karl Marx*. London: Routledge.
Gundle, S. (2008). *Glamour: A History*. Oxford: Oxford University Press.
Heidegger, M. (1978). *Being and Time*. London: Blackwell.
Illouz, E. (2007). *Cold Intimacies: The Making of Emotional Capitalism*. Oxford: Polity Press.
Jackson, T., and D. Shaw (2007). *Mastering Fashion Marketing*. In the Palgrave Master Series. New York: Palgrave Macmillan.
Jurgenson, N. (2011). "Digital Dualism and the Fallacy of Web Objectivity." Cyborgology, accessed September 13, 2011. http://thesocietypages.org/cyborgology/2011/09/13/digital-dualism-and-the-fallacy-of-web-objectivity/.
Kempf, H. (2007). *Comment les riches détruisent la planète*. Paris: Éditions du Seuil.
Lanier, J. (2014). *Who Owns the Future?* United Kingdom: Penguin Books.
Lipovetski, Gilles (1987), *L'empire de l'éphèmere*, Paris: Gallimard.
Morin, E. (1962) *L'Esprit du temps* [*The Spirit of the Time*]. Paris: Grasset.
Negroponte, N. (1995). *Being Digital*. New York: Alfred A. Knopf.
Pine, J. B., and J.H. Gilmore (1999). *The Experience Economy*. Boston, MA: Harvard Business Review Press.
Polhemus T., Proctor (1984). *Fetish Fashion in Fashion 1985*, New York, St. Martin's Press.
Reich, Robert B. (2008). *Supercapitalism: The Transformation of Business, Democracy, and Everyday Life*. New York: Vintage Book.
Rifkin, J. (1994). *The End of Work: The Decline of the Global Labor Force and the Dawn of the Post-Market Era*. New York: Tarcher/Putnam.
Rivoli, P. (2006). "The Travels of a T-Shirt in the Global Economy: An Economist Examines the Markets, Power, and Politics of World Trade." In *Foreign Affairs* 85 (no. 2).
Scheffer, Michiel (2009). "Fashion Design and Technologies in a Global Context." In E. Paulicelli, and H. Clark, eds. (2008), *The Fabric of Cultures: Fashion, Identity, and Globalization*. 1st ed. Oxford: Taylor & Francis.
Schmitt, B. H. (1999). *Experiential Marketing: How to Get Customers to Sense, Feel, Think, Act, Relate to Your Company and Brands*. New York, NY: The Free Press.
Steele, Valerie, *Fetish. Sesso moda e potere*, Roma, Meltemi, 2005.
Steele, Valerie, *Fifty Years of Fashion: New Look to Now*, London-New Haven, Yale University Press, 1997.
Stegemann, Nicole, "Unique Brand Extension Challenges For Luxury Brands," *Journal of Business & Economics Research*, Vol. 4, October 2006.
Veblen, T. (1912). *Theory of the Leisure Class*. New York: B. W. Huebsch.
Welters, L., and A. Lillethun, eds. (2011). *The Fashion Reader*. 2nd ed. New York: Berg Publishers.

MORE THAN THE EVENING NEWS B-ROLL
THE POTENTIALITY OF HEALING FOR QUEER AND TRANS PEOPLE OF COLOR (QTPOC) THROUGH COMICS

KATLIN MARISOL SWEENEY

Pulse, a popular LGBTQ+ nightclub in Orlando, Florida, became the site of national and global attention when a gunman entered the club during "Latin Night" on June 12, 2016 at approximately 2 a.m. and proceeded to fire into the crowd, resulting in the death of forty-nine people and the injury of fifty-three others. This tragedy received substantial media attention by a variety of mainstream news and smaller media outlets, particularly in the days immediately following the attack. However, much of this coverage failed to acknowledge that the victims and survivors of the Pulse nightclub shooting were not only queer, but QTPOC—queer and trans people of color—who were primarily Latinx, Black, and Afro-Latinx. Relatedly, the majority of the 2016 media coverage alternated between offering up-to-the-moment information about the shooter and looking at the tragedy from the perspective of those who were inside of Pulse that night. [229] The coverage of the Pulse nightclub shooting by mainstream news outlets did ensure that the tragedy was made visible to the public on a global scale, but the film and still images produced during this time demonstrated the media and public's appetite for seeing the victims and survivors of the shooting as graphic carnage in the form of bleeding limbs, pain-stricken faces, and cries of anguish, rather than as people deserving of adequate privacy.

In response to the Pulse nightclub shooting and the tone of the media coverage that happened on and immediately after June 12, 2016, IDW Publishing and DC Entertainment copublished a collection of multiauthored comics titled *Love Is Love*. The project was originated by Marc Andreyko and IDW Publishing. It was edited by Sarah Gaydos and Jamie S. Rich, assistant edited by Maggie Howell, and designed by Annie Brockway-Metcalf, all of whom are affiliated with DC Comics. Unlike the media coverage of the time, which primarily functioned as a way for this tragedy and its aftermath to be made visible to the global public, *Love Is Love* is invested in QTPOC's feelings about the tragedy to a degree that goes beyond fighting for temporary representation in the media. Instead, the anthology is invested in QTPOC's process of dealing with this tragedy in the long-term by making their pain visible vis-à-vis comics focalized by QTPOC and their allies. In this way, *Love Is Love* offers an alternative space to what was created by mainstream media's coverage of the shooting in June 2016. By centering the feelings and experiences of

[229] The shooter's name and background will remain unnamed in this piece out of respect for the victims and survivors of the attack.

LOVE is LOVE
A comic book anthology to benefit the survivors of the Orlando Pulse shooting
I LOVE MY LESB DAUGHTE
My Dads Rock
My Moms Rock
TRANS MEN ARE MEN
LOVE IS LOVE
Elsa 2016.

QTPOC and their allies in these comics, *Love Is Love* makes a healing space narratively possible for characters and readers. This healing takes two forms: first, these narratives explore how QTPOC survivors, victims, and their loved ones reacted to the shootings (these stories intentionally refuse to dwell on who the gunman was or to offer any rationale for his actions, unlike the mainstream media); second, the anthology was published to make a real-world difference in the lives of those affected by the tragedy, as all proceeds raised from this anthology were donated to help those impacted by the shooting.

With this combination of narrative and financial goals focused on healing, *Love Is Love* emerges as a collection deeply invested in the potentiality of healing from pain for QTPOC. I argue that the individual comics in *Love Is Love* visualize opportunities for QTPOC's healing in three ways. First, many of the comics feature the recurring image of the living room television delivering the news of the Pulse nightclub shooting, which indexes the June 2016 media coverage to engage with the limitations of the television news media format in contributing to the healing of QTPOC. Second, narratives that show the living room television delivering the news coupled with a meaningful conversation between the viewer and their loved ones emphasizes the real-world impacts of media coverage in QTPOC's daily lives, especially in coming out and family acceptance. Third, the open-endedness of these comics resists the trope in fictional media of all queer partnerships and lives ending in death by representing QTPOC's lives as having the capacity "to go on" after grappling with tragedy. These three characteristics of the narratives included in the anthology, combined with the paratextual elements of the physical edition that highlight its charitable contributions, coalesce to inform my reading of *Love Is Love* as a comics anthology "for a cause" that is effective in generating a site of narrative and real-world healing for queer and trans people of color rather than perpetuating their erasure in the media.

THE "CONSTRUCTEDNESS" OF HISTORICAL TRAGEDY ON THE PAGE

IDW Publishing and DC Entertainment's decision to collaborate on and publish a comics anthology about QTPOC survivors and their loved ones' reactions to the Pulse shooting reflects how comics as a storytelling mode are equipped to visually and narratively explore questions of tragedy, identity, and futurity in ways that are unique to graphic narrative/sequential art. How these comics look and how the reader looks at them matter in terms of how these histories are recorded and remembered by the reader, who may or may not have previous knowledge of these histories prior to reading the comic.

Related to but not identical to the process of watching a film or reading a prose novel (during which a viewer or reader may forget that they are receiving these images through a particular narrative form), the material qualities unique to the comic book and graphic novel remind the reader holding these texts that they are engaging with a narrative that is mediated through the creator/creators' perspective on how it should appear in print. The materiality of the comic book and graphic novel is not only visible in its status as a physical object, but also through the markings on the comics page that hint at the substantial amount of time and work that went into producing this "constructed" form. As Jared Gardner writes in "Storylines" (2011), we approach the comics page much differently than we do the page of a prose novel, given that

> we never look at the printed book and imagine that the font gives us access to the labor involved in the scene of writing ... [while] on the other hand, we *cannot* look at the graphic narrative and imagine that the line does *not* give us access to the labored making of the storyworld we are encountering (p. 64).

Characteristics of comics such as the line, shading, color depth (or the lack of color), and the size of panels allow the reader to engage with works like the individual comics in *Love Is Love* in a way that goes beyond reading, analyzing, and trying to imagine the dialogue. Readers are able to hear and see these conversations in action, which allows them to visually access additional details about the storyworld that would otherwise be unknowable without verbal acknowledgement, such as facial expressions, body language, and seemingly "unimportant" objects scattered throughout the mise-en-scène.

Additionally, as Hillary Chute contends in *Why Comics? From Underground to Everywhere* (2017), the comics page can represent multiple moments in time at the same time, while also visually commanding that the reader embrace what she calls its "all-at-onceness" or "symphonic effect" in which "one's eye takes in the whole page, even when one decides to start in the upper left corner and move left to right" (p. 25). The ability to see "everything" at once, even if one attempts to restrict later panels to their peripheral view, impacts how the reader interprets the meaning of earlier panels, especially when these earlier panels are building to later explorations of significant tragedy. This is of particular importance to how one reads the stand-alone comics in *Love Is Love*, given that each of these comics are brief in length and typically show the beginning, middle, and end of the story within one to two pages. Additionally, the reader enters these individual storyworlds with the prior knowledge that they will all address the Pulse shooting or QTPOC's feelings during the post-June 2016 period in some way, which influences how they may interpret the meaning of some of the details of the page as mentioned above.

Scholars of comics studies have also underscored the unique potential of comics to narratively represent communities, historical tragedies, and experiences that may otherwise remain unacknowledged or purposefully erased. As Kate Polak describes in *Ethics in the Gutter* (2017), integral to the storytelling mode of comics and graphic novels is "the constructedness of the page" that allows for "a space of imaginative possibility" to flourish (p. 14), in order for "graphic narratives [to] fictionalize dark episodes in history" (p. 36). Chute argues in *Disaster Drawn: Visual Witness, Comics, and Documentary Form* (2016) that the process of representing historical tragedy on the comics page is an act of "'materializing' history ... [which] creates it as space and substance, giv[ing] it a corporeality, a physical shape—like a suit, perhaps, for an absent body, or to make evident the kind of space-time many bodies move in and move through" (p. 27). Similarly, as Frederick Luis Aldama writes in his introduction to *Redrawing the Historical Past: History, Memory, and Multiethnic Graphic Novels* (2018), "graphic novelists from a wide variety of ethnic planetary experiences use visual-verbal formats to enrich our understanding of individuals weighed down and destroyed by the past along with those who overcome histories of racial oppression" (p. ix). Consistent among Polak, Chute, and Aldama's arguments is the emphasis that the power of comics and graphic novels is significant in part due to their ability to "make space" for characters and stories whose living counterparts have experienced marginalization in mainstream media and in the real world. Comics and graphic novels, particularly those written by and for people who experience different forms of marginalization, effectively offer a kind of media representation distinct from what is produced by mainstream US media, which often fails to put these stories on-screen or on the page at all. Understanding the comics page as bearing markers of "constructedness" that effectively "materialize" history is a useful framework for examining how the comics form is an effective, alternative storytelling mode for creators wanting to represent the Pulse shooting differently than how it appeared in mainstream news media in 2016.

MAINSTREAM HISTORY: CABLE NEWS COVERAGE OF THE PULSE SHOOTING IN JUNE 2016

In order to examine the kind of intervention that is being made by the comics in *Love Is Love* that feature the living room television delivering news of the shooting, it is crucial to first assess the trends of how mainstream media portrayed survivors and victims of the Pulse shooting on the day of and the days immediately after June 12, 2016. While the Pulse shooting was well represented in both print and television news media, this essay specifically tracks how television news segments reported on the tragedy.

This dedicated focus on television news is not meant to discount the many articles that were written during this period from both mainstream and smaller media publications, the content of which varied widely in terms of how the survivors and victims were represented. In part, the purpose of looking exclusively at how mainstream televisual media reported on the shooting is to identify some of the problems with cable news coverage that have arisen in the post-9/11 and post–Operation Desert Storm media age. As Jared Gardner explains in “Time under Siege” (2015),

> as the new twenty-four-hour cable news station, pioneered by Ted Turner’s CNN in the 1980s, merged with the rise of the World Wide Web in the 1990s ... satellite television and the Internet ... have combined to make the world feel “smaller,” collapsing distances and creating the demand for “real time” access to news and information (p. 23).

The inundation of the US public with television content in the early 2000s, combined with the onset of technological expediency in the post-2010 era, has produced a national news ecosystem comprised of more: more content, more access, and more speed. While the viewing public demands to know what is happening as soon as it happens—to the point that cell phones are now programmed to deliver this information in the form of a notification—television, print, and online news publications clamor to be the first to report on these events as they happen. As a result, the US news business now faces the dilemma of not only having to deliver the news as quickly as possible, but having to offer content to its viewing public at all hours of the day.

Part of what this project is invested in is problematizing how the bodies of injured QTPOC, specifically those that were bleeding or in distress, were repeatedly filmed and shown on-screen as the “b-roll” as news anchors spoke about the latest updates on the shooting. B-roll can be described as the video footage shown on-screen during a television news segment, typically without any sound, while the newscaster narrates the events that are taking place or speaks about a related topic. This b-roll is related to the news segment’s focus in that it is meant to present us with images that visually explain what is being verbally expressed to us. The inclusion of b-roll footage is common on cable newscasts as a way to hook hearing viewers—especially those in public spaces like bars or restaurants—who may have their television set to mute. By including these images on the screen in addition to the breaking news ticker and headline, viewers who may be glancing at the screen may be motivated to turn on the sound so that they can hear the “full story.”

The television news coverage of the event was consistent in two regards: repeating the same information about the event while showing different b-roll each time and downplaying that it was QTPOC who were most affected by the tragedy. These consistencies can be tracked based on how the coverage produced by television news media in 2016 primarily covered the shooting by showing four images: footage of victims running for their lives, interviewing loved ones of the survivors and victims, uncovering details about the gunman’s motivations and history, and airing brief memorials for each of the forty-nine victims.

Consistent across the news coverage offered by mainstream news channels such as CNN, ABC, FOX, and NBC on and after June 12, 2016 were the following trends: camera crews following or chasing after injured QTPOC, zooming in on injured QTPOC from afar to better film their distress, and sustaining view of their injured bodies on-screen. Many of these stations often showed the same or similar footage of injured QTPOC, in addition to interviewing similar people on the night of the attack. For example, in the news broadcasts released by ABC, b-roll footage that did not show the survivors and victims alternated between the emergency vehicles and personnel, crowds gathered around the site, and the Pulse nightclub sign. Alternatively, the b-roll footage that focused on survivors and victims showed the following: a person whose waist was covered in what is either blood or a red pattern (it is unclear) and whose white pants had fallen down, revealing their bare buttocks, while they are limping away from the club with the help of two other people; an individual being carried by two people, whose right knee is bleeding profusely and had been bandaged with what appears to be a bandana; an injured individual laid out on the ground with two people kneeling above him; another person whose pants are starting to fall down while they struggle to walk while leaning on another person as they seek out help.

In each of these instances, the person operating the camera rapidly zooms in on the injured person's body to get a better look at the individual's wounds. The camera moves particularly quickly when blood is visible on the person's body, when pain is visible on their face, or when an article of clothing is falling off. In the case of the person whose white pants are falling down, the camera begins with a wide shot and quickly narrows to a closer view to focus on the individual's waist where the pants are beginning to slip. In the case of the individual who is laying on the ground, the clip is relatively brief, with their face and shoulders are the only parts of their body visible to the camera, which is at a distance from the scene. Once the camera is able to focus on the individual's face in distress, the distanced shot zooms in to give the viewer a closer look at the person's pained expression. Additionally, in the case of the person whose knee is bleeding, the two people carrying this individual are rushing them along to get them help from the emergency personnel. The person operating the camera quickly moves it to follow the movement of the three people as they rush past them, with the camera zooming in closer and closer to the bloody knee until the person's face is no longer visible.

These images are shown repeatedly on-screen in different news segments—such as in the ABC Special News Report, where these images function as the b-roll to a survivor's interview, and in the breaking news segment reported by David Muir, where these images are the b-roll that supplements Muir's report. These filming techniques reflect how the decision of major news networks to constantly show new images of "bodies in pain" allows them to heighten their profit margins through increased viewership and more advertising sold. These images cater to the instinct that viewers have to gaze upon images of crisis—whether a car accident or a shooting—by continuing to show these images on-screen whenever the story is being covered.

The video coverage that survivors and victims of the Pulse shooting received is consistent with how cable news covered other tragedies of the time: the Manchester Arena bombing at an Ariana Grande concert in Manchester, England (May 2017) and the Route 91 Harvest music festival shooting in Las Vegas, Nevada (October 2017). Though these tragedies did not explicitly impact the QTPOC community, like the coverage of the Pulse shooting, these events were talked about in the news by showing a continuous stream of images of people running for their lives, screaming, and being scared. This visual representation of tragedy as one that is predicated on giving the viewing public the chance to gaze upon pain and suffering is something that *Love Is Love* pushes back against by showing QTPOC as subjects with agency and humanity, not simply bodies in pain to be consumed as part of the mainstream news's daily scoop.

THE "SHELF-APPEAL" OF PARATEXT: COMICS "FOR A CAUSE"

The format of *Love Is Love* is that of a comics anthology: a collection of works that has been assembled by a primary organization or creator around a central theme, with each of the collected works functioning as stand-alone comics. Similar to formats such as zines and literary magazines, the comics anthology format brings together a multitude of voices and presents the possibility for a reader who may not be familiar with one or all of the authors to engage with their work. In doing so, the comics anthology allows for readers to learn more about a topic that they otherwise may know only generalized information about. Additionally, the spaces where an anthology like *Love Is Love* is available matters in terms of its potential success in selling copies to the public. This genre, while sometimes available in both print and digital forms, typically relies on the materiality of the anthology for its distribution to be far-reaching. Temporary sites such as zine fests, tabling events, and fundraisers allow for anthologies like *Love Is Love* to be made available and sold to specialized crowds that are more likely to have an interest in purchasing and reading this work. However, given that many potential readers for this anthology were already engaging with the Pulse shooting through online news media, much of the success of the collection is due to the fact that it was well marketed in online publications such as the *New York Times*, *NPR*, and *Washington Post*—spaces where readers were already going to read related articles about the shooting.

The comics anthology genre that *Love Is Love* is part of also features a more specialized subgenre that can be identified as the comics anthology "for a cause." This genre consists of multiauthored works that originate with an organization or main creator proposing a call for submissions in which creators will submit content to be included in an anthology and sold to raise funds for a charitable cause. Recent examples of the comics anthology for-a-cause genre include *Puerto Rico Strong* (2018) and *Ricanstruction* (2018), which each donated one hundred percent of proceeds to disaster relief and recovery programs for the people of Puerto Rico, and *Little Heroes* (2018), a UK-based organization that uses proceeds from the anthology to purchase comic-making kits for chronically ill children who spend a lot of time in hospitals.

Love Is Love participates in this specialized subgenre of "for a cause" in its move to donate all proceeds from purchased copies of the anthology to support the Orlando-based charity Equality Florida Institute. It is worth noting that after *Love Is Love* was released in December 2016, it raised $165,000 by March 2017, which, through the support of Equality Florida Institute, was donated to the OneOrlando Fund to support the survivors and loved ones of victims of the Pulse shooting.

While the anthology was released as both a digital and physical collection, some of the paratext that emphasizes its charitable purpose relies on a degree of "shelf-appeal"—that the anthology be seen in person at a bookstore and physically held or flipped through—for a potential buyer to be made aware of this aspect of the project. For digital readers, the possibility exists that they may overlook these details given that these paratextual elements tend to blend into the page as it appears on-screen.

The comic book industry comes together to support the survivors and honor those killed at the Pulse nightclub in Orlando, Florida, on June 12, 2016.

Writers and artists from across the globe have created exclusive new material expressing their sorrow, compassion, frustration, and hope, all inspired by the tragic events. In doing so, they celebrate the victims, survivors, and their families while also spreading a message of peace and inclusion.

Organized by **Marc Andreyko** and **IDW Publishing** with assistance from **DC Comics**, **LOVE IS LOVE** also includes many beloved characters from across the comics field, including **Supergirl, Hack/Slash,** Will Eisner's **The Spirit, Southern Bastards, Batwoman,** Archie Comics' **Kevin Keller, Alters,** and **Harley Quinn,** to name a few.

INTRODUCTION BY
PATTY JENKINS
(DIRECTOR OF *MONSTER* AND *WONDER WOMAN*)

Cover by **Elsa Charretier**
with **Jordie Bellaire**

$19.99 USA ISBN: 978-1-63140-939-4
Suggested for mature readers.
Printed and bound in Canada.
idwpublishing.com
dccomics.com

IDW

INCLUDES WORK BY
PHIL JIMENEZ
CAT STAGGS
STEVE ORLANDO
GAIL SIMONE
MATT BOMER
PAUL DINI
MORGAN SPURLOCK
MING DOYLE
JOE KELLY
JAY EDIDIN
JONATHAN HICKMAN
BRIAN MICHAEL BENDIS
MARC BERNARDIN
MARGUERITE BENNETT
ED LUCE
MATT WAGNER
JASON AARON
JASON LATOUR
JAMES ASMUS
MARK MILLAR
PAUL JENKINS
SINA GRACE
PHILIP TAN
G. WILLOW WILSON
JOSÉ VILLARRUBIA
TARAN KILLAM
JIM LEE
ANEKE
JUDD WINICK
MIKE CAREY
P. CRAIG RUSSELL
BRAD MELTZER
PATTON OSWALT
DONNA BARR
CARLA SPEED MCNEIL
AND MANY MORE

EDITORIAL AND RELATED SERVICES PROVIDED BY

All proceeds from this and future editions of Love is Love will go to LGBTQA charities.

The physical edition of *Love Is Love* resists this possibility that its charitable cause can be overlooked by repeatedly putting readers into physical contact with this message when they hold the anthology. The cover, title page, spine, back cover, and inside back cover of *Love Is Love* all bear notations that state the anthology's purpose of raising funds for those impacted by the Pulse shooting. The spine, inside back cover, and back cover all specify which foundation the proceeds will be donated to, with the back cover stating that "all proceeds from this book will be going to the victims, survivors, and their families via Equality Florida." Equality Florida Institute, as described on the inside back cover and on the organization's website, "is the largest civil rights organization dedicated to securing full equality for Florida's lesbian, gay, bisexual, and transgender (LGBT) community." The cover and title page offer a generalized description of the anthology's charitable cause by describing the collection as "a comic book anthology to benefit the survivors of the Orlando Pulse shooting." This notation about *Love Is Love*'s purpose to fundraise for Equality Florida Institute makes it clear that this collection can be designated as not only a print comics anthology, but as part of the genre that is the print comics anthology for a cause.

Relatedly, the anthology also opens to an in-memoriam page printed on the inside cover, which contains a list of the names and ages of the forty-nine QTPOC who were murdered in the Pulse Nightclub shooting. The in-memoriam page's location in the text also literally comes first, which symbolizes how these victims should also be remembered before and eventually instead of the gunman. The inclusion of this list frames *Love Is Love* as a visual media that seeks to center QTPOC rather than the gunman, who was the focus of much of the news coverage in 2016.

The physical version of the anthology allows for the reader to take most notice of its charitable cause given that wherever the reader touches before reading the individual comics—the front cover, the back cover, the spine, the title page, or the inside back cover—they will be reminded of the anthology's purpose of supporting survivors, victims, and their loved ones in the long term. In total, all of the paratextual elements that the reader engages with prior to reading the individual comics inform the reader's understanding of this text as one that is invested narratively and financially in helping QTPOC survivors heal from the pain caused by the Pulse Nightclub shooting.

COMICS AS A MODE OF HEALING

The more than thirty-five pieces of art and stories featured in *Love Is Love* memorialize the forty-nine victims and stand in solidarity with the fifty-three survivors of the shooting by indexing the 2016 media coverage of the Pulse shooting to explore how these images produced real-world impacts to those watching from their living rooms. The eleven comics in the anthology that represent the television in the living room are often left open-ended to reflect how the lives of these characters, particularly those that are QTPOC, can and will continue on after the moment of tragedy. By resisting conclusion or a stable ending, these comics demonstrate how the lives of QTPOC are not restricted to the moment of tragedy and are literally and figuratively left open to the possibility of a future after the shooting. Additionally, these narratives alternate between showing what can be understood as the reaction moment and subsequent inaction, and the meaningful interaction produced by the breaking news. When the physical version of the comics anthology is opened to show two pages, these television news comics are often placed near stand-alone comics by creators such as Emma Houxbois and Alejandra Gutierrez (p. 51), Brian Michael and Olivia Bendis (pp. 52–53), Dave Crosland (p. 64), Tim Seeley and Mark Englert (p. 85), B. Alex Thompson (p. 87), and Ed Luce (p. 129), which portray QTPOC in various states of love, joy, and fulfillment. The inclusion of these joyful comics alongside the comics that index the 2016 media coverage demonstrates how the anthology as a whole is invested in the possibilities for narrative healing beyond the Pulse nightclub shooting by showing alternate storyworlds that are not centered on the day of the tragedy itself.

The comics in *Love Is Love* that feature the reaction moment and subsequent inaction are typically structured as narratives with a definitive ending, which reflects how the narrative addresses that the narrator has failed to act in any way that will produce healing for QTPOC and their loved ones. "Thoughts and Prayers: A Confession," written by Jeff Jensen, illustrated by David Lopez, and lettered by Dezi Sienty, tells the story of a father and son getting ready to go to church when they see the news on their living room television that the Pulse shooting has happened (pp. 8–9). The father hurries his son away from the breaking news segment that bears the CNN logo in the bottom right hand corner and proceeds to brainstorm in a pain-stricken mode over what he can contribute to activist efforts. In the second to last panel, the father sits at the dinner table with his family, hands clasped together in prayer, and he thinks to himself, "But the day gets away from me, as it always does. And so all I do is bring it before the Lord" (p. 9). In the final panel, the father is seen laying in his bed with his arms outstretched on either side of him and his face vacant of any expression; above his head, it reads, "And I leave it there" (p. 9).

"Thoughts and Prayers: A Confession" symbolizes the failings of the television media form in contributing to the healing of QTPOC through its depiction of all of the "almost" that the father fails to act on. For example, the father pulls up an article by *People* magazine that offers strategies for demanding gun law reform, but does not make the call. The father also considers donating to a GoFundMe and sharing his donation on Facebook, but ultimately chooses not to make a contribution. Additionally, he considers calling his queer and QTPOC friends to check on them, but does not follow through on this either. The father's failure to follow through on any of these possible actions to support QTPOC in a meaningful way demonstrates how the mainstream media coverage that occurred during this time failed to direct viewers to ways that they can contribute in ways that smaller media outlets did. Based on the final panel's last look at the father lying awake in his bed, the comic leaves the reader with the sense that the father will never change his ways. The comic ending on the line "And I leave it there," which completes the sentence from the previous panel, also acts as a conclusion to the narrative in its completion of the storyline.

The comics in *Love Is Love* that feature the meaningful interaction between two characters because of the breaking news coverage of the Pulse shooting on the living room television offer an open-endedness that suggests infinite possibilities beyond the moment of encounter with the news. The untitled comic written by Jeff King, with art by Steve Pugh, lettered by Todd Klein, and colored by Quinton Winter is a coming-out narrative in which the breaking news is the catalyst for the queer child to talk about their sexuality with her father (p. 16).

The comic opens with the daughter sobbing in front of the living room television, with her text conversation with a friend floating above the speech bubble of her father, who is standing at the foot of the steps and uttering homophobic remarks. These negative comments from the father prompt the daughter to ask how he would feel if she were gay, with the following conversation in which her father struggles to accept and support her taking place in front of the living room television showing the Pulse nightclub vigil on-screen. The final panel of the single-page comic shows the daughter running to her father to hug him, with the father's arms beginning to raise in reciprocation as the narrative ends. The hug, which was prompted by the father's remarks that he would like to accompany his daughter to the Orlando memorial for the Pulse victims, visually represents an open ending in that the hug does not guarantee that this newfound connection will be sustained. Rather, it hints at a possible outcome in which the father may literally and figuratively embrace his daughter and her sexuality. The presence of the living room television delivering the breaking news makes the space possible for this conversation to take place, effectively creating the opportunity for healing to occur between this distanced father and daughter.

Similarly, the comic written by Rob Williams, with art by Mike Dowling, and letters by Corey Breen titled "Helpless/Not Helpless" also shows a father-child relationship potentially strengthened or recovered after the parent witnesses the breaking news about the Pulse shooting (p. 17). The reader receives little context for this single-page panel, which shares visual space with Jeff King's unnamed comic on the previous page. Rather, the only voices the reader hears is that of the newscaster on the living room television, who is stating the death toll of the shooting, and that of the father when he goes upstairs to check on his child in their bedroom,

THEY'RE CONSOLING EACH OTHER, DAD.
HOW WOULD YOU FEEL IF I WAS THERE?
HOW WOULD YOU FEEL IF I WAS GAY?
I'M GAY.
IF YOU BELIEVE WHAT HE BELIEVES, YOU'D DO WHAT HE DID. THINK ABOUT IT. PEOPLE ACT BASED ON HOW THEY SEE THE WORLD. EVERY SINGLE TIME. UNDERSTANDING SOMEONE ELSE'S STORY, THAT'S HARD. BUT IT'S WORTH IT. EVERY SINGLE TIME. THEN WE DON'T HAVE SH*T LIKE THIS HAPPEN.
LOVE IS LOVE
THERE'S A MEMORIAL TOMORROW IN ORLANDO.
I'M GOING.
LET ME COME WITH YOU.
PLEASE. I'M THE LAST PERSON YOU WANT TO BE WITH BECAUSE WHAT I SAID JUST NOW WAS THOUGHTLESS AND CRUEL.
BUT...IF WE LEAVE NOW AND DRIVE ALL NIGHT, WE CAN JUST MAKE IT BY MORNING.
I...I NEED YOU TO KNOW THAT NO MATTER WHAT I SAID ABOUT THOSE MEN, I WILL NEVER, EVER FEEL ANYTHING OTHER THAN LOVE FOR YOU. WHATEVER YOU SAY DOESN'T CHANGE THAT.

during which he simply says, "I love you." This relatively silent comic offers no conclusions or additional details about the state of the relationship between the parent and child, but makes it clear that hearing the news of the shooting is what prompts the father to express their love to their young child.

When the father clicks off the television, he picks up the teddy bear that the child has left on the ground and takes it to them, effectively placing it on the child's pillow in a way that gestures at the child being given a "bear" hug. The open-endedness of the comics suggests that the intimacy of the shooting vis-à-vis the living room television prompts a reaction—though seemingly small—that alters the parent's behavior with their child. Additionally, this moment suggests that something is beginning to open up and potentially heal in this relationship based on its attention to the father's desires to be with his sleeping child to protect them in the dark.

The consistency in which these comics portray a change in the relationship between parent and child is also present in the untitled comic by Nunzio DeFilippis and Christina Weir, illustrated by Emma Vieceli, colored by Christina Strain, and with letters by Neal Bailey (p.15). In this comic, the living room television is barely visible in the first panel, with the speech bubble from the screen being the reader's strongest clue that this is what is being obscured from view. The following panels show the viewer, who is a person of color, sobbing as he watches the screen and hears the rising death toll. His viewing is interrupted by a phone call from his parents, during which they beg him to be careful but also out and proud. The single-page comic ends with the viewer clutching the phone to his chest while his eyes are closed, tears continue to stream down his face as he smiles. The tears on the protagonist's face at the beginning and end of the comic are connected in that they visually mark the shift from pain to healing in the comic. The tears are produced during his reaction to the news segment and change during the course of his conversation with his parents. The protagonist attempts to hastily wipe away his tears as he answers the phone call, but upon hearing from his parents that they accept him as he is, he resumes crying in the final panel. However, while these tears initially marked the significant pain he is feeling at learning of the shooting, the tears in the final panel show how the television coverage—which his parents presumably have also seen—created space for he and his parents to have a conversation of acceptance. Additionally, the glow of his cell phone that illuminates his chest after the call ends visualizes how he feels loved and accepted rather than hated as he did originally.

CONCLUSION

The failings of mainstream news' treatment of QTPOC is something that narrative media like comics can confront and unpack by creating an alternate visual space that allows for the nuances and diversity of queer and trans people of color to be seen and heard from, especially for those members of the queer community who are often most marginalized. In particular, the print comics anthology format is one of the most effective models for this work to be done in, given that these anthologies typically ask contributors to submit content that addresses a mutual theme, with the intent of placing these comics back-to-back in the collection so that readers are exposed to works by artists familiar to them and those who are not.

If we consider how the mainstream news coverage of the Pulse Nightclub shooting often times situated QTPOC as b-roll, what these comics do instead is resist this secondary, b-roll status by presenting QTPOC as the main characters of these narratives, as figures we have a declared investment in beyond simply hearing the latest tragic news of the day. Through reading these individual, memoir-style comics, what occurs is that readers, even if they do not identify as QTPOC, have the opportunity to develop empathy for those who are affected by this tragedy and who are trying to heal from this pain by listening to these narratives. Additionally, this empathy develops while readers who purchase this anthology also make a financial contribution to those affected that can produce a real-world, positive impact beyond simple awareness.

References

Aldama, Frederick Luis (2018). "Coloring a Planetary Republic of Comics." In Martha J. Cutter and Cathy J. Schlund-Vials, eds., *Redrawing the Historical Past: History, Memory, and Multiethnic Graphic Novels*. Athens, GA: The University of Georgia Press.

Chute, Hillary (2016). *Disaster Drawn: Visual Witness, Comics, and Documentary Form*. Cambridge, MA: The Belknap Press of Harvard University Press.

——— (2017). *Why Comics? From Underground to Everywhere*. New York: HarperCollins.

Gardner, Jared (2011). "Storylines." In *SubStance* 40 (no. 1).

——— (2015). "Time under Siege." In Daniel Worden, ed., *The Comics of Joe Sacco: Journalism in a Visual World*. Jackson, MS: University of Mississippi Press.

Gaydos, Sarah and Jamie S. Rich (2016). *Love Is Love*. San Diego, CA: IDW Publishing.

Lopez, Marco et al., eds. (2018). *Puerto Rico Strong*. St. Louis, MO: Lion Forge.

Miranda-Rodriguez, Edgardo, ed. (2018). *Ricanstruction: Reminiscing and Rebuilding Puerto Rico*. Somos Arte, LLC.

Polak, Kate (2017). *Ethics in the Gutter: Empathy and Historical Fiction in Comics*. Columbus, OH: The Ohio State University Press.

RACIALIZED COMEDIC SIDEKICKS IN DISNEY'S *CINDERELLA* AND *THE PRINCESS AND THE FROG*

BONNIE OPLIGER

Disney's issues with racial representations, especially in the company's earlier films, are fairly well known. However, it is interesting to consider why so many of these problematic racial stereotypes are reproduced again and again in the form of the comedic sidekick. Often depicted as anthropomorphized animals or objects, these sidekicks are dependent, goofy, unquestionably loyal, and perhaps most significantly, racialized. I argue that as happily "othered" subjects in these films, comedic sidekicks offer important insight into the most pernicious aspects of racial inequality. These racialized comedic sidekicks emphasize the authority and legitimacy of the central protagonists according to mainstream Western values because of their accentuated difference. Of course, these values have shifted significantly within the eras of classical Disney, Renaissance-Disney, and present-day Disney films. As popular cultural artifacts, these films can be placed in conversation with specific historical events during their respective time periods, but also as reinforcing larger, hierarchal formations of power.

Various sidekicks from classical Disney, Renaissance-Disney, and present-day Disney films. Specifically: *Snow White and the Seven Dwarfs* (1937), *Cinderella* (1950), *Sleeping Beauty* (1959), *The Little Mermaid* (1989), *Aladdin* (1992), *Pocahontas* (1995), *Mulan* (1998), *The Princess and the Frog* (2009), and *Tangled* (2010). http://mitchj.info/new/disney-movie-animal-sidekicks.html.

Building off of Lee Artz's work on capitalist ideology in popular Disney films, I will focus my examination on Disney sidekicks in *Cinderella* (1950) and *The Princess and the Frog* (2009), two fairytale films released almost sixty years apart and representing the Classical and Contemporary Disney eras. Further research on this subject would ideally address a film from the Renaissance Disney era.[230] I have selected fairytale films both because of their enduring popularity in Disney's formula for success across all three time periods, but also because of the explicitly gendered and classed dynamics at work in these adaptations, which I see as adding another layer of depth to the hierarchies being enforced. I have given particular attention in my argument to the way the characters are introduced, primary because I believe these introductions most clearly impact the viewer's understanding of these minor characters.

Jaq and Gus look up adoringly at the title character in Disney's 1950 film *Cinderella*.

[230] For instance, Sebastian in *The Little Mermaid* (1989) as a Jamaican servant to King Triton, the enchanted servants of *The Beauty and the Beast* (1991) and their European rather than American accents, or Mushu from *Mulan* (1998), voiced by Eddie Murphy.

Louis singing "When We're Human" with Tiana and Naveen.

Through my analysis, we can see how dominant ideals are constructed through representations of difference. Ania Loomba discusses the "Manichean allegory" in which a "binary and implacable discursive opposition between races is produced" (1998, p. 104). Disney has a pattern of using the same dichotomous relationships to appeal to a mainstream audience. Consider how Disney's villains are often gender-inverted, animalistic, and socially isolated individuals. We see this reflected in such characters as the widowed and age-shifting Evil Queen of *Snow White*, the octopus-hybrid sea witch Ursula of *The Little Mermaid*, or the effeminate tower-dwelling (darker-skinned) Jafar of *Aladdin*. These villains are interesting in their own right, that is, as culturally stigmatic, but powerful, monstrous others. But perhaps more insidious is the noble savage—the colonized comedic subject who is portrayed by the oppressor as joyously accepting of their own subjugation.

Various villains pose for side by side in this poster from the "Disney Villain Invasion"(2015).

We should be careful to note that diversity of racial representation in classical Disney was notably negative, as in *The Song of the South*, or largely absent, as in almost all popular Disney animation films before the 1990s. But what about a more progressive, modern Disney that features more central characters of color? Both Steinberg and Artz acknowledge the increasing diversity within recent Disney films but argue that these aspects are only used to retain the same hierarchy of power in an increasingly diverse society under the guise of inclusivity. Steinberg argues, "[w]hite heroes are frequently provided with a nonwhite or female sidekick to overtly signify the value of diversity—a strategy that covertly registers the need for white male control of a diverse society" creating a "safe, common-culture type of diversity that sanitizes and depoliticizes any challenge to the harmony of the status quo" (2011, p. 44). This neoliberal sentiment uses a sanitized form of diversity for economic prosperity rather than social change. While Steinberg focuses her studies on kinderculture TV heroes, and Artz on four Disney Renaissance-era films, I am interested in how Steinberg's and Artz ideas can be traced through Disney's comedic sidekick characters.

Sebastian tries to advise Ariel unsuccessfully in this scene from *The Little Mermaid* (1989).

Mulan in her matchmaker outfit in this scene from *Mulan* (1998).

Disney's *Cinderella* had been in production for two years before its release in 1950. Politically, the United States was experiencing a post–World War II economic boom but also a rise in conservatism as fears of communism were exacerbated by the Cold War (Gilderhus). In her analysis of *Saludos Amigos* (1943) and *The Three Caballeros* (1944), Karen S. Goldman examines the influence of Hoover's policy as made popular by Roosevelt in 1933. Similarly, Prajna Parasher examines the influence of McCarthyism on problematic portrayals of Native Americans in *Peter Pan* (1953). Both Goldman and Parasher argue that in these films the racialized others are rendered childlike through unthreatening caricatures. Turning to the characteristics of the comedic sidekick characters in *Cinderella*, we may note that many parallels exist between both Parasher's observations regarding the usefulness of racialized others given the rise of McCarthyism, and Goldman's analysis of the enduring influence of Good Neighbor policies on Disney films depicting Latin American characters.

Donald Duck dons a sombrero in this scene from *The Three Caballeros* (1944).

Three unnamed house mice "draw straws" with Jaq and Gus in *Cinderella* (1950).

The Chief smoking alongside Tiger Lily and Peter in *Peter Pan* (1953).

The comedic sidekicks of *Cinderella*, particularly the mice, are racialized.[231] The mice's dialect and outsider status mark them within the first scene of the film. After Cinderella's morning routine, in which blue birds and mice help bathe her and prepare her clothing for the day, Jaq anxiously tries to grab Cinderella's attention saying, "New mouse in the house. Brand-new. Never saw before." After getting over her initial fixation on finding the correct gendered clothes for the new mouse, Cinderella goes to release the trapped foreign mouse. She says, "Jaq, maybe you'd better explain things to him." To which Jaq replies "Zuk-zuk, Cinderelly. Zuk-zuk. Now, now, now. ... Look-a, little guy. Take it r-easy. Nothin' to worry 'bout. We like-a you. Cinderelly like you too. She's nice. Very nice." Cinderella then fashions the new mouse in a rather tight shirt and hat and names him Octavius (Gus for short). Cinderella is obviously being situated in this opening scene as the caretaker and master of the anthropomorphized mice. These little brown mice are clothed in handmade garments to mark their acceptance into the household, and therefore more civilized status from other mice. In exchange for clothes, food, and shelter, the mice serve Cinderella. But other than this exchange of services, the mice are likewise marked as outsiders by their disjointed English. Jaq speaks in sentence fragments (for example "Never saw before"), mispronounces words (for instance "Cinderelly"), and makes up other words (the notable "Zuk-zuk" is not colloquial French or any identifiable language). Naomi Wood points out that in the film the mice

231 While there are other minor characters: Bruno (the dog), Lucider (the cat), Major (the horse), and the blue birds that help the mice. These animals exhibit human characteristics; however, I have chosen not to focus on them because they cannot speak as the mice can.

"speak in a high-pitched pidgin, suitable to their status as colonized subjects"(1996, p. 32). These mice are characterized as outsiders—because of their dialect, their exclusion from the main areas of the house, and their constant habit of working for Cinderella.

However, this exchange of services is romanticized and race is somewhat obfuscated by the anthropomorphized status of the mice. Perhaps the mice are Latin American, remnants of Disney "Good Neighbor" film days. Or perhaps they are representative of a more generalized immigrant population—either way, they are marked as deviating from the American accents of the admirable and benevolent protagonist Cinderella.

Gus is stuck in a mouse trap in this scene from *Cinderella* (1950).

Arguably, this opening scene also demonstrates how *Cinderella* infantilizes the mice in order to highlight their deficiency and/or nonthreatening status as subjects whose purpose is to serve Cinderella in her transition from child/servant to woman/queen. For instance, when Jaq and his other mouse friend rush to tell Cinderella about the "visiting" mouse stuck in the trap, Cinderella says, "Now wait a minute, wait a minute, one at a time please." Or when she finds Gus stuck in the trap she says, "The poor little thing's scared to death." Cinderella's commentary emphasizes Gus's, and all the mice's, greater vulnerability and deference to Cinderella. This characterization echoes Parasher's observation regarding the infantilization of the Indians in *Peter Pan* as functioning to reinforce Peter's dominance; however, here the mice are serving Cinderella's interpellation into the status quo. The mice emphasize Cinderella's mothering skills and enforce gender roles. Cinderella herself is imbued with romanticized notions of individual hard work, honesty, and upward mobility, Western values which are in fact complicated by her upper-class status, the overemphasis on her physical appearance and the film's predictable resolution predicated on heteronormative marriage. Likewise, Cinderella may love the mice, but their racialization as rodents highlights the more insidious power dynamics at work. We can see how comedy functions to racialize the mice through dialect and comedic stereotypes that naturalize their subjugated position, further obfuscating social constructions based on gender, class, and race. Certainly, we can see the emergence of more enlightening Disney films in recent decades, but how progressive are these films, really?

Cinderella is distracted by love ... and her own reflection in this scene from the film.

Over a half century later, Disney's *The Princess and the Frog* was released in 2009. Based very loosely on the "Frog Prince" fairytale, Tiana's story is set in New Orleans and features Disney's first animated, African American princess. Politically speaking, the United States had just elected the first African American president to the White House the previous year. The election would inspire in many ways a more open political discussion about race, but also the erroneous idea that we were somehow "postracial." Still, Disney's decision to reveal the first African American princess at this time was inevitably tied to developing cultural discussions about race—most insidiously claims of a "postracial" Obama era.

Barack Obama's *TIME* magazine cover from October 20, 2008.

Additionally, the setting of the film in New Orleans has obvious political implications especially when aligned with 2005's Hurricane Katrina. While obviously not all residents were African American, this population was in many ways more greatly affected because of gaps in pay and greater rates of poverty. Neal A. Lester notes that the film's physical and temporal settings do not escape valid critique. He surmises whether selecting this place "expressly reinforces the vision that with persistence, resources, and patience, the traumatized city of New Orleans will rise from the flood of destruction stronger and better than before Katrina" (2010, p. 301). Likewise, Moon Charania and Wendy Simonds are critical of the setting, stating,

> It's disturbing that this American dream has been set in a city where recent ruination so clearly highlighted the division between haves and have-nots. How ironic is the film's recognition that, "freedom

> takes green," when considered against the backdrop of the real losses the people of New Orleans have suffered? (2010, p. 71).

Indeed, Tiana's success does depend on money, and it seems that it is entirely her responsibility to work hard enough to reach that dream.

This issue of historical erasure, even while attempting to humanize blackness through a hyperfocus on individuality that Lester, Charania, and Simonds critique, is at times especially evident in the problematic portrayals of Alligator Louis and Firefly Ray. The majority of the film takes place in the swamp, in Ray and Louis's world, where race might be represented but is not directly addressed. Charania and Simonds note that "[s]pecies bond together here [the swamp] against adversity; maybe it's a metaphor for multiculturalism, Disney's way of valuing diversity" (2010, p. 71). A closer examination of the sidekick characters reveals why this somewhat covert way of valuing diversity remains problematic and echoes a pattern from Disney's earlier eras. Louis and Ray's characterization harkens back to Artz's argument regarding the strategic use of diversity in recent Disney films in order to naturalize hierarchy, defend the elite, promote hyperindividualism, and undermine democratic solidarity (2004, p. 126). Both Ray and Louis are meant to elicit laughs because they reside in the bottom levels of this hierarchy. Louis demonstrates a comedic disjoint between the stereotype of the dangerous black male body and a harmless music-loving alligator, while Ray both accentuates cultural difference for laughs and suffers the death of a noble "savage."

Ray is exasperated in this scene from *The Princess and the Frog* (2009).

Louis is introduced by a pair of eyes creeping up on Tiana and Naveen rowing down the bayou after the pair are transformed into frogs. Drawn to the strumming of Naveen's improvised ukulele, Louis immediately brings up his horn to play music with Naveen. Naveen says enthusiastically, "Play it brother!" while Louis responds in kind, "Where you been all my life?" The scene plays with the idea of predator and prey, a more oblique reference to the culturally stigmatized black male body. Jazz music is shown as reconciling a threatening out-group (Louis) to a privileged in-group (Naveen). This reconciliation is achieved through the humanizing power of music. Louis is racialized by his enthusiasm for jazz, but also because of his threatening alligator body, rendered unthreatening here because of his comedic sidekick status.

Louis makes his grand introduction to frogs Tiana and Naveen.

In this way, Louis contrasts well with several other characters within the film—for instance, Dr. Facilier, whose sexualized black body remains threatening, or even Tiana's father, who dies early on in the film. Dr. Facilier in particular is problematically sexed as dubious—he is both feminized (with his purple attire, feather in hat, and thin stature) and tied to the image of a pimp (with his top hat/cane/manipulative cunning). Much has also been made about Prince Naveen's racially ambiguous status and his royal lineage as the "Prince of Maldonia," a made-up kingdom never thoroughly explained in the film. This peculiarity is especially pronounced given the very specific location and time period of New Orleans during the 1920s in which the film takes place. Naveen's racial ambiguity, Tiana's father's very early death, and Facilier's hypersexed villain status indicate Disney's struggle to represent black masculinity within the film. While Louis might be the best indicator of a positive black masculinity, he still remains an alligator who, no matter how much he might insert himself into the human world, can never be fully human.

Dr. Facilier, a.k.a "the shadow man," gets some help from his "friends on the other side" after making a deal with Prince Naveen.

James lovingly takes care of Eudora and Tiana and dreams of one day owning his own restaurant during this clip from the film. It is revealed that James was later killed in combat before getting to see this dream become a reality.

This is the problem with Disney's insistence on coding race into anthropomorphized subjects. Louis is never able to be become human, and the film's issues with stigmatized black masculinity (Louis is big and intimidating), while somewhat subverted by Louis's sweet and rather timid characterization, is most predominantly used for laughs rather than serious social critique. Artz echoes this sentiment when he argues: "[b]eing true to one's self is reliant on social position. Orphans, merchants, zebras, baboons, sailors, warriors and workers face no dilemma. Their true selves are patently, graphically obvious. They remain in the background unless needed in the elite narrative" (2004, p. 134). While Tiana is not initially an aristocrat, this royal order is restored in the film's conclusion. Louis's ability to play the trumpet whenever he wants has to be allowed by a royal order—not because he "found himself" in anyway, but because Tiana has graciously sanctioned it.

Louis plays his trumpet to a dismayed band in this scene from the film.

Following traditional formulations of racial difference, Ray's white Cajun racialization is used to both accentuate Tiana and Naveed's normality and produce laughs. Ray is introduced after stumbling upon Naveen and Tiana's entanglement via their frog tongues. After untangling them, Ray says, "'Bout time I introduce myself. My name Raymond, but everybody call me Ray." To which Naveen responds, "Pardon me, but your accent ... it's funny, you know?" Ray explains, "Oh I'm a Cajun brotha, born an' bred in the Bayou." Maria Herber-Leiter identifies how this interaction highlights racial difference, arguing, "Not only does Naveen speak directly to the different, thus funny, sound of Ray's accent but he also demonstrates the point through his own accent, also influenced by his native French language. In contrast to Ray's thick, folksy speech, Naveen's is more polished, standard French-influenced English" (p. 793). Additionally, Ray's appearance is noticeably marked as lower-class; he has disheveled hair as well as crooked and missing teeth. Similarly, Ray's love for the star he has named Evangeline is mocked by some characters (like Naveen) and permissively protected by others (like Tiana) who view it as endearing although delusional. Ray's dialect, visual lower-class signifiers, and perceived stupidity mark him as an othered subject. In this way, Ray can be compared to other white characters such as Lottie La Bouff, Lawrence, or the backwoods frog hunters who are marked for similar character flaws in intelligence tied to gender (as in Lottie's case) or class (as in Lawrence and the frog hunters cases).

Ray's death, unprecedented among other fairytale comedic sidekicks, is the catalyst for Tiana and Naveen's acceptance of their frog status and decision to marry. Ray's death legitimizes his love for Evangeline as his bright star appears next to hers in the sky but as Herber-Leiter points out, only after Tiana doubts him "confirming Ray's difference and subsequent inferiority for the audience. ... Moreover, Ray must die for Tiana and Naveen to realize they are not as different from each other as they think" (p. 970). While sweet, the destruction of this divide has to come through the death of a blatantly marginalized character. Ray is literally sacrificed in the narrative for the protagonists, thereby reinforcing their normality through his difference. Even while Ray is revealed to be correct about Evangeline in the end, the stereotypes persist throughout the film and culminate in a romanticized "noble savage" image.

Ray shows off his light in this scene from the film.

Ray's final moments before death after being stepped on by Dr. Facilier.

Tiana and Naveen are allowed to occupy the animalistic, more racially/difference-inflected world of the bayou for a short period of time before ascending through the traditional ceremonial marriage—this time with some extra emphasis on the power of entrepreneurialism. While Disney does seem to more self-consciously use stereotypes in the anthropomorphized comedic sidekick characters to address racial bias, a definite improvement from early films like *Cinderella*, ultimately the classic Disney formula remains: using difference for laughs and to further the primary characters' storylines.

Tiana and Naveen are married by Mama Odie
in the bayou as frogs.

Tiana and Naveen are transformed back into humans after their marriage is made official, and the couple holds a second wedding celebration in New Orleans in this scene near the end of the film.

Undoubtedly related to the company's desire to appeal to a developing sense of multiculturalism, this film provides physical displays of blackness but continues to relegate the more difficult social commentary to the sidelines. And if we don't explicitly recognize difference and all the social implications that identity carries in a real historical context, then we are rewriting history and reproducing whiteness. Drawing attention to the biases of our thinking has never been so integral. Of course, the struggle over power structures and racial divides will not be resolved through corporately funded animation films. However, I do believe recognizing how issues of race are cagily relegated to the sidelines, and historical truths ignored, will foster the continued progression of Disney entertainment. We must demand more direct representations of historically subjected groups, grounded in specificity, that challenge dominant narratives of success and happiness. Never underestimate the power of Disney or the intelligence of its young viewers.

REFERENCES

Artz, Lee (2004). "The Righteousness of Self-Centred Royals: The World According to Disney Animation." In Critical Arts: A Journal of South-North Cultural and Media Studies 18 (no .1), pp. 116–146. Print.

Charania, Moon, and Wendy Simonds (2010). "The Princess and the Frog." In Contexts: Understanding People in Their Social Worlds 9 (no. 3), pp. 69–71. Print.

Clements, Ron, and John Musker, directors (2009). The Princess and the Frog. Walt Disney Studios Motion Pictures. Film.

Geronimi, Clyde, and Wilfred Jackson, directors (1950). Cinderella. Walt Disney Productions, 1950. Film.

Gilderhus, Mark (2006). "The Monroe Doctrine: Meanings and Implications." In Presidential Studies Quarterly 36 (no. 1), pp. 5–16. Print.

Goldman, S. Karen (2013). "Saludos Amigos and The Three Caballeros: The Representation of Latin America in Disney's 'Good Neighbor' Films." In Johnson Cheu, ed., Diversity in Disney Films: Critical Essays on Race, Ethnicity, Gender, Sexuality and Disability, pp. 23–37. North Carolina: McFarland & Company. Print.

Lester, Neal A. (2010). "Disney's the Princess and the Frog: the Pride, the Pressure, and the Politics of Being a First." In The Journal of American Culture 33 (no. 4), pp. 294–308. Print.

Loomba, Ania (1998). Colonialism/Postcolonialism. New York: Routledge. Print.

Parasher, Prajna (2013). "Mapping the Imaginary: The Neverland of Disney Indians," in Johnson Cheu, Diversity in Disney Films: Critical Essays on Race, Ethnicity, Gender, Sexuality and Disability, pp. 38–50. North Carolina: McFarland & Company. Print.

Steinberg, Shirley (2011). "Introduction." In Shirley Steinberg, ed., Kinderculture: The Corporate Construction of Childhood, 3rd Edition, p. 44. Westview Press.

Trousdale, Gary, and Kirk Wise, directors (1991). Beauty and the Beast. Buena Vista Pictures, 1991.

Wood, Naomi. "Domesticating Dreams in Walt Disney's Cinderella." In The Lion and the Unicorn 20 (no. 1) (June 1996), pp. 25–49.

DIGITAL

Collections · Explore · Exhibitions · Blog · OUR SITES · LANGUAG

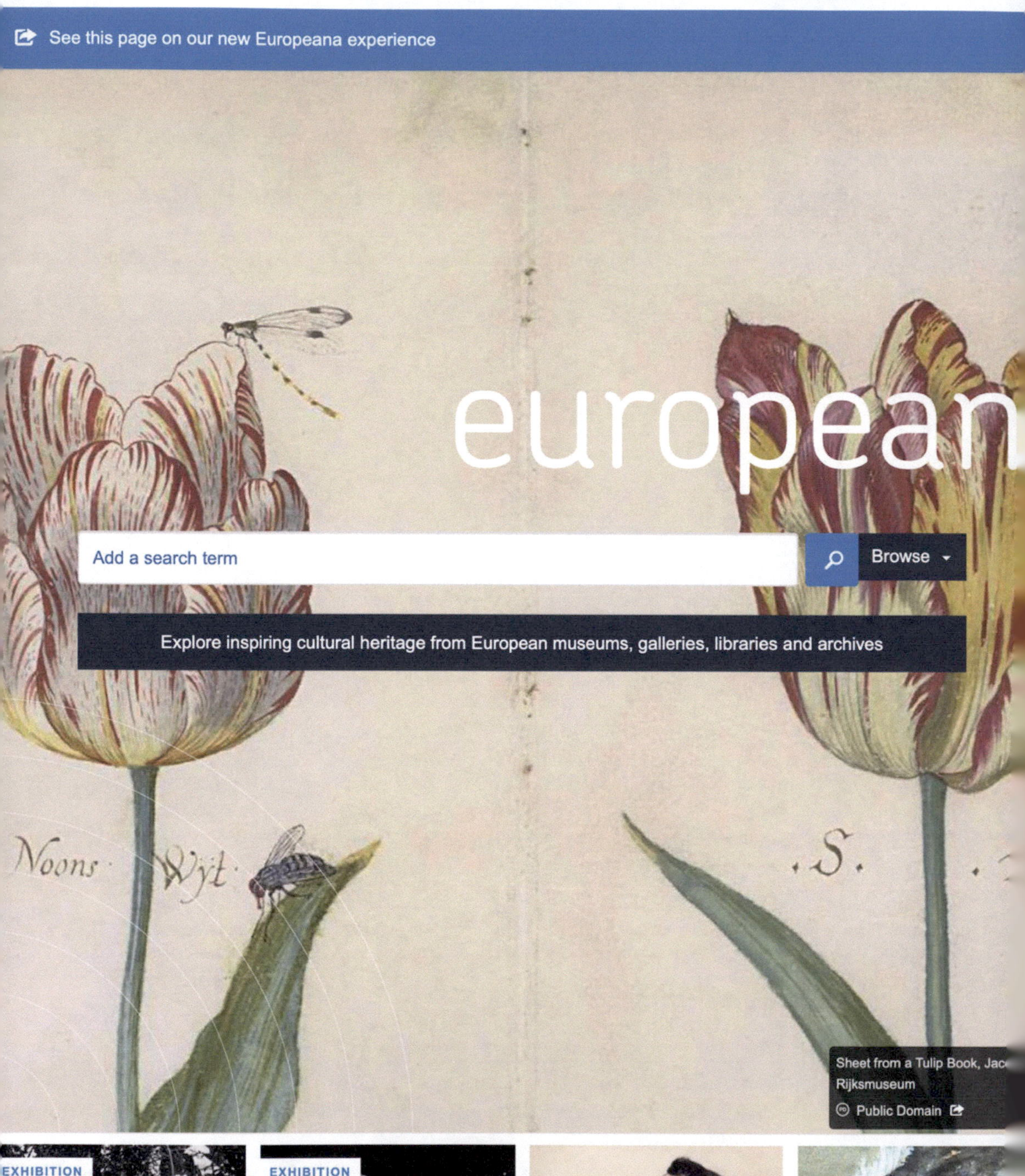

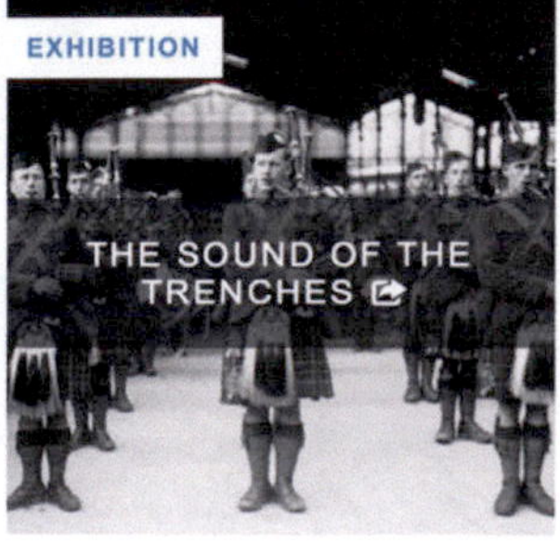

THE DOMESTICATION OF THE PAST
DIGITIZED HERITAGE ONLINE AND THE ADMINISTRATION OF TIME

MATTEO TRELEANI

> The desire of contemporary masses to bring things "closer" spatially and humanly ... is just as ardent as their bent toward overcoming the uniqueness of every reality by accepting its reproduction. Every day the urge grows stronger to get hold of an object at very close range by way of its likeness, its reproduction.[232]

In 1936, according to Walter Benjamin, the mechanical reproduction of art was the reflection of a social circumstance: the need to make things more readily and efficiently available. The photo of a painting in a magazine, for instance, is a reproduction that makes the image easier to access. The same trend could be observed today in the phenomenon of online heritage publication. To bring *archive material* closer spatially and humanely is a major issue for heritage politics. The full availability of documents online seems to be the aim of institutional and private policies in the field of heritage.

Digitization seems to allow cultural production to float in the social space regardless of its location—everywhere and all the time. We can read an article, watch a film, and contemplate a painting in its digital version in any circumstance thanks to the internet and the devices that give access to it. Cultural contents circulate in the mediascape, and the evolution of media seems to live under the aegis of immediate accessibility from a distance. This phenomenon, also called the age of access, from an economic point of view, impacted our access to works of art and media content, and our relation to the past has also been shaped.[233]

We will argue that digitization has contributed to a form of domestication of the past. This domestication is part of a more general tendency toward the administration of time, which is evident when looking at the evolution of media and cultural industries.

The past, having passed, is no longer reproducible, quantifiable, or rationizable. The digitization of archives and the dissemination of documents can be seen as yet another step toward crossing the last frontier of industrialization—that of something that is no longer there. What may be seen as a main problem is that this form of domestication seems to

---

232 Walter Benjamin, "The Work of Art in the Age of its Technological Reproducibility: Second Version," trans. Edmund Jephcott and Harry Zohn, in Michael W. Jennings, Brigid Doherty, and Thomas Y. Levin, eds., *The Work of Art in the Age of Its Technological Reproducibility, and Other Writings on Media*, Cambridge, MA, Harvard University Press, 2008, pp. 19–55.

233 Jeremy Rifkin, *The Age of Access: The New Culture of Hypercapitalism*, TarcherPerigee, New York, 2000.

erase the specificity of the past. In appropriating it (from Latin *domus*—house—to make it ours), we *presentify* it (make it present): all the interest of the past contained precisely in its difference from the present. The past is a foreign country, where they do things differently. ...

HERITAGE AND ACCESSIBILITY

Recent decades have seen a large number of projects of digitization and publication of archival documents online. In Europe, Europeana is the largest institutional platform where libraries and archives publish much of their content. In the United States, a private initiative like the platform Archive.org, conceived by the Internet Archive foundation, gives access to digitized documents and recorded digital records, such as the web and other online circulating content. Conserved and transmitted to future generations, these documents are considered to be "heritage content" because of their historical value.

Accessibility is one of the main reasons for digitizing archived material. Stocking contents digitally is probably the less durable way for their preservation (even if things may probably evolve in the future[234]): migrations are needed every five years because of the risk of demagnetization. But the efficiency of access that the digital makes possible through the web is unparalleled. More specifically, digitization is useful in order to make things accessible immediately and from a distance. We have always been able to access documents in a library or in the archives, but through digitization, we can consult them without having to move.

Digitization therefore implies publication: making things available to the public. This is obviously good news—one has access, easily and quickly, to a large quantity of documents which were previously difficult to access. Nevertheless, the fact that digitization implies, *de facto*, publication is rarely questioned in heritage policies, as if the idea of publishing and therefore putting online of archive documents, which means broadcasting from certain points of view, was self-evident (whereas it has almost never been done before—or at least not to this extent). Accessibility is defended by the Charter on Preservation of Digital Heritage published by UNESCO in 2003. Article 2 clearly affirms that "the purpose of preserving the digital heritage is to ensure that it remains accessible to the public."[235] Another supranational institution, such as, for example, the European Union, recommended the Member States to make digitized

[234] Migration every five years at least is necessary because magnetic supports are not reliable; demagnetization implies the loss of all data. As a term of comparison, in the audiovisual domain, a 35-mm film can last seventy-five years in good conditions, and paper lasts centuries.
[235] UNESCO, Charter on the Preservation of Digital Heritage, 2003.

heritage accessible online (see the Commission's Recommendation 711 of October 27, 2011).[236]

The accessibility of archival documents requires digital interfaces for their consultation. Europeana, for instance, designs interfaces that allow an extremely complex consultation of its archives: several points of view or entry keys are possible, and our way of reading or viewing documents will therefore depend on the choice of parameters that we will have configured beforehand. The document is thus edited or editorialized in this way,[237] but more often this process is automatic; its contextualization is delegated to the machine.

We will thus question the meaning of a domestication of the past, which passes though these interfaces. Interfaces have an epistemic role. In other words, they are the instrument of a digital mediation of memory, and their authority on the meaning of content is thus strong.

Before we continue, a short digression is required. We adopt here a critical approach to a cultural issue, the issue of cultural heritage, linked to a technological evolution. Even if digitization is described as a tool for domesticating memory, our perspective is far from technological determinism. In 1964 in *Apocalypse Postponed*, Umberto Eco wrote very modern words—if we apply them to our actual relationship with digital logic.[238] Eco affirms that the attitude of the man/woman of culture, facing the era of industrial machinism, should not aim for a return to nature, but should rather ask himself/herself under what circumstances is man / woman dominated by the cycle of production and how to elaborate a new image of man / woman in relation to this system; a man/woman who would not be free from the machine but who would be free in relation to the machine. In the same spirit, we wish to highlight some problematic aspects, in order to understand how this technique can become a tool of emancipation. The goal is not to challenge digitization (to emancipate from the machine), but rather to highlight a form of power that manifests itself through technology. Technology is not the target of criticism, but technology is where some social tendencies actualize themselves. By analyzing it, we can thus look at these tendencies and thereby understand power relations. In other words, technique is a revelatory: it can be interrogated in order to discern power relations, neither a cause

---

236 "Digitization is an important means to ensure greater access to and use of cultural material." Commission Recommendation of 27 October 2011 on the Digitisation and On-Line Accessibility of Cultural Material and Digital Preservation, *Official Journal of the European Union*, L 283/39.

237 For a conceptutalization of "editorialization," see Marcello Vitali Rosati, *On Editorialization: Structuring Space and Authority in the Digital Age*, Institute of Network Cultures, Amsterdam, 2018.

238 Umberto Eco, *Apocalypse Postponed*, Indiana University Press, Bloomington, 1994, p. 17.

nor an effect but a case study that makes visible some subjacent tendencies. As the philosopher Gianni Vattimo says, the problem is not technique, it is domination.[239]

THE DOMESTICATION OF TIME

How can we describe the idea of domesticating time? The evolution of media can be seen as a tendency to invade and administer every moment of daily life. This does not begin with television, which administers "available human brain time" to sell to advertisers, according to the famous formula of Patrick Le Lay, former CEO of the French private channel TF1. From Jonathan Crary's perspective, for example, techniques can be seen as a means of monetizing people's daily time, at least since the invention of electricity (with night lighting, which finally made it possible to make half the human day profitable, making evening shopping, working in blast furnaces and so on possible).[240]

According to Crary, capitalism has monetized everything (hunger, leisure, sexual desire ...), and the only rampart against total profitability was sleep, a moment that is by definition unproductive (and which, according to the studies cited by Crary, only diminishes: we slept an average of ten hours a day at the beginning of the century; they have become seven today).

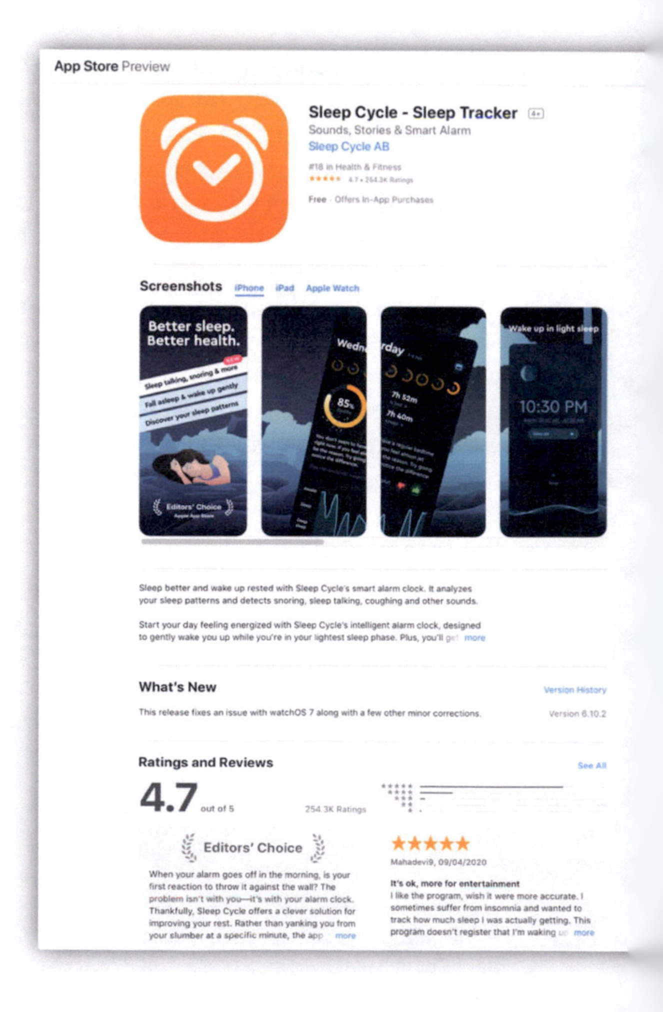

Smartphone applications now aim at administering our sleep as well, with measurement and, therefore, the setting of cycles, for example. The tools of mediation of any activity are therefore essential if we want to have control over every moment of an individual's day.

---

[239] Gianni Vattimo, "La technique n'a pas tenu ses promesses de liberté. Entretien avec Gianni Vattimo de Matteo Treleani," *Ina Global,* no. 4, March 2015, pp. 78–82.

[240] Jonathan Crary, *24/7: Late Capitalism and the End of Sleep*, Verso, New York, 2014.

Let's take the smartphone as an example. According to Maurizio Ferraris, having one means to be subject to what he calls "total mobilization."[241] With a mobile phone, you can always be spotted and therefore available for work. You can reply to an e-mail in the evening or on weekends, thus removing the boundaries between domestic and work space—limits that have finally weakened quite a bit if one takes into account the notions of digital labor or, more generally, any form of entertainment of cultural industries that make the available moments of our life profitable or potentially profitable. Technology obviously plays an important role in the task of helping the domestication of private time.[242]

In the past, certain moments of daily life were not necessarily profitable—at night, for example, as without illumination, it is very difficult to do anything.[243] This time management can also be related to a form of control, as demonstrated by the constant digital traceability of any activity. Online work and leisure produce traces that will be archived and preserved. The notion of the "right to be forgotten"—the claim to be able to erase our traces from Google, for example—falls within this framework. From the domestication of daily activities, we then move on to the administration of its traces. Traces are recorded in order to better measure and question human behavior.

THE DOMESTICATION OF THE PAST

Can we say that the digitization of cultural heritage is part of this general tendency toward an administration of time? It would no longer be a question of managing the daily life of individuals, but rather of acting on historical time, the domain of history and collective memory. Domestication is not only about monetizing the past and making it economically available.[244] It is about administering it as a consumable commodity—in other words, being able to manipulate it, use it, and manage it.

It is interesting to emphasize the temporal nature of the phenomenon described by Jonathan Crary. It is not just a question of trying to find collective, profitable, and productive occupations for community and economic purposes. It is the diachronic dimension of the question that seems particularly fascinating. For Crary, this unrestricted continuity of

---

[241] Maurizio Ferraris, *Ame et iPad*, Presses de l'Université de Montréal, Montréal, 2013, p. 82.

[242] Historian David Landes has shown that passing from the idea of obeying time to that of disciplining it, and not the contrary, was the main change that brought to the invention of clocks. David S. Landes, *Revolution in Time: Clocks and the Making of the Modern World*, Harvard University Press, Cambridge, MA, 1983.

[243] About the industrialization of light: "Each evening, the medieval community prepared itself for dark like a ship's crew preparing to face a gathering storm," Wolfgang Schivelbusch, *Disenchanted Night: The Industrialization of Night in the Nineteenth Century*, University of California Press, Berkeley, 1995, p. 81.

[244] Media archaeologist Wolfgang Ernst rightly emphasizes this aspect: "the audiovisual memory of postwar Germany remains with companies that might sell these media archives to private investors. Memory will be commodified." Wolfgang Ernst, *Digital Memory and the Archive*, University of Minnesota Press, Minneapolis, 2013, p. 203.

productive activities implies some form of instant flattening. We are heading toward a twenty-four hours a day, seven days a week productivity. The 24/7 would thus be a nontime: a time without duration and without history, a "hallucination of [...] presence" (2014, p. 47).

Now, it should be said that the assumed interest of society for heritage and its dissemination through digitization that we describe here seems to contradict this trend. Unlike the nontime 24/7, the passion for heritage should bring us closer to the past. The excess of "making heritage" or *patrimonialization*, however, could paradoxically fit into a form of presentism. Archiving everything implies the desire to make the past present—bringing it closer to us, as Walter Benjamin asserted. Making it accessible means to make it present and thus erase its specificity.

According to François Hartog,

> the heritage wave in tune with that of memory, has become more and more extensive to the point of reaching the limit of "all-heritage". Just as all things are announced or claimed to be valuable, so everything could be or become heritage. [...] "Patrimonialization" or museification has gained, becoming ever closer to the present.[245]

Heritage inflation would in other words be linked to a system of historicity that can be called "presentist." Emmanuel Hoog translated this tendency into an inversion of the traditional archival logic: "In the modern adage, we no longer preserve because it is important, but something becomes important because it had been preserved. Or rather, we allow it to become important."[246] In other words, we record what is technically recordable and then decide on its value. The semiologist Claudio Paolucci summed up this process with a formula: previously, the memorable was memorized if memorizable. Today, the memorizable is memorized and therefore becomes memorable.[247]

Domestication thus acts on the mass, the fact of "archiving everything" is a way of presentifying the past, making it materially available in the present, but it also develops in the singularity of the relationship with the past. Many projects are based on the 3D reconstruction of old towns: thanks to a gadget in Place de la Bastille in Paris, we can see through a virtual-reality immersive lens the square in 1789, for instance (the project is called TimeScope, the same as the start-up which developed it). The

---

[245] François Hartog, *Regimes of Historicity: Presentism and the Experience of Time*, Columbia University Press, New York, 2016, p. 97. Also see Rudi Laermans, "Paradoxes of Patromonialization," *Open Platform for Art, Culture and The Public Domain*, Open!, September 30, 2004, www.onlineopen.org/paradoxes-of-patrimonialization.

[246] Emmanuel Hoog, *Mémoire année zéro*, Seuil, Paris, 2009, p. 121.

[247] Claudio Paolucci, "Archive, patrimoine et mémoire. Un regard sémiotique sur la tiers-mondialisation du savoir," in Valentine Frey and Matteo Treleani, eds., *Vers un nouvel archiviste numérique*, L'Harmattan, Paris, 2013, pp. 75–103.

past is thus materially domesticated: we can see it and almost touch it. It is even industrialized: we obviously pay for the experience, and the tool is exportable to other squares and towns. The main problem is that there is no distance between the viewer and what the tool shows her/him. There is no way of having a critical gaze on this version of the past that is shown as the only one conceivable, without any reference to data and information that have been used to reconstruct the square in that way. In a transparency society, as Byung Chul Han calls it, what's lacking is distance, which means the respect for the other (the alterity being here the past).[248]

OBJECTIVATION

The role of digital technology can only accentuate this phenomenon. In Bruno Bachimont's words, "digital technology reifies cultural contents into manufactured objects that can be subjected to automated and industrial processes."[249] The aim is of course to make this content profitable and monetize it. However, this is not the only effect: it is in particular an epistemological change that leads to an objectification of the past.

The objectification is achieved through reification, the idea of seeing heritage as a stock where objects are placed in order to preserve them from time and opened so that people can see them. The "all heritage" requires us to think of archive documents as things. The European Commission's 2011 Recommendation on Digital Preservation (711/2011), for example, recommends that Member States publish "objects"—as they are called—on Europeana, even defining the desired quantity for each state. In the UNESCO World Library, we find the same attitude toward archive documents. The interface shows us first of all how many documents each institution published online (an attitude that seems normal today but is not: it would be strange to see the quantity of books on each shelf of a library instead of having the names of the authors). It's thus a reification, which involves quantification. A recent marketing campaign of the French National Library, the BnF, affirmed "Pour tout lire comptez 150000 ans" ("To read all, you will need 150,000 years"). Again, it is the symptom of a quantitative way of seeing heritage: works of literature and of the human mind have to be reified as objects and as singular elements in order to quantify them.

If the document becomes an object, what's materialized is also the relationship with the reliability of the past. The archival material has to be reliable without the need to interpret it and contextualize it within a more general discourse about the past. The document acquires a form of objectivity—we have to trust the document—and this objectivity needs

---

[248] Byung Chul Han, *The Transparency Society*, Stanford University Press, Redwood, CA, 2012, p. 13.

[249] Bruno Bachimont, "Le nominalisme de la culture. Questions posées par les enjeux du numérique," in Bernard Stiegler, ed., *Digital Studies. Organologie des savoirs et technologies de la connaissance*, Fyp Editions, Limoges, 2014.

the lecture to bypass discursive mediations (the mediations of historical discourses that build our image of the past). Making the past present means building absolute trust in the archival object: in other words, the archive document must become the past (and not an inscription that has to be interrogated and interpreted in order to build a historical discourse that talks about the past). The content of archive materials thus becomes "objective."

Indeed, publication is based on a criterion of transparency—everything must be public and available. The idea would be precisely that of going beyond the intermediary in order to give direct access to knowledge (a knowledge that would, therefore, be reified in a digitized object, in this case, the document). It is therefore quite fascinating to note that this transparency must be based on a delegation to the technical mechanism. A mediation, in order to be visible, must disappear and fade away, like a window that is effective when it cannot be seen: mediation hides. The device also becomes transparent, therefore. Its epistemic role is no longer evident. It becomes neutral, not meaningful—as if mediation, for the mere fact of being digital, was not cognitively important.

The objectification thus passes by a reification of the mass and an occultation of the mediation. The delegation of observation to the technical instrument is one of the principles of the concept of mechanical objectivity for Daston and Galison (who write a history of the concept of objectivity).[250] Putting the subject between parentheses through the intervention of the instrument (technical and therefore neutral) implies a sense of objectivity. The atlases at the end of the nineteenth century begin to replace the drawings of the leaves, for example, by daguerreotypes, where the detail is much less precise and defects are added to the image (such as black and white effects or stripes) preventing its good visibility, but which, for the reader, seem to give a sense of major objectivity because photography is the product of a technical and nonhuman (and therefore not subjective) activity.

Some kind of mechanical or digital objectivity seems to be applied in the consultation of archives online. In 2010, the French National Audio-Visual Institute published a story from 1964 where an engineer explained why they were moving the Eiffel Tower about a hundred meters. The report is actually an April Fools' Day joke, as you can tell by looking at its date (1st of April 1964). But many bloggers at the time thought it was true and asked themselves how was that possible.[251] Digital interfaces have here an important epistemic role in making the user think that what he finds is actually trustworthy. The user finds the document and manipulates it. The archive material here obtains by force its most performative status that of "proof" of an event: bloggers who believe the reality of the

---

[250] Lorraine J. Daston and Peter Galison, *Objectivity*, Zone Books, Cambridge, MA, 2010, p. 115.
[251] For a developed analysis of this case study, see Matteo Treleani, *Mémoires audiovisuelles*, Presses de l'Université de Montréal, Montréal, 2014, pp. 15–20.

reportage claim to have “found an archive document” from 1964. They found it: the digital interface is only a technical and therefore nonsignificant mediation—a tool that is used to manipulate and that does not need to be questioned. They do not look at the date and do not question the fact that this could be a joke. The document online becomes reliable; you do not have to interpret it in order to trust the representation of the past that it evokes. Antoinette Rouvroy calls this phenomenon “digital truth regime” (by borrowing from Foucault the notion of truth regime as a mode of categorization of reality): it’s about believing in the truth of data visualized in infographics and digital interfaces.[252]

More generally, the tendency is to move away from these forms of strangeness that give meaning to the past—a meaning that would reveal its specificities precisely in the non-accessibility. Being able to control and manage everything seems to be the trend implemented by digital technology as the culmination of an evolution that was already underway at the beginning of the last century. It is not a question here of asserting that the possibility of reading in Paris a manuscript preserved in Venice would be problematic—quite the contrary. This is a major development for research and for our media regimes. However, it is a phenomenon that is not insignificant that must be questioned. Now, the point is not to defend the idea that we should stop digitazation so as not to capitalize on the past, which is obviously absurd, but rather not to take as evidence processes that impact our relationship with the past.

These questions become glaring in some cases of reuse of historical sources. We can describe them as misuse of archives.

MISUSING ARCHIVES

In the field of film studies, a case of misusing archives is particularly well known: the documentary series *Apocalypse* about World War II, produced by France 2 and directed by Clarke and Costelle in 2009. Several commentators have highlighted this “manipulation of history.” Here, the digital medium is not preeminent. A montage of 35-mm films could have had the same effect. The series is a reuse of audiovisual archives with a voiceover commentary: a rather traditional practice for historical documentaries. Let’s take a look at the most glaring fact: the black-and-white

[252] Antoinette Rouvroy and Bernard Stiegler, “Le régime de vérité numérique,” in *Socio* no. 4 (2015), 113–140.

images of the past have been colored. Georges Didi-Huberman, in an editorial for the newspaper Libération, affirms that colorization has the same function of makeup, and its goal is to disappear the signs of the time.[253] To erase black and white means to presentify the images, to make them closer to the present and to the audiovisual genres we are used to seeing on television today. It is a form of domestication, therefore, which aims to erase the forms of strangeness/distance of documents of the past.

Misuse is not just digital, therefore. Rather, the digital acts as a revealing phenomenon that makes a certain attitude toward the past visible.

STRANGENESS AND OBLIVION

> We have partly domesticated that past, where they do things differently, and brought it into the present as a marketable commodity. But in altering its remains we also assimilate them, ironing out their differences and their difficulties in the process.[254]

If misuse cases are obviously problematic, the question is what is lost from a domesticated, systematized, and digitized past? Having access to all the cultural works of our time should be a dream for any civilization. Two points can be interesting: strangeness and oblivion—the strangeness of a past time, which is no longer there and can no longer be. The archive evokes in us—or ought to—the feeling of a lost world that we cannot understand because of the impossibility of getting rid of our present; and oblivion: forgetting, an essential act for the functioning of our memory. "The oblivion of oblivion" as Milad Doueihi put it: the era of all-

---

[253] George Didi-Huberman, "Apocalypse. En mettre plein les yeux et rendre l'histoire irregardable," in *Libération*, March 22, 2009.

[254] David Lowenthal, *The Past Is a Foreign Country Revisited*, Cambridge University Press, Cambridge, 2015, p. xxv.

heritage has forgotten the act of forgetting, which is nevertheless a constituent part of all mnemonic activity.[255]

Presentification aims to erase the signs of time in order to tame the past—to make it understandable and close to us. The example of colorization is particularly effective. As the historian Carlo Ginzburg says, the interest of the past is precisely where it is other than the present, where we are asked to make an effort to rethink our current categories by determining their historicity. Precisely where it is difficult to grasp, the past gives itself: in its difference from the present.[256] As soon as we realize that the document refers to a world different from ours, where we see things differently (Jean Pierre Vernant affirmed, regarding the representation of cities in Greek vases, that we will never be able to see them as we did at the time; we would have to get rid of the notion of perspective that shapes our way of looking at any representation and which formed the idea of landscape as we know it).

We could call "strangeness" this form of distance that separates us and at the same time allows us to grasp a world other than our own—"the past as foreign country" which is given where the gaps that make up this distance are irremediably present. The decomposition of media, its degradation, and the difficulty of vision that these degradations often involve is a revealing factor: take burns and scratches in images of films of a century ago, for example. These defects refer to a void, a sense of vertigo, which seems to circumscribe the document, the void of the world that surrounds it and which is no longer there and which can only be awkwardly evoked, since to get rid of our present gaze is the only way to effectively plunge into this otherness. All this does not necessarily concern times far from ours. Just think how different the internet was only fifteen years ago. Without Facebook and Twitter, reading a page on the New York Times website in 2000 should take into account a considerable difference in everything around it. The writing is no longer the same, nor the choice of titles, which are nowadays very often linked to the possibilities of sharing in social networks, for example. The strangeness is then the opposite of the proximity that immediate remote availability seems to guarantee us.

The second loss is forgetting. Forgetting is the unthinked and unthinkable aspect of digital technology and culture for Milad Doueihi. In order to think about the past and the future, forgetting is necessary. Memory is based on a process of selecting, filtering, and highlighting certain elements. For Doueihi, forgetting does not mean only and exclusively losing. It means being able to see and question new things and territories, to explore novelties while leaving aside known or memorized things.[257]

---

[255] Milad Doueihi, *Digital Cultures*, Harvard University Press, Cambridge, MA, 2011.

[256] Carlo Ginzburg, *Wooden Eyes: Nine Reflections on Distance*, Columbia University Press, New York, 2001.

[257] Milad Doueihi, *Pour un humanisme numérique*, Seuil, Paris, 2011, p. 155.

The recording society, where every document is saved and every object archived, seems to have difficulty managing this essential element. We can glimpse a world of digital traces, impossible to erase.

However, to assert that the age of the all-archive would be an era where oblivion is unthinkable may seem paradoxical. If the selection is not carried out at the material level, since all documents would be logically preserved, it should be assumed that it will be carried out elsewhere. But thinking that the all-archive would actually be a lossless memory is the real problem. Nothing tells us, for example, that we are not losing something—everything that is not digitally recordable, for example.

Strangeness and oblivion are therefore words that do not seem to fit with a more general form of administration of the domain of history. The usefulness of a digital humanism undoubtedly resides in the reminder of a certain way of seeing things that does not forget its human side and where technology is not external to man but one of its dimensions, where strangeness and forgetfulness have a place because they are not excluded from reasoning.

In Umberto Eco's novel *The Mysterious Flame of Queen Loana*, an amnesiac character, Yambo, has forgotten all his life and remembers only his readings or the notions he learned.[258] All of his personal experiences, his emotions, and what allows him to make the connection between things seem to be gone. Where Funés the Memorious [from the short story of the same name by Jorge Luis Borges] can't forget and suddenly locks himself in a room to stop memorizing, Eco's Yambo, on the contrary, remembers only notions, having lost the humanity of his memory—another metaphor that we should probably consider to understand how we approach notions of heritage in the digital age.

[258] Umberto Eco, *The Mysterious Flame of Queen Loana*, Harcourt, San Diego, 2005.

WEB STARDOM
A NEW PHENOMENON
VANNI CODELUPPI

From around the mid-1990s, people in contemporary societies began to transform from purely passive recipients of media messages into active creators of various kinds of messages. Since then, an array of different tools have become widely available for this purpose, including websites, e-mail, digital photos, smartphones, technologies for the production of high quality videos, and so on. In parallel with these changes, people have developed an awareness of their growing power within the world of media.

In actual fact, signs of change had already become clear before that time. When the black cab driver from Los Angeles Rodney King was violently beaten by policemen on March 3, 1991, for example, the incident was recorded on video and broadcast over and over again on American television. The video was very misleading (it didn't show the audience that the event was preceded by a long car chase at 110 miles an hour, that the cab driver was driving under the influence of drugs and alcohol, and that he had failed to heed the policemen's warnings not to move), but it was very effective at the communication level. So much so that, when the policemen in question were acquitted at the trial, violent disorders erupted in the streets of Los Angeles, leaving fifty-four dead and two thousand wounded.

But it was with the advent of the web, particularly when it shifted to the so-called 2.0 phase, that people were truly transformed into effective creators of media messages. A key driver of this process was the introduction and large-scale diffusion of social networking sites and, more recently, also of services like Twitter's Periscope and Facebook Live, which make live-streaming possible. Obviously, stars were the first to take advantage of all this. But as we shall see in due course, ordinary people, too, have learned to use these tools in the same way as stars.

STARS AND SOCIAL MEDIA

Stardom has probably always been a participatory phenomenon. Efforts to actively engage audiences go right back to the early forms of cinematography. Back then, however, the tools used for the purpose were rudimentary and included such things as letters to members of fan clubs or photographs signed by the actors. Modern-day digital media, by contrast, as we well know, have greatly enhanced the participatory dimension of stardom and have also significantly improved the quality of the relationship that stars are able to establish with their fans. Yet many of the activities that fans used to practice in the past persist to this day. With digital media, today's participant fans become creators of content, just as the fans before them created scrapbooks out of clippings and star-related items and wrote letters or fantasy stories (Reich and O'Rawe 2015, p. 53). The real change has to do with the new distribution and dissemination opportunities now available. These have grown considerably and, through social networking sites, fans can play an even greater role in boosting and publicizing the image of a given star.

At the same time, the new digital media give stars greater ability to control the image of themselves that they wish to deliver. This is illustrated by the episode that took place during the Oscars 2014 ceremony, in which the hostess Ellen DeGeneres invited all the actors and actresses sitting in the front rows to take a photo together with the aim to break the record as the most retweeted selfie in the world.

Featuring the likes of Brad Pitt, Meryl Streep, Julia Roberts, Kevin Spacey, Bradley Cooper, and Jennifer Lawrence, the target was easily reached as the selfie was retweeted over three million times in a short space of time. This highlighted what had long been well known, namely that the advent of social networks has enabled celebrities to enhance their media presence and, above all, to independently manage their relationship with fans and create a greater feeling of closeness and intimacy with them, giving fans the sense that they have "backstage" access to them.

In the past, in order to communicate, stars had to employ a huge staff (assistants, agents, public relations specialists, and so on), have their photos taken by professionals, and sometimes also had to defend themselves from the latter's aggressive behavior—as shown by Princess Diana's tragic death in Paris while trying to escape some intrusive paparazzi. Today's celebrities, by contrast, tend to view the role of these professionals as anachronistic because, thanks to social media, they now have the opportunity to create forms of communication addressed directly to their fans.

Obviously, this comes more naturally and easily to younger stars, who have grown up in the era of the internet. Sometimes, these young celebrities have no special qualities or abilities but manage to achieve wide public visibility thanks to their particular skills in using social network sites very effectively. A case in point is American web star Kim Kardashian, who can neither sing, dance, nor act, but has garnered tens of millions of followers on Instagram and Twitter.

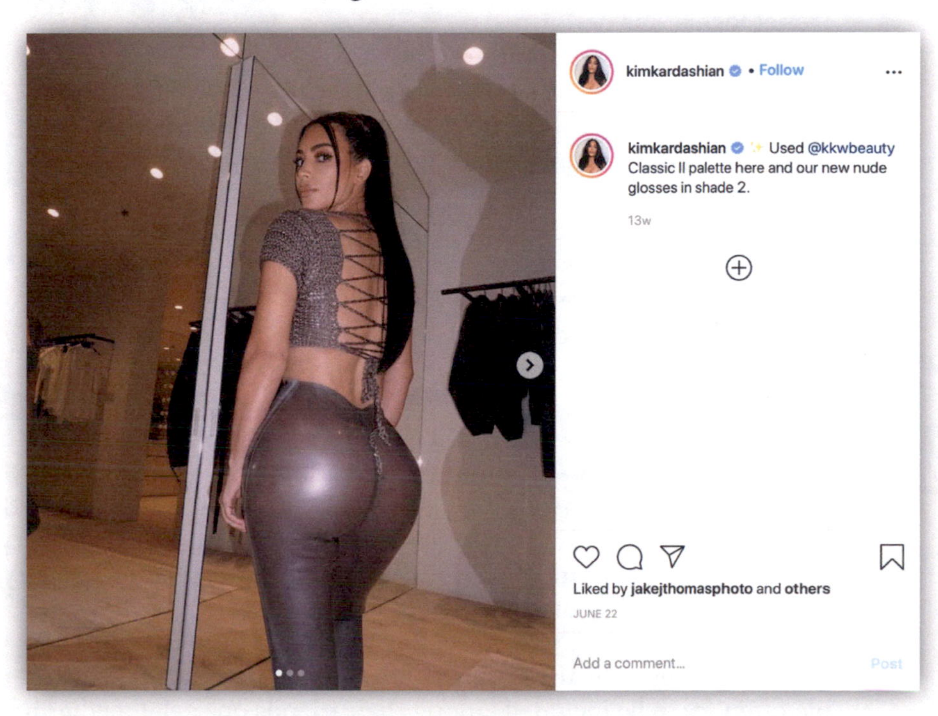

All she does is use various media to talk about what she does in her everyday life with family and friends, and, most of all, she constantly exhibits her body. This, despite the fact that, as some have pointed out, it is an absurd body, consecrated to the collective desire, with huge Bambi eyes, hips that would throw anyone off balance, and a famous B-side insured for fifteen million dollars (Salemi 2014, p. 131).

A similar case is that of Paris Hilton, the rich heiress of the American hotel empire who in recent years has become the focus of the gossip world, but who also has the power to influence the fashion trends and consumption behaviors of droves of adoring teenage girls who ape her garish, childlike, and trash style. Her media career was launched by a leaked video showing her, nineteen years old at the time, having sex with her former boyfriend Rick Salomon, then thirty-two years old. It was no coincidence that the video came out just before her debut as the star in *The Simple Life*, a reality show that lasted for five years. In the show, alongside her friend Nicole Richie, the adopted daughter of singer Lionel Richie, she performed all the unsavory chores that have to be done in a country farm. For several years, Hilton led a life of excesses, going to drink and drug parties with friends like Britney Spears and Nicole Richie, and was arrested on various occasions for driving with an expired license and under the influence of alcohol. In 2007, having served her sentence, she came out of jail looking as if she had never been there. She emerged smiling, surrounded by a flock of photographers and video cameras, and that same evening she gave an exclusive interview to the leading American journalist Larry King on CNN. And today, she very successfully endorses numerous product brands (clothes, shoes, jewelry, as well as hotels and resorts).

The fame enjoyed by these kinds of celebrities means that ordinary people often try to be included among their friends on social media sites. As

Hilton herself stated, "People *need* to believe your life is better than theirs" (Hilton and Ginsberg 2005, p. 8). Film director Sofia Coppola picked up this theme in her 2013 movie *Bling Ring*, showing the real-life exploits of a gang of Hollywood teenagers obsessed with celebrities. They burgled the homes of famous people to steal not just valuable clothes and jewelry but items that had actually been worn by the stars, and they sometimes even naively posted images of their "trophies" on social media. The film director had this to say:

> To write the film script I met the "real" protagonists of the story. They didn't believe they had done anything wrong. They broke into the houses because the keys were actually hidden under the doormat, and there was no burglar alarm. [...] Their idols, too, love to exhibit themselves don't they? And those houses were like impregnable fortresses. The kids were looking for luxury, but most of all they wanted to be recognized; they wanted to be famous [Piccino 2013, p. 13].

Thus, there are no significant differences between the behavior of stars and that of their fans. In contemporary societies, all individuals are equally anxious to exhibit themselves as well as they can.

THE STAR IS ALWAYS THE STAR

As we said, nowadays stars and fans behave in a similar way. Yet the social role of the star and that of the fan remains essentially different. This is illustrated by the proliferation of so-called fan fiction in recent years, namely stories featuring celebrities as the lead characters and usually written by their fans to stage a chance meeting with their idols (Mazzucco 2016). They consist of novels, like those in the *After* series created by American writer Anna Todd, for example, recounting from the fans' perspective and in fairy tale–like language the personal exploits of various kinds of stars. In the case of *After*, the stars are the members of the music band One Direction, but in this literary genre, written by teenagers for teenagers, the stars can also be soccer players like Messi and Higuain, or YouTube celebrities. These stories are currently disseminated chiefly via Wattpad, an online platform featuring over 250 million narratives with over forty-five million users worldwide.

The fact that there is still a significant difference in social status between stars and their fans today is also clear from the way celebrities use social media. The scholar Tiziano Bonini (2012) conducted a quantitative study analyzing the activities of stars within these digital spaces and found that "they post their personal reality show on their twitter profile twenty-four hours a day, seven days a week, and their fan audiences just watch them." His findings suggest that fans are essentially passive recipients of the messages posted by stars, while the latter do nothing but talk about themselves, their whereabouts, their actions, and their opinions at

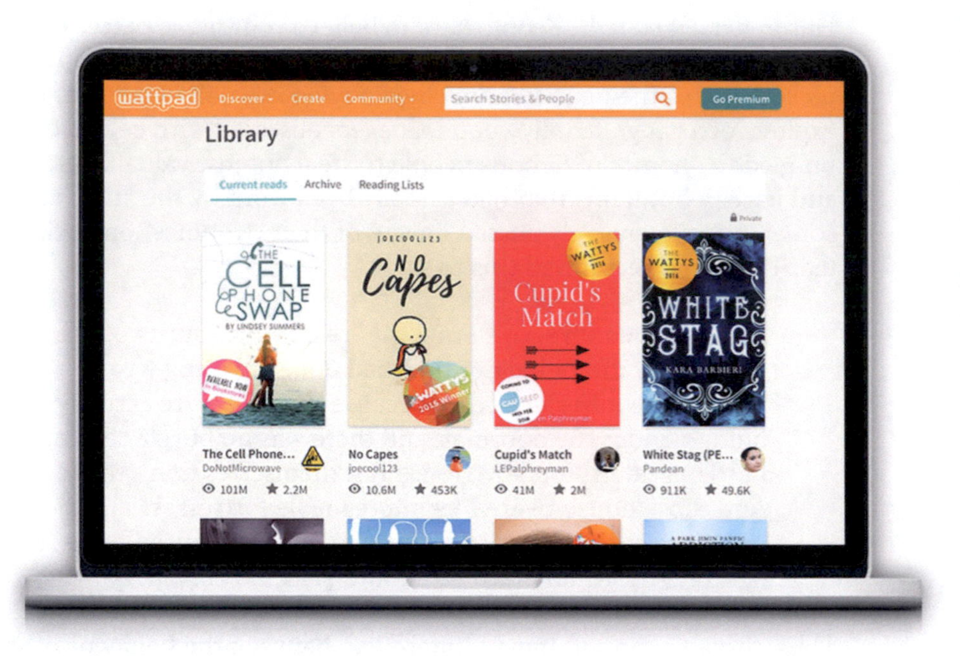

the time of writing. In other words, stars refuse to behave as social media users normally do, that is they do not retweet news or messages posted by other users, they do not make political comments, and, most notably, they do not follow anyone, or hardly anyone. They merely describe or discuss what they do in their daily life in order to promote their personality. In theory, the web provides the opportunity for total interaction among all those involved in the communication process, but in practice there are significant human and social limits preventing this from happening. Celebrities achieve their fame through their ability to create a wide audience, but they are then unable to respond to all the messages they receive. The celebrity "can't talk to even a fraction of a percent of her audience, ever, because she is famous, which means she is the recipient of more attention than she can return in any medium" (Shirky 2008, p. 70).

It should be noted that when social media first appeared, stars were able to deliver a certain image of themselves and maintain a distance from their fans fairly easily. But since the selfie has become a mass phenomenon cutting right across society (Codeluppi 2015), stars feel a growing need to differentiate themselves from ordinary people. This explains why they sometimes also try to communicate by making an original and provocative use of the images concerning their lives.

But in the era of the selfie, the fans' greatest ambition is to be able to take a picture of themselves alongside the stars they admire and then share it with their friends on social media, while Hollywood stars, for their part, are constantly trying to escape from this veritable modern-day obsession that has taken hold of their fans. Some celebrities have even left their luxurious Los Angeles mansions, immersed in sunny palm-fringed settings, to go and live in New York, a far less friendly and appealing metropolis, but which offers a better chance to protect their private lives by blending into the anonymous crowd. It also makes it easier to escape from the growing number of cases of fans who become intrusive and obsessive because they regard the celebrity as their own exclusive property.

At the same time, in an era dominated by social media networks, stars are inevitably required to have greater media exposure. Fans use the web to find out their whereabouts in both their private and professional lives, and often follow them to chic restaurants or fashion shows and use their smartphones to take pictures of their intimate moments, which they then regularly post on social media. These images bear testimony to the fact that the fan has been able to share a moment in the life of the star, and, as such, they become precious trophies to display or to share with others. This activity is generally known as "VIP watching," using the analogy with the traditional practice of bird watching. To cater to it, special events are organized where fans have access to their favorite stars. In the past, there used to be fan clubs, but nowadays these are no longer sufficient.

In conclusion, celebrities today can no longer keep away from their audience. And, as is the case with reality shows, in which they often happen to star, they are frequently subjected to the merciless judgment of their fans. This also happens in specific areas, like the artistic sphere. One such example is *The Mercury Theatre*, a French show featured on Instagram involving a competition in which thirty international artists are invited to compete for the public's "likes," and where the public cruelly decides who to eliminate.

THE STARS NEXT DOOR

The widespread growth of media exposure affects the lives not only of stars but also of ordinary people. Indeed, today it is actually possible to be filmed by the watchful eye of a video camera from the moment we are born, as is the case in the maternity ward of the Macedonio Melloni hospital in Milan, where the images of the newborns are sent to friends and relatives wherever they happen to be. And as individuals grow up, they become aware that they are constantly at the center of other people's attention and feel distressed if that level of attention declines. It is only natural, then, that all individuals try to present themselves in their social media profiles as well as they can. They reduce physical contact and try to manipulate from the "backstage" of a protective digital display window the aspects of their personal identity that they wish to showcase. Thanks to the technical possibilities offered by social networking sites, they take great pains both to create the personal image they wish to deliver and to actively manage it over time. They are also concerned to have an audience of their own and to nurture it successfully. So, just like stars, their goal is to achieve personal fame, which in this case has been termed "microcelebrity" (Senft 2008), that is, fame shared by a selected group of individuals through the communication tools available on the Web. To manage this fame as effectively as possible, people try to imitate the communication strategies for building one's image and reputation usually adopted by stars, who in turn imitate the communication behaviors adopted by company brands (Brognara and Codeluppi 1992).

Individuals thus exhibit themselves daily in these digital display windows because, to the extent that these make them feel as if they are onstage in front of an audience, they offer gratification. Although out of the hundreds of people who call themselves their "friends" a large proportion are in fact perfect strangers, those who communicate feel as if they are standing before a huge audience that is constantly listening to them. This provides a sense of reassurance to people, whose social condition today usually involves high levels of uncertainty. And digital display windows offer yet another possibility that is no doubt also reassuring, namely to give one's identity stability over time, in other words to attain a sort of "virtual immortality."

One of the reasons for this behavior is that people believe they have the ability to select and control everything they want to publicize about their identity. And indeed, the technology currently available also makes it possible to create a private dimension that can only be shared with a limited number of people. This is what Giovanni Boccia Artieri, Manolo Farci, Fabio Giglietto, and Luca Rossi (2014) have called "connected intimacy." This type of intimacy with restricted access is achieved by using specific communication strategies whose meaning is inaccessible to the wider audience, but not to the few who are aware of the relevant codes of interpretation. Such strategies include, for example, posting photos, which are more ambiguous than written words, or links, lines from songs, poems, and music videos. And whatever their nature, they always involve a method of expression that can only be understood by a specifically selected group of individuals.

That, however, raises the question of how compatible all this is with today's prevailing idea that people should show themselves on the web as they really are and totally transparently. On this subject, Lee Siegel (2009) has argued that, with the advent of the internet, people's private space has shifted into society's public sphere, but at the same time it has led to the development of two different types of private spaces: one that is artificially constructed for the purpose of self-presentation on the web so as to make it effective at the communication level, and one that is real, inaccessible, and jealously guarded. In other words, people are aware that in a world where social networks are deemed to play an essential role for the development of social relations it is impossible to avoid being constantly under observation. Yet, it is possible to try to divert other people's attention by showing them a private image of oneself calculated to prevent them from focusing on the private side that one is seeking to protect and restrict to a few selected individuals, as mentioned earlier.

This, however, does not prevent the recipients of "exclusive" messages from posting them in turn on the web. And, most importantly, it does not prevent people from adopting a dedicated strategy of self-promotion within the purpose-built private space they have created, in other words, a "window display" strategy (Codeluppi 2007) designed to enhance and promote oneself and one's life as much as possible.

We should consider, moreover, that every feature we choose for our "window display" today also ends up telling others what we would not wish to be known. But everything we do online inevitably tends to generate a huge mass of information concerning each and every web user, who generally forgets about it or is hardly aware of it.

WEB STARS

The practice of ongoing self-exposure that people adopt on the web has consequences on the use they make of their bodies. In social life, people have always used their bodies as the primary tool to present themselves to others. All individuals thus seek to present their specific bodily capital, namely the sum of all their physical features, as well as they can. Today, people frequently do this by exhibiting their bodies online, with digital platforms such as YouTube providing the ideal medium of expression in that regard. YouTube has the potential to reach around one billion unique visitors per month, having established itself over the years as a sort of constantly available television that people can watch from anywhere and, most importantly, in which anyone can freely enter. Its very nature as a platform overflowing with images and sounds, and with such a wealth of features as to literally submerge individuals, offers everyone the chance to have their own exhibition space. While this space is usually limited, it is nevertheless both available and appealing.

It all started a few years ago with the occasional posting of videos showing dancing cats or the most hilarious bloopers by politicians and television personalities. But it soon became clear that, in some cases, these videos were able to achieve surprising results in terms of visibility, with hundreds of millions of viewings worldwide. This was true of Psy, the South Korean rapper with his hit single "Gangnam Style," and of video clips featuring singers like Jennifer Lopez, Rihanna, and Lady Gaga. That was how web professionals, known as YouTubers, were born. YouTubers try to make the most of this medium's prevailing communication standards, where the idea is to deliver short, simple, and possibly entertaining messages. They usually create videos featuring funny gags, for example, or instructions on the best way to dress and put on make-up, or how to win in video games. The videos are usually created and presented by ordinary people whose very "ordinariness" makes their viewers identify with them, and turns them into veritable celebrities with millions of followers regularly watching them on YouTube or similar platforms like Vimeo, Twitch, Vine, and TikTok.

One such example is Felix Kjellberg, better known as PewDiePie, a twenty-year-old Swedish internet celebrity who is regularly watched by over forty million viewers on YouTube and has the largest following in the world. He specializes in producing short and entertaining commentaries on video games, and, as with other YouTubers, the secret of his success probably lies in his ordinariness and strong similarity to his viewers, who can therefore easily identify with and relate to him.

Today, there are a great many YouTubers. British-born Zoe Elizabeth Sugg, best known as Zoella, started out by uploading video tutorials on fashion and makeup as well as on how best to behave in challenging situations. But once she became famous, she started talking mostly about herself and her life, posting videos of her having breakfast, going shopping, and organizing parties with friends.

All these personas show little talent for doing anything at all. They don't usually dance or sing; they are not particularly witty and don't even know how to act. But, precisely because of this, they are living proof to their fans that despite their lack of skills, they have still managed to achieve great success and popularity. Hence, their fans, too, can take up the challenge and try to make something of their lives.

These celebrities usually embody a deeply conformist cultural model. It is a model with an entirely positive focus and no scope for complaint or criticism: everything has to be presented as happy and carefree. A prime example of this is the video showing Nash Grier, an American Vine celebrity, looking extremely pleased with himself after brushing his teeth. The world of YouTubers is also strongly based on consumerism, where what matters most is to achieve enough popularity to secure profitable contracts with companies to promote their brands. Companies in fact often seek out these kinds of individuals. This is especially true of fashion companies, which use YouTubers as influencers with the ability to persuade masses of consumers to buy their products. In Italy, the blogger Chiara Ferragni, who has droves of young followers, is a typical example of this (Keinan et al. 2015). And it is not surprising to find that, in 2016, the English brand Burberry decided to commission the campaign for its new fragrance to Brooklyn Beckham, the seventeen-year-old son of soccer star David and singer and fashion designer Victoria Beckham. The youth was given preference over leading professionals because he was thought to have greater appeal for younger consumers based on his success as an Instagram celebrity—as evidenced by his six million followers in this particular social platform. And for their fashion shows and advertising campaigns, fashion companies today frequently choose makeup artists, hairdressers, and stylists based on the size of their social media following.

In actual fact, the number of YouTubers who have achieved considerable financial success and manage to make a living out of this work is extremely small. And all of them have invariably had to fully accept the fact that, in order to earn money in the digital universe of the web, it is to some extent necessary to “sell” one’s life by associating it with brands that enjoy varying degrees of success. All this, however, does not happen by chance. YouTubers are not communication whiz kids who rise to fame single-handedly by sending out messages from the confines of their bedrooms. This is the deceptive and romantic image of them usually presented to the world. But in reality, behind the most successful YouTubers, there are specialist companies that have trained them and have carefully designed their image. And these companies naturally expect to reap the fruits of their labor, including by featuring the YouTubers in television shows and films. It is no wonder, then, that in the new digital world, success is achieved first and foremost by those who are most willing to become goods to promote other goods.

References

Boccia Artieri, G., M. Fraci, F. Giglietto, L. Rossi (2014). "Intimità connessa. Intimità e amicizia tra gli utenti italiani di Facebook," in G. Greco, ed., *Pubbliche intimità. L'affettivo quotidiano nei siti di Social Network*. Milan: FrancoAngeli.
Bonini, T. (2012). "Appunti per un'anatomia della celebrità su Twitter." Doppiozero. http://www.doppiozero.com/materiali/Web-analysis/appunti-un%E2%80%99anatomia-della-celebrita-su-twitter.
Brognara, R., and V. Codeluppi (1992). *Imagineering. Costruzione dell'immagine e strategie di comunicazione*. Milan: Guerini.
Codeluppi, V. (2007). *La vetrinizzazione sociale. Il processo di spettacolarizzazione degli individui e della società*. Torino: Bollati Boringhieri.
——— (2009). *Tutti divi. Vivere in vetrina*. Rome-Bari: Laterza.
——— (2015). *Mi metto in vetrina. Selfie, Facebook, Apple, Hello Kitty, Renzi e altre "vetrinizzazioni"*. Milan-Udine: Mimesis.
Gabler, N. (1995). *Winchel: Gossip, Power and the Culture of Celebrity*. London: Vintage.
——— (2001). *Toward a New Definition of Celebrity*. Los Angeles: The Norman Lear Center, Annenberg School of Communication, University of Southern California.
Hilton, P., and M. Ginsberg (2005). *Your Heiress Diary: Confess It All to Me*. New York: Fireside.
Keinan, A., K. Maslauskaite, S. Crener, and V. Dessain (2015). "The Blonde Salad." Faculty & Research, Harvard Business School. http://www.hbs.edu/faculty/Pages/item.aspx?num=48520.
Marshall, D.P. (1997). *Celebrity and Power: Fame in Contemporary Culture*. Minneapolis-London: University of Minnesota Press.
Marwick, A.E (2013). *Status Update: Celebrity, Publicity, and Branding in the Social Media Age*. New Haven: Yale University Press.
Mazzucco, M. (2016). "Come riscrivere Vacanze romane e conquistare i lettori ventenni." In *La Repubblica*, June 7, 2016.
Morin E. (2005). *The Stars*. Minneapolis, MN: University of Minnesota Press.
O'Rawe, C. (2014). *Stars and Masculinities in Contemporary Italian Cinema*. London: Palgrave Macmillan.
Piccino, C. (2013). "Quell'ossessione di celebrità che attraversa il nostro tempo." In *Il manifesto*, September 18, 2013.
Redmond, S. (2013). *Celebrity and the Media*. London: Palgrave Macmillan.
Reich, J., and C. O'Rawe (2015). *Divi. La mascolinità nel cinema italiano*. Rome: Donzelli.
Rojek, C. (2001). *Celebrity*. London: Reaktion.
——— (2011). *Fame Attack: The Inflation of Celebrity and Its Consequences*. London: Bloomsbury Academic.
Salemi, R. (2014). "Scandalosa Kardashian." In *L'Espresso*, December 4, 2014.
Senft, T. (2008). *Camgirls: Celebrity and Community in the Age of Social Networks*. New York: Peter Lang.
Shirky, C. (2008). *Here Comes Everybody: The Power of Organizing without Organizations*. London-New York: Penguin Press.
Shingler, M. (2012). *Star Studies: A Critical Guide*. London: BFI.
Siegel, L. (2009). *Against the Machine: How the Web Is Reshaping Culture and Commerce—and Why It Matters*. New York: Spiegel & Grau.
Turner, G. (2004). *Understanding Celebrity*. London: Sage.
——— (2010). *Ordinary People and the Media: The Demotic Turn*. London: Sage.

IMAGOCRACY
THE POLITICAL IMAGINARY IN THE DIGITAL AGE

GUERINO BOVALINO

INTRODUCTION

One of the most famous images in the collective imaginary is that of an elegant Obama posing in front of a man-height Superman statue. That is to be considered the apex of the American *messianism*, and it is in that mythological frame that the President of the United States consolidated his prophetical image as the superhero who would be able to make the world population converge through his words, and as the one who would save and change the world.

Speaking of suggestive political imaginaries, we could recall how in most recent times, Marine Le Pen publicized a pamphlet by Laurent Obertone, an unknown French intellectual, to support her political thesis. The pamphlet was called *La France Orange Mécanique*,[259] and she renamed it

[259] Laurent Obertone, *La France Orange Mécanique*, Ring, France 2015.

“the orange booklet.” According to the author, France has become a fertile environment for unpunished crime, allowing for a parallel with the cult movie from 1971 *A Clockwork Orange*, which denounced the banality of juvenile ultraviolence. The idea of Le Pen to mention the book as the “orange booklet,” as well as its original title, creates a clash of different types of imaginaries such as Mao’s red booklet, a cornerstone of a traditional political theory, or the cinematic masterpiece by Stanley Kubrick, which is the symbol of a postmodern, postalphabetical, and pop culture.

Moreover, it has become publicly known how terrorist movements appraise the effect of their actions in the public imaginary. We could recall ISIS’s communication strategy, which uses the publication of the slaughtering of the “unfaithful” on the web, using special effects and the editing of the videos that parody Hollywood movie productions and reality shows through the spectacularization of death in world vision.

The so-far cited examples of the postpolitical dimension of our times suggest that, in order to reflect on contemporary political forms, it is necessary to explore the mediologic[260] aspect of it, meaning to focus on the mythical and media imaginary,[261] which is the territory where politics has been constituted.

POLITICS VS POLITICAL?

According to our view, in fact, politics is not to be intended in its administrative form, or rather politics as the art to govern the world. It is to be intended in order to examine “the political,” meaning politics in its essence through the symbols and images on which the substantial dimension of the *individual* and of *society* is constituted.[262]

Single individuals and society are indeed the smallest units of more complex constellations from which different imaginaries come from.

The “political” can be interpreted and analyzed as one of the many aspects of everyday life, going beyond its historical-ideological dimension. It can be considered as a mix of suggestions and images, words, and sentiments, which are all elements that evoke the irrational and emotional aspects of politics. Every act that is considered political is created, lives, and is alimented by myths and imaginaries and cultural references that are rooted in the society.

---

[260] Mediology is a newly coined term which indicates the study of the means of mass communication, which are understood as a place of a sort, which influences, includes, and puts in relation the social, institutional, and ideological aspects of society.

[261] Media imaginary is to be intended as ensemble of images and symbols which, once filtered by media, constitute the core of society and individuals.

[262] The distinction between politics and “the political” can be connected to the distinction between “pouvoir” and “puissance,” as highlighted by Maffesoli.

THE MEDIUM IS THE MESSAGE
THE MEDIUM IS THE MESSAGE
THE MEDIUM IS THE MESSAGE

The "political" is included completely in the dynamics of cultural industry. The crisis of ideologies, as well as the end of the left-right dichotomy, for instance, are concepts that are now rooted in a contemporary political analysis and can no longer be overviewed by analysts.

The crucial cultural and political point for an incisive research is without any doubt the analysis of how the political consensus is gained by political leaders today. What are the ideas that shape the new political fractures that push towards the polarization of political choices?

Some intellectuals have outlined new theories that go beyond the classical left-right dichotomy, pointing out other solutions, but not limited to: liberal-communitarian, radical- reformist, populist-elitist. [263]

THE POWER OF MEDIA

Our vision explicitly refers to the intuition of the Canadian comparative literature professor Marshall McLuhan, who had already understood in the 1960s the way in which new electronic media had reconfigured our culture and society in all of their aspects: political, social, and economic.[264] McLuhan was the first one to recognize the potential (and the power) of the media in creating and recreating stories and worlds, creating needs and fears, and feeding wishes. He highlighted the power of media to ignite innate and primordial aspects of men, therefore becoming agents of anthropological change themselves.

Media makes itself explicit by reaffirming the coexistence of archaic and ancestral factors with other ultramodern aspects. New communication technologies and biotechnologies have profoundly modified our society by penetrating the everyday life of all individuals. The changes that they have brought about have touched every aspect of our existence, including the political aspect.

Deciding who will govern our existence requires an analysis of the context in which we live and a clear decision on how we would want to live. It requires defending and promoting the idea that we have of the world and contrasting those who do not agree with us. The mentioned idea of the world is not limited to the ideological field anymore. We could say that it is not possible to define and justify our actions through classical concepts, such as "political parties" and "leaders," which nowadays often do not fully embody the response to our wishes and needs. The inefficacy of the traditional forms of representation makes it hard for the parties and their leader to maintain a public consensus, and therefore to promote a political program. In this sense, it becomes useful to try to use a different way to interpret "the political." According to Michael

[263] Cf. Marcello Veneziani, *Alla luce del mito*, Marsilio, Venezia, 2017; Alain de Benoist, *Droite-gauche, c'est fini !: Le moment populiste*, Éditions Pierre, Guillaume de Roux, 2017.
[264] Cf. Marshall McLuhan, *Gli strumenti del comunicare*, Il Saggiatore, Milano, 1964.

Maffesoli's idea in *La Transfiguration du politique : La tribalisation du monde postmoderne*, the concept of politics as competition between parties (based on the conflicts determined by cleavages) has been substituted by a new modality of being together: tribes of people that get together to share passions; humans who recognize each other as a community based on symbolic images that relate to immutable archetypes and myths that are embodied today in new stereotypes, which are crystalized in precarious but recognizable figures and forms.[265] This idea should be propaedeutic to every reflection on the matter.

Politics is therefore transformed: it goes from being a medium of persuasion to a means of seduction. We could say that the programs and ideologies themselves do not create followers anymore. Indeed, it is the ability of the political leaders of embodying said programs and ideologies and representing them in their symbolic and communicative sphere that can stimulate the irrational side of the individual who votes. As effectively explained by De Kerkhove and Susca in *Transpolitica (Transpolitics),* politics has to take the images of the public imaginary and makes them its own. That is because politics itself has to be defined as a powerful show that represents the society depicted in the media and not as an aseptic institution of power.

VIDEOCRACY, COMMUNICRACY, AND IMAGOCRACY

A premise to our discourse on new forms of the "political" in the digital era necessarily highlights the necessity to reconfigure both ideological dichotomies and political language: the words we use to talk about politics.

We consider it necessary to underline how terms such as *politics, community, and person* are conceptually obsolete. In fact, politics went from being an instrument of warrantee to a means of control and oppression. The concept of *community* is always balanced between the necessity to keep one's *humus* united, and the contrast with the external factors that threatens its integrity. And finally, the basic molecule of the "political," the person, should be interpreted today in its Latin meaning of *mask*, a *fragmented and destructured figure* which is divided into different subjectivities, impersonating different lives.

Exploring the instruments that are necessary to clear out the critical aspects of the issue allows us to set a mediologic discourse to take into account the changes of society and of the individuals in front of the new communication technologies and digital language. The mediologic approach allows us to investigate the future perspectives of the political dimension in its constant shifting from democracy to inedited forms of *videocracy*, *communicracy*, and *imagocracy*.

---

[265] Cf. Michel Maffesoli, *La Transfiguration du politique*, La Table Ronde, 1992.

Videocracy is the kind of politics related to mass media in a way that is still vertical and with a vertex. In Italy its maximum interpreter was Silvio Berlusconi.

Communicracy constitutes the subsequent passage, which is connected to personal media and the social media network, through which we all become consumers and producers of a message. We specifically refer to the passage from telepolitics to the politics of *tweets* and *selfies*, all the way to new movements such as the German Pirate Party and the Italian 5 Stelle Movement.[266]

Imagocracy derives from the theory of Michael Maffesoli that states: "the surface is the true depth of things." This said surface is the place where our relation to the environment around us and to others is negotiated—it's the atmosphere where ideas, books, visions, movies, advertisements, everyday life, comics, culture, philosophy, and banality mix. The imaginary is filtered, produced, and reproduced by media, and in particular by new digital technologies. The *political* has metabolized the consumption dynamics and lives entirely inside these Babelic multiform imaginaries, and could not survive otherwise. The true challenge is to be able to get the constant interaction and intersection of traditional forms recognizable as politics with new instruments of consensus and debate. We need to assimilate new forms that inhabit our everyday life, and to understand these forms as non-divisible from the *political*. Studying said forms is useful to fully understanding the mix of impressions, suggestions, and passions that are at the basis of the actions in the *polis* of each citizen (influenced in his/her electoral functions as well as its function of creator of common sense).

The political has metabolized the dynamics of consumption. Nowadays, TV stars have become opinion leaders and have the same amount of influence of renowned intellectuals. Pornography is considered a cultural form equal to literature or poetry. The *political leader* has become a brand like Zara or Nike, and Steve Jobs is recalled as a Messiah who came on Earth to bring the Apple-Bible, revolutionizing our existence.[267]

Politics is fed by the public imaginary, which actualizes some archetypes filtered by media. By filtering them, the media reconfigure said imaginaries into images and digital figures, like all the other areas of society. These archetypes, despite being disguised in new forms, continue working as compasses for the individuals in terms of representation of the people and the media themselves.

---

[266] Cf. Angelo Romeo, Posto, taggo, dunque sono? Nuovi rituali e apparenze digitali, Mimesis, Milano, 2017.

[267] [Editor's note: As this collection of essays goes to press Apple's Market Cap value is at 2 trillion dollars.]

There is a gap between the man and the world that it inhabits that has to be filled: human beings, in fact, need to give a sense to their everyday life. Said gap was filled in the centuries through a *dispositif* that was able to transform the human being into an entity able to act in the world and to relate to it as part of a larger network.[268]

The *dispositif*[269] that allowed the human being to become a historical subject has changed forms over time: Godly figures, mythological, religious, or ideological figures. Today the gap is filled by the media in all of their expressions, as well as by the imaginaries that come from the mix of old and new symbols read through the lenses of new technologies. As Alberto Abruzzese said "Christ today is re-embodied into a phone."[270]

PROMETHEUS, DIONYSUS, AND ORPHEUS

Trying to analyze the new fractures of society, as well as the visions that come from it and the imaginaries that give it sense, it is possible to create a trinity that is constituted by three existential mythologies, three archetypes that determine just as many macroimaginaries. Each one of them is a dispositif used by a certain portion of society to represent themselves and their lives, both at the individual and societal level.

While these dispositifs are only useful as paradigms, they can help us highlight three visible mythologies that can work as new schematization and interpretation categories of politics in the society and in the imaginary: so we are not talking about the ideological dispositifs such as left and right division, but about the myths that are the dispositifs of Prometheus, Dionysus, and Orpheus. Each one of these mythologies constitutes a particular dispositif that allows us to interpret reality and to live following specific trajectories. The mythological imaginary that each of these mythical devises can, when added to contemporaneity, clarify the new clash between political positions since it isn't limited to a mere political view, but it expresses forms that contemplate aesthetical, emotional, and irrational dimensions as well, since these are proper to each vision of the existence.

---

268 *Dispositif* is a term used by the French intellectual Michel Foucault, generally to refer to the various institutional, physical, and administrative mechanisms and knowledge structures which enhance and maintain the exercise of power within the social body. See M. Foucault, *L'archeologia del sapere*, Milano, Rizzoli, 1969.

269 Cf. Gilles Deleuze, *Che cos'è un dispositivo*, Napoli, Cronopio, 2005; cf. Giorgio Agamben, *Che cos'è un dispositivo?*, Roma, Nottetempo, 2006.

270 Alberto Abruzzese, "Sul disprezzo che abbiamo per la vita che siamo," series of papers published in *Gli Altri* newspaper in September 2009.

The myth of Prometheus includes the eagerness of power that is present in Obama's *messianism* as well in the dystopias of technical and bureaucratic control of the society, and therefore in all forms of progressivism and fideism and in the revolution in the hands of the hero and superhero (whether that is a liberal revolution or techno-utopist one).

The American *Neocon*, with their mission to export democracy to the world, is one of the epiphenomena of this project as much as the cyber democratic utopias of the 5 Stelle Movement, and as the universalistic and postpolitical projects of the neohumanist manifesto of Zuckerberg.

The myth of Dionysus[271] talks about the refusal of every paradigm at the foundations of modernity, about a critic of progressivism in favor of the emotions and sensuality of the being. This is the communion with life itself and the indulgence to any intimate desire, while individuals usually are stuck in their digital existence as avatar that consume themselves and the others. When answering to the needs of the flesh (both real flesh and the mystical virtual flesh, which are to be considered indivisible), one is ignited with pure life without any filters.

Finally, the myth of Orpheus is characterized as a tension toward the return to earth. It is a look straight to the eyes of apocalypse that Orpheus's followers see in the present time, which is invaded by technology and gender theories, to be considered as an apex of a general attack to anything traditional and spiritual of the communitarian and naturalistic ideal of life. Orpheus represents the supremacy of the concrete superhuman act over the virtual dimension of our digital lives.

Neoidentity representatives, anti-Europeans, and anti-multiculturalists fight for the safeguard of the traditional family and of the marriage, praising the role of religion as a social glue. The followers of the Orphic vision have a renewed criticism toward capitalism, a transversal and postideological position that is against the so-called strong powers, which according to them motivate injustices because they are fruit of the only political form that can maintain the best possible world.[272]

Analyzing the forms that embody different mythologies is an arbitrary exercise that does not crystalize people or events in a preordered scheme. It works as an indicator of tendencies. It can be very fascinating to investigate the new cleavages in which political scenarios and the existential visions are being reshaped. It is necessary to keep the attention high in order not to fall into banal schemas, and it is more than ever urgent to maintain a curious eye while staying away from the prejudice of predefined political schemes, which offer a sneak peek of a world to come and at the same time the world we live in.

---

[271] Michel Maffesoli, *L'Ombre de Dionysos*, Le Livre de Poche, 1982.

[272] Cf. Alain Finkielkraut, *Nous autres, modernes: Quatre leçons*, Paris, Ellipse, 2005.

THE USE OF ONLINE SEXUAL CONTENT BY YOUTHS IN THE ABSENCE OF SEXUAL EDUCATION

AGNESE PASTORINO

A snapshot of young people's online practices in European countries, particularly regarding to the use of sexual contents, is provided by the 2010 survey EUKidsOnline. This survey has been carried out by the London School of Economics and Political Sciences in twenty-five European countries, funded by the Safer Internet Programme (European Commission) and aimed at investigating online risks encountered by children between the ages of nine and sixteen. The research teams identify four types of risks: aggressiveness, sexuality, values, and commercial. Further, these are classified into three additional categories: content (where the child is a mass production receiver); contact (where the child participates in activities initiated by adults); and conduct risks (where the child is an actor as responsible or victim; Hasebrink et al. 2009).

Among others, sexual risks have the main diffusion, in the forms of pornographic contents (14%, content risk), sexting (15%, contact risk), and communicating with unknown people (30%) and then meeting offline (9%, conduct risk). Audiovisual contents are involved in pornography and sexting, which respectively correspond to content and conduct risks. The survey shows how sexual images are related to the child's age. One-third of children between fifteen and sixteen years old (36%) report seeing these images much more than do those nine to ten years old (11%). Use of sexual contents takes place mostly online (14%). Girls watch them less than boys do (12% vs. 16%), but they may remain shocked from this vision (39% vs. 26% of boys). The risk of watching them is prevalent among children aged fifteen to sixteen. The younger children are, the more they might be shocked by the use of sexual contents (56% for children of nine to ten years, 24% for children of fifteen to sixteen years). The most exposed children to online sexual contents reside in Northern countries (Norway, Denmark, Sweden, the Netherlands, and Finland) and Eastern Europe (Czech Republic, Lithuania, Estonia, and Slovenia). A reliable exposure of children concerns Germany, Italy, Spain, Ireland, and the United Kingdom. Children use sexual material on video-hosting sites (such as YouTube—5%), adult/X-rated websites (4%), social networks (3%), other sites (3%), video game sites (2%), and peer-to-peer file download sites (2%). Sexual material reproduces naked people (11%), explicit sex (8%), genital organs (8%), or violent acts (2%) (Livingstone et al. 2011).

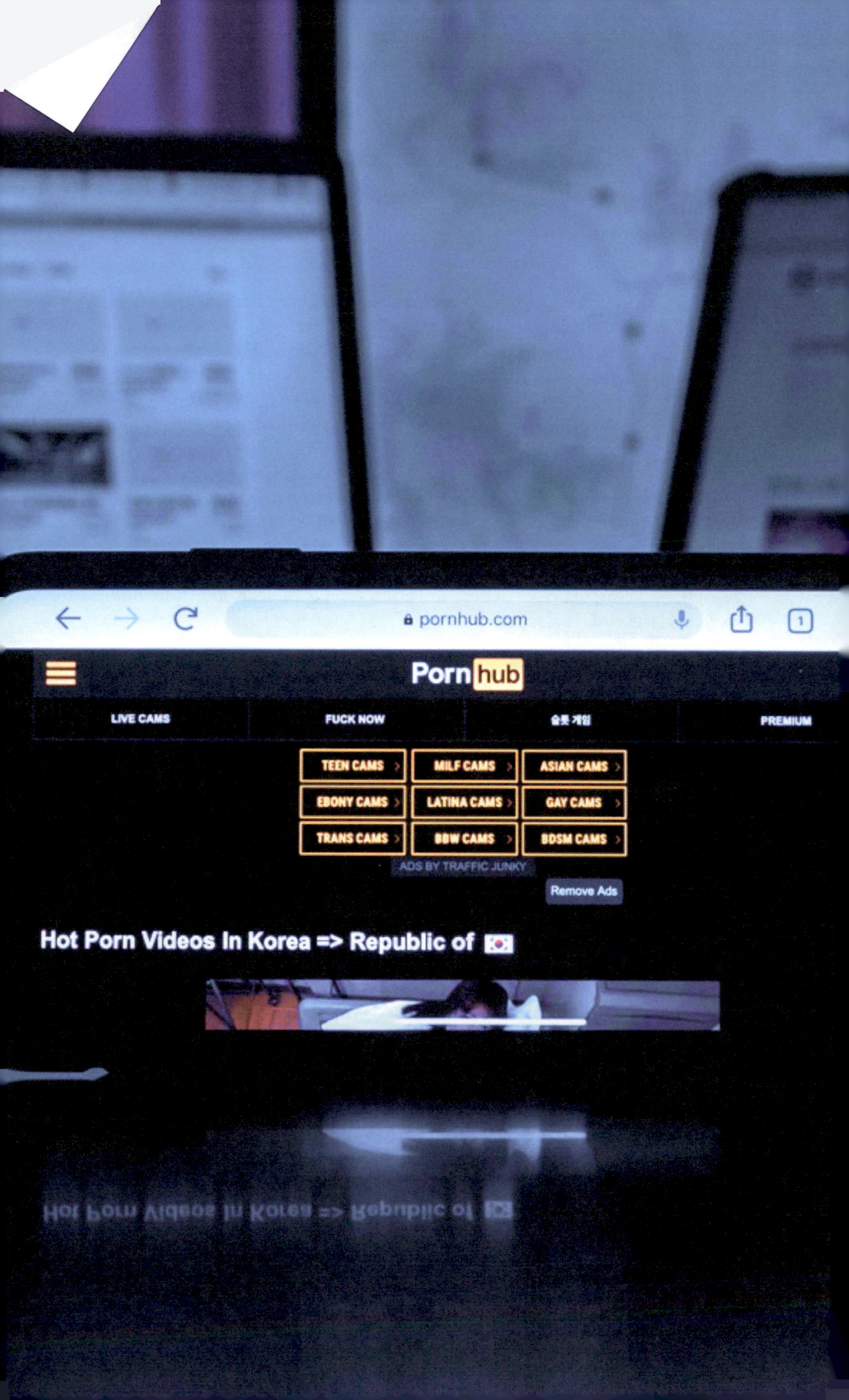
pornhub.com
Pornhub
LIVE CAMS
FUCK NOW
슬롯 게임
PREMIUM
TEEN CAMS
MILF CAMS
ASIAN CAMS
EBONY CAMS
LATINA CAMS
GAY CAMS
TRANS CAMS
BBW CAMS
BDSM CAMS
ADS BY TRAFFIC JUNKY
Remove Ads
Hot Porn Videos In Korea => Republic of

Key to all of this is the research contained in Sonia Livingstone's and Leslie Haddon's *Children, Risk and Safety on the Internet: Research and Policy Challenges in Comparative Perspective*—in particular within Antonis Rovolis & Lisa Tsaliki's chapter dedicated to the use of pornography (2012), and Livingstone's & Anke Görzig's chapter focused on the exchange of online sexual images (2012). These findings represent the starting point for our analysis of the literature on the subject.

The scientific tradition on media effects has mainly looked at the impact of pornography on children as harmful, by highlighting the need of children's protection, especially from online sexual contents. Nonetheless, these theories have been criticized for being too deterministic, in comparison with liberal positions which emphasize positive sides of exposure to sexual images; these latter perspectives defend the potentiality of breaking out repressive or normative expressions of sexuality (Flood 2009). Another major issue, pointed out by several researchers, is ubiquitous access to pornographic content (McKee, Albury, and Lumby 2008; Ey and Cupit 2011). Limits of research results on these topics are due to ethical implication and lack of empirical evidence of harm (Helsper 2005). Also, the relation between sexual risk and psychological or physical harm strongly varies (Livingstone 2009).

Recent research on the use of pornography by young people shows how they can express personal opinions on this subject (Buckingham and Bragg 2004; McKee, Albury, and Lumby 2008) and they are aware of the falseness of media representations. Ethnographic research has demonstrated how exposure to pornography is rarely traumatic but is part of sexual initiation or education and the possibility of experiencing pleasure (McKee, Albury, and Lumby 2008; Buckingham et al. 2010; Roviolis and Tsaliki 2012).

Beyond pornography, new devices and social networking platforms have facilitated the diffusion of other modalities of exchange of audiovisual sexual contents, including self-produced material. "Sexting" is the term used to indicate the exchange of sexual content through the internet, which increases with online anonymity if compared to other forms of sexual communication via other media (Subrahmanyam and Smahel 2011). From the perspective of adolescents, based on the results of focus groups, sexting is described as a modality of filtration (Lenhart 2009). The scientific debate on sexting is controversial especially with regard to the risky effects of this practice among young people; in specific cases, sexting may have harmful effects, for example when images are sent to accidental receivers or are modified (Barak 2005; Ybarra, Mitchell, and Wolak 2006). Other arguments focus on "hyper-sexualization" of the female body (Ringrose 2010, p. 179), sexualizing children (Greenfield 2004), or grooming through sexual images (Davidson and Gottschalk 2010). Concerns toward online risk have less explored the relation with sexual development in adolescence (Stern 2002) or sexual education. Other authors distinguish criminal activity from sexual experimentation (Wolak 2011). Despite the fact that sexting is not surveyed in depth within EUKidsOnline, some researchers of the network highlight the importance to pay attention to cross-cultural analysis with particular attention toward interreligious diversity in European countries (Livingstone and Görzig 2012)

In media representations, discourse on risks prevails, for example in European countries (Haddon and Stald 2009). Besides, media and social representations of risk change across countries and periods (McKee, Albury, and Lumby 2008). Attention to risks contributes to the proliferation of public anxieties and moral panic (Critcher 2008; Tsaliki 2011). News has often transmitted narratives on the diffusion of illegal sexual contents produced and diffused by adolescents themselves (Arcobascio 2010; Sacco et al. 2010). Public debate tackles three sexual risks: child abuse material, grooming, and online pornography (McKee, Albury, and Lumby 2008).

Based on these international findings, on March 2012, I conducted a quali-quantitative sociological survey in Italy focused on the main online risks, pornography and sexting, under the supervision of Monique Dagnaud. The sample was composed of 200 young Italian students aged between sixteen and twenty years old, belonging to diverse sociocultural backgrounds.[273] The sample was selected in three high schools in Southern Italy (Salerno), chosen according to sociocultural and gender composition. By surveying their practices and opinions, I wanted to answer the following questions: how do youth receive online sexual contents? How do these contents transform the activities of their daily life? How do they modify their social life and relationship with others?

My survey shows how the use of sexual contents online is aimed at satisfying youths' needs, curiosity, and desires; moreover, it reveals a relation to sexual practices, relational dynamics, and processes of identity development during adolescence.

The methodological tools adopted included an online questionnaire and focus groups. Quantitatively, the survey analyzed online practices, especially the use, exchange, and production of sexual contents. Qualitatively, it surveyed the opinions of youth on the link between pornography and love, female body, and children. Focus groups were composed based on sex (either female, male, or mixed ones) and sociocultural background of participants. Many methodological precautions were applied, so that young people could openly talk about their online "sexual" practices, which are mainly solitary from parental control. Several challenges were related to the topic, such as the difficulty to talk

[273] The survey was carried out within the masters in sociological research program of the EHESS Ecole des Hautes Etudes en Sciences Sociales in Paris in 2012.

about sexuality with young people and to propose the research topic to schools' directors; these methodological difficulties are present also in other pieces of research on children, which highlight vulnerability, anonymity, confidentiality, consensus, agency (Lobe, Livingstone, and Haddon 2007).

Online surveys were developed through the LimeSurvey software and submitted in the computer rooms at school. A ratio of 53% boys and 47% girls answered online questionnaires. All boys in the quantitative survey declared having watched sexual contents (106 boys, compared to 16 girls). Among all the users, three different groups may be distinguished based on two variables: willingness (voluntary or involuntary) and frequency of use (32% of users). These groups were composed of accidental voluntary users (20%) and inadvertent accidental users (9%). Most of the surveyed population (52%), mainly boys, declare they voluntarily use sexual contents either usually or occasionally. The profiles of these different types of users:

> *Usual users:* the use of pornographic sites takes place mainly at home (98%), through the computer (82%), in the afternoon (64%), most when their parents are at home (20 out of 33). The presence of parents at home doesn't influence the use of pornographic sites, which confirms it to be a solitary practice (96%). The duration of visits of pornographic sites is quite limited (between a quarter to half an hour—80%). The most visited websites are YouJizz.com (42.86%) and the network of pornographic sites PornHub Network (34.94%, among which mainly are YouPorn.com—27.87%, Tube8.com, and PornHub.com). The contents of these websites present many common aspects. For usual users, the use of pornographic websites is mainly aimed at sexual excitement and pleasure.
>
> *Occasional voluntary users:* the majority of occasional voluntary users (74%) declare they visit pornographic sites alone; but, differently from voluntary usual users, 26% declare to visit them with friends. They have emotions either of excitement (18), pleasure (4), and curiosity (10).
>
> *Involuntary accidental users:* most of the times involuntary users receive porn images by chance during a normal navigation on the internet (16, both through a research browser, in particular Google—8, and through an email—4, or through advertising on a website—4); therefore, accidental use of porn websites online is the effect of a form of advertising that may touch both adults and the youngest ones. A minority phenomenon (4) is the exchange of links among friends.

Open answers to online questionnaires explored young people's opinions. Although visiting pornographic sites has been a common practice among boys, the general opinions about sexual contents are mostly negative (42.72%) or negative and positive (26.70%). Only a minority expresses a positive opinion. Most of those expressing a negative opinion consider that pornography does not respect fundamental values of society, including human dignity, love, and family; in their view, practices shown are immoral (for example, the fact that it shows sex on payment) and may degrade common morality (13 out of 45). They also find that sexual life, which should be a fundamental activity in society, is represented through pornography as a form of vulgar transgression;[274] sexuality is trivialized and underestimated (6 out of 45). This twisted vision of sexuality can have a negative effect, especially on children (6 out of 45). Some young girls perceive pornography as disgusting, embarrassing, disagreeable, and revolting (7 out of 45), when it involves sexual acts with little girls, trans, or gays. Most of the respondents who express either a positive or neutral opinion on pornography are online voluntary users of pornography. The group which considers its use as neutral, meaning both positive and negative, recognizes that while online pornography is a source of excitement (13 out of 37), it also has negative sides (such as lack of feeling, risk of dependence, immorality, and above all possibility of conveying a twisted vision of sexuality to children—9 out of 37). Only 7% of surveyed youths express a positive opinion on pornography. They find these images exciting (6 out of 16); others find them educational (3 out of 16) due to the absence of sexual education at schools.

The majority of respondents do not share their opinions about online pornography with anyone (53%). Those who shared their opinions (29%) mostly chose a particular friend (36 out of 59), a group of friends (35), or their boyfriend/girlfriend (28). Among family members, they mainly choose their peers (i.e. sisters, brothers, cousins, etc.—11). Sharing opinions about online pornography takes place mostly in the peer group (i.e., best friend, friend group, boyfriend/girlfriend, sister, brother, etc.); parents are almost absent (7—especially the mothers).[275]

Regarding the perception of pornographic images as a reproduction of real sexual life, the majority of young people interviewed find that pornography doesn't replicate real sexual life faithfully (70%).[276] One of the main reasons for this common perception is the clear distinction between pornography, as a show of standardized sexual acts, and sexuality as an intimate expression of individual feelings and emotions; in addition, young people find that sexual activities represented are out of limits of normality (24), for example in terms of physical performances and bodies of actors. The only realistic aspect of pornography is the act of penetration, the realism of which is irrefutable (5).

The nonrepresentability of sexual acts through pornography is confirmed within focus groups, where young people define pornography as a

---

[274] This negative opinion was confirmed by those who found pornography "disgusting."
[275] This paragraph refers to the qualitative analysis of the question "Have you ever shared your opinions with anyone? With whom?"
[276] This paragraph refers to qualitative analysis, relative to the question: "According to you, do pornographic images reproduce real sexual life?"

representation of sex without love, without feelings, without emotions, a reproduction of a physical relationship without any mental, sentimental, spiritual connection. Young people point out three main differences between love and pornography: intimacy, non-remuneration, and spontaneity of feelings.[277] They clearly distinguish women, who make love, from prostitutes, who have sex for payment. On the one hand, boys accuse pornography of being able to confuse their idea on women and inducing them to believe that all women are prostitutes; on the other hand, they keep an image of their girlfriends as women to respect, with whom they cannot do the same activities watched in pornographic videos (such as anal or oral sex). Young people also report sexual objectification of the female and male body.

Finally, young people highlight possible negative consequences of the premature exposure to porn on children, including a high probability of experiencing early sexual experiences, a desire to emulate pornographic actors, the risk of creating a false idea of sexuality, and the possibility of developing a self-referential sexuality that arises from masturbation and dependence on pornography. Among the causes of children visiting pornographic sites, young people consider the lack of parental control and sexual education either within families or schools.

The common perception of all groups is that pornography cannot be considered as a means for children's or adolescents' education, since it may confuse ideas about the relationship with others and it does not offer information about sexual precautions or diseases. In any case, young people recognize an informative use of pornography that compensates the lack of institutional sources of information in family or at school.

Most relevant results of our research derived from six focus groups. Some young people have defined sexual arousal and masturbatory practices related to the use of these contents as an activity having a beneficial effect on their psychophysical well-being and satisfying their needs. They consider it as a natural and spontaneous activity, which concerns their sexual instincts and their attraction for women. Exchange or production of sexual contents was described during focus groups; they are at the center of more complex and articulated dynamics of filtration and sexual exploration. These dynamics take place in three fundamental moments: first, boys look at profiles of female users on social networks (especially Facebook), with whom to talk in chat possibly on erotic themes; secondly, they propose to pass from online Facebook chat to MSN to exchange private photos; third, they propose to open the webcam either directly on MSN or through Skype, to strip and masturbate. Additionally, boys online develop a progressive contact with girls through the web, of which some of them are at the same time animators, through chat, and actors, by sending sexual and naked images

[277] This last opposition also corresponds to the difference tie/relationship and body/soul.

and using the webcam. The transition from one social network to another is linked to the user experience and design of these social networking sites: Facebook is used to select the potential partner through real-life pictures and chat, but it is considered controlled by adults; then, on MSN, perceived as less controlled, they exchange sexual or naked images; finally, for the higher quality of the webcam, Skype is used for sexual proximity. As opposed to the pornographic sites, the webcam is characterized in the narratives of young people as a tool offering a greater possibility of intimate interaction in online sexual practices. During focus groups, boys declared they appreciated this activity, whereas girls described their experiences as victims of unknown users, who were either sending naked images or using webcams. Feeling of girls about these practices of online interaction are mainly of shame and embarrassment, whereas on the contrary boys express pride for their masculinity. The erotic narcissism of boys is also strongly evidenced in the production of pornographic videos. In fact, they said they produced these videos to show them to their classmates. Girls, on the other hand, declare to be almost never consenting,[278] except in the exceptional case where they want to attract the attention of classmates or family.

Proceeding from these results, our research has offered interesting reflections on the association between young people and online sexual contents and more generally on the condition of youth in contemporary society. Psychological isolation of youth from family is a central feature of our research; young people do not talk to their parents about their sexual and sentimental activities, although they are fundamental aspects of their development. This distance from parents is compensated by internet browsing and online activities, which allow young people—isolated in their rooms—to communicate with the outside world; adult life, a source of curiosity for young people, is thus known and explored through audiovisual media in a way completely free from parental control. Therefore, the web incentivizes a "room culture."[279] Through social networks, the room, which was a place of isolation, becomes an opportunity to meet with friends and strangers; through the internet, the bedroom also becomes a space for exploration and knowledge of external world.

Sexual contents and women: young people emphasize the distinction between a woman, with whom to have a sentimental relationship, and a prostitute (or female), with whom to have an exclusively sexual relation for money or for interest. The boys emphasize the difference between girlfriend and prostitute and how pornography can make them believe that all girls are prostitutes. They admit liking a more sexually liberated woman, with whom they can do the acts seen in sexual contents; at the same time, they do these acts only with girls with whom they have sex. On the other hand, with their girlfriends, they do not reproduce the acts

[278] Sometimes they are not aware to be filmed or that the videos will be shown.
[279] M. Dagnaud, Génération Y, Sciences Po Presses, Paris, 2011, p. 24.

seen in the pornography (like anal or oral sex). Women are represented for their physical appearance, not considering their psychological characteristics (such as intelligence, sympathy, etc.). The female body becomes a sexual object. In this context, young people consider that the man himself is represented as an object. Body reification is attributed by young people either to women or men. Pornography influences the daily lives of young people, especially girls, who on the one hand are inspired by fashion to dress attractively; on the other hand, they feel judged by their comrades (either boys or girls) as easy girls, especially when their clothes are attractive.

Sexual contents and children: The relationship between pornography and children has undoubtedly been a central topic of qualitative analysis. The first access of children (eleven to thirteen years old) to sexual contents takes place through word-of-mouth among grandchildren at the beginning of the school; they start very early to exchange these websites and images. Access to these sites is also facilitated either by the ease of use or the ability of children to use the internet and digital devices. Psychological reasons for access are especially the emulation of comrades and curiosity toward sexuality. The possible consequences are having early sexual experiences, wanting to imitate pornographic actors, creating a false idea of sexuality, developing a self-referential sexuality that arises from masturbation, and lastly becoming dependent on pornography. Among the causes, young people consider the absence of parental control and the lack of sex education either within families or schools. The common opinion of all focus groups is that pornography cannot be considered as a means for the education of children or adolescents. In any case, young people recognize an informative use of pornography that offsets the lack of institutional sources of information in family or school.

Children rarely talk about sex in family or school, although it is a topic of great interest to them. Therefore, sexual contents become a source of information especially for the youngest ones; these images can confuse their ideas about social relationships, by considering the other person as an object of sexual desire and love; in addition, they don't provide information about precautions or sexual diseases. Young people recognize that the lack of information from families and schools may be considered as one of the main causes of teenage use of sexual contents; therefore, they recognize that it may be difficult to find a solution, due to shyness regarding talking about sexuality with adults. In this respect, boys propose a major parental control through blocking these contents on computers or even censoring the most violent, transgressive, or forbidden ones (for example, pornography which reproduces young girls). On the other hand, girls, in female focus groups, consider two solutions, which emphasize the centrality of families and schools in educating children about sexuality. The first solution proposed is the institution of a sexual education, which—from primary to high schools—

teaches young people what is sexuality. The second proposal is a change in the ways to talk about sexuality, from a vertical communication—where young adolescents are "forced" to listen to an institutional source—to a horizontal one—which allows young people to interact with people who can help them to understand their problems with respect to sex and love. Furthermore, these new modalities to communicate are also associated with the need to change the sources of information; in fact, they find that neither teachers nor parents are persons with whom they can speak about this subject, but that it is necessary to find an intermediate and neutral figure between parents and children, who, on the one hand, has the maturity, the right distance, and the skills to guide them and, on the other hand, has the mental openness and ability to listen to them, to understand their experiences and personal issues.

As a conclusion, this essay reveals that children's psychological isolation in families and schools impacts their views on the subject of sexuality. The daily use of the internet represents a revolutionary change in family life characterized by the "growing empowerment" of children from educational instances and parental control (Martin and Lelong 2004). Also, the use of mobile devices (i.e., phones, iPads, iPods, tablets, etc.), inaccessible to parents either for their characteristics of portability or difficulties encountered by "digital immigrants," increases the possibility for children of escaping parental control. In the so-called "bedroom culture" (Bovill and Livingstone 2001), peer communication through ICTs becomes central to relational dynamics and processes of identity development. First, audiovisual media and then the internet give rise to a true "childhood transformation," where social norms of interaction—whether among children or between children and adults—change remarkably. The concept of "child" is radically transformed by the spread of the internet (Meyerowitz 1985). Children are no longer excluded from the information of adult life, and their contact with the external world may no more be managed by parents. Already, the advent of television and cinema had introduced speeches and images which were at first forbidden in the intimate space of the house; today, the internet makes even less controllable the access to social contacts, information, and content available for children (Boyd and Hargittai 2010).

Several authors have analyzed the "transformation of childhood" in contemporary society, including Neil Postman, within *The Disappearance of Childhood*, successively quoted by Joshua Meyrowitz in *No Sense of Place, The Impact of Electronic Media Behavior* and finally revised in a different way by David Buckingham in 2000 in his book *After the Death of Childhood*. All these authors inquire on the origins of childhood. First, Postman and then Meyerowitz support the theory of technological determinism, rooted in McLuhan's ideas (McLuhan 1968). The two authors relate the emergence of the notion of childhood to the birth of

printing and the spread of literacy. Children, who lack the knowledge of writing and reading, are socially excluded from adult life and require education provided by schools. Additionally, David Buckingham takes distance from this theory, highlighting how the notion of childhood has existed (and evolved) since previous times; therefore, printing cannot be considered as a phenomenon directly linked to the formulation of the notion of childhood. Also, Buckingham considers that several factors "concur to sweep or shorten the phase of childhood." All these authors question the notion of childhood subsequent to the emergence of electronic media, which expose children to an early knowledge of the adult world.

To conclude, we find a whole generation (called Y ("Why") by Monique Dagnaud) that grows with an imaginary distinct and distant from those of previous generations—the internet thus transforms cultural practices and inaugurates new ones. With an approach that crosses the sociology of youth and sociology of the web, Dagnaud describes the characteristic traits of the Y generation. Through this term, the author defines digital natives (Prensky 2001). By analyzing data on internet use, a strong online presence of young people is immediately evident. Dagnaud considers internet use by children as an "anthropological mutation." Generation Y, from adolescence, is strongly determined by the external world, and young people's choices depend on "peer approval," rather than on parent education. The model of behavior is particularly conformist with peers. At the same time, young people unite to this tendency an exhibitionist behavior, defined by the media sociologist as a "therapy of transparency and autofiction." Consequently, young people expose their intimacy in online public space, living between "concrete experience and imaginary projection," "here and elsewhere" (Dagnaud 2011).

References

Arcabascio, C. (2010). "Sexting and Teenagers: OMG R U GOING 2 JAIL???" In Journal of Law & Technology 16 (no. 3), pp. 1–43.

Barak, A. (2005). "Sexual Harassment on the Internet." In Social Science Computer Review 23 (no. 1).

Bovill, M., and S. Livingstone (2001). "Bedroom Culture and the Privatization of Media Use." In S. Livingstone and M. Bovill, eds., Children and Their Changing Media Environment: A European Comparative Study, pp. 179–200. Mahwah, NJ: Lawrence Erlbaum Associates.

Boyd, D., and E. Hargittai (2010). "Facebook Privacy Settings: Who Cares?" In First Monday 15 (no. 8).

Buckingham, D. (2000). After the Death of Childhood. New York: Wiley.

Buckingham, D., and S. Bragg (2004). Young People, Sex and the Media: The Facts of Life? Basingstoke: Palgrave Macmillan.

Critcher, C. (2008). "Moral Panic Analysis: Past Present and Future." In Sociology Compass 2 (no. 4), pp. 1127–1144.

Dagnaud, M. (2011). Génération Y. Paris: Sciences Po Press.

Davidson, J., and P. Gottschalk (2010). Internet Child Abuse: Current Research and Policy. London: Routledge.

Ey, L.A., and C.G. Cupit (2011). "Exploring Young Children's Understanding of Risks Associated with Internet Usage and Their Concepts of Management Strategies." In Journal of Early Childhood Research 9 (no. 1), pp. 53–65.

Flood, M. (2009). "The Harms of Pornography Exposure among Children and Young People." In Child Abuse Review 18 (no. 6).

Greenfield, P.M. (2004). "Inadvertent Exposure to Pornography on the Internet." In Journal of Applied Developmental Psychology 25 (no. 6), pp. 741–750.

Haddon, L., and G. Stald (2009). "A Comparative Analysis of European Press Coverage of Children and the Internet." In Journal of Children and Media 3 (no. 4), special issue, pp. 379–393.

Haddon, L., and S. Livingstone (2012). "EU Kids Online: National Perspectives." EU Kids Online, The London School of Economics and Political Science, London, UK. (This version available at: http://eprints.lse.ac.uk/46878/.)

Hasebrink, U., S. Livingstone, L. Haddon, and K. Ólaffson (2009). "Comparing Children's Online Opportunities and Risks across Europe: Cross-National Comparisons for EU Kids Online." EU Kids Online, LSE, London (Deliverage D3.2, 2nd edition).

Helsper, E. (2005). R18 Material: Its Potential Impact on People under 18. An Overview of the Available Literature. London: Ofcom.

Lenhart, A. (2009). Teens and Sexting: How and Why Minor Teens Are Sending Sexually Suggestive Nude and Nearly Nude Images via Text Messaging. Washington, DC: Pew Internet and American Life Project.

Livingstone, S. (2009). Children and the Internet. Cambridge: Polity Press.

Livingstone, S., and A. Görzig (2012). "Sexting: The Exchange of Sexual Messages Online." In S. Livingstone, L. Haddon, and A. Görzig, eds., Children, Risk and Safety on the Internet: Research and Policy Challenges in Comparative Perspective. Bristol, UK: Policy Press.

Livingstone, S., L. Haddon, A. Görzig, and K. Ólafsson (2011). Risk and Safety on the Internet: The Perspective of European Children. Full Findings. London: EU Kids Online Network.

Martin, O. and B. Lelong (2004). "Internet en famille (dossier thématique)." In Réseaux 22 (no. 123).
McKee, A., K. Albury, and C. Lumby (2008). The Porn Report. Melbourne: Melbourne University Press.
McLuhan, M. (1968). Understanding Media: The Extensions of Man. New York: McGraw-Hill.
Postman, N. (1994). The Disappearance of Childhood. New York: Vintage Books.
Ringrose, J. (2010). "Slut, Whores, Fat Slags and Playboy Bunnies: Teen Girls' Negotiations of 'Sexy' on Social Networking Sites and at School." In C. Jackson, C. Paechter, and E. Renold, eds., Girls and Education, 3–16, pp. 170–182). Maidenhead: Open University Press.
Rovolis, A., and L. Tsaliki (2012). "Pornography." In S. Livingstone, L. Haddon, and A. Görzig, eds., Children, Risk and Safety on the Internet: Research and Policy Challenges in Comparative Perspective. Bristol, UK: Policy Press.
Sacco, D.T., R. Argudin, J. Maguire, and K. Tallon (2010). Sexting: Youth Practices and Legal Implications." Cambridge, MA: Berkam Center for Internet and Society.
Stern, S. (2002). "Sexual Selves on the World Wide Web: Adolescent Girls' Home Pages as Sites for Sexual Self-Expression." In J. Brown, J. Steele, and K. Walsh-Childers, eds., Sexual Teens, Sexual Media, pp. 265–285). Mahwah: Lawrence Erlbaum Associates.
Subrahmanyam, K., and D. Smahel (2011). Digital Youth: The Role of Media in Development. New York: Springer.
Tsaliki, Liza (2011). "Playing with Porn: Greek Children's Explorations in Pornography." In Sex Education 11 (no. 3), pp. 293–302.
Wolak, J., and D. Finkelhor (2011). "Sexting: A Typology." In Research Bulletin (March 2011). Durham, NH: Crimes Against Children Research Center, University of New Hampshire.
Ybarra, M., K. Mitchell, and J. Wolak (2006). "Examining Characteristics and Associated Distress related to Internet Harassment." In Pediatrics 118 (no. 4), pp. 1169–1177.

PS4
HELLBLADE
SENUA'S SACRIFICE

BAFTA
WINNER
GAMES
GANHADOR DE 5 PRÊMIOS

THE
GAME
AWARDS
WINNER - 2017
GANHADOR DE 3 PRÊMIOS

16

BUILDING EMPATHY IN VIDEO GAMES THROUGH DIGITAL *MESTIZAJE* IN *HELLBLADE: SENUA'S SACRIFICE*

CARLOS KELLY

The twenty-first century has brought rapid changes to video games, making the medium a highly innovative mode of reading and storytelling. These advances bring about faster speeds, vivid visual spectacles, a constant stream of artistic feats, more possibilities for immersion, and the potential to spur empathy, just to name a few. In 2006, Sony released the PlayStation 3 (PS3), marking the beginning of a new age for faster speeds and better graphics in gaming consoles. The PS4 came next in 2013, followed by the PS4 Pro in 2016.[280] With the PS5 release rumored for 2020, we can see the emphasis on upgrading console experiences for players; thus, it is no wonder video games contain the potential for creating unrivaled immersive experiences. Video games offer an immersive space where players may traverse wide-ranging representations of race, gender, and sexuality—a world of subjectivities. According to the dialogue created by Philip Penix-Tadsen between scholars Anna Everett and Vít Šisler, we need to "seriously analyze not only those games that obviously participate in the 'redeployment and reification of specious racial difference' (Everett 120) but also the ways in which game designers are undertaking 'genuine attempts to transcend the simplifying patterns of representation in video games' (Šisler 205)" (Penix-Tadesen 2013, p. 180). Thus, I will situate my study of Ninja Theory's *Hellblade: Senua's Sacrifice* (*Hellblade*) in the tensions evident in the dialogue above.

In order to properly delve into the representative experience that is playing *Hellblade*, I will turn to the methodological framework of Gloria Anzaldúa's liminal third space. Anzaldúian theory serves as a window allowing me to examine the significance of Senua's journey and Ninja Theory's decisions behind the creation of *Hellblade*. As part of this chapter, I will examine the implications around race, gender, and disability as I explore how *Hellblade*'s narrative provides the potential for players (through gameplay) to empathize with Senua's life and her experience with mental illness. In addition, I will argue playing *Hellblade* and playing video games in general place players into Digital *Mestizaje*, a liminal space allowing for continual cultural fusion(s) arising from the mixture of game players' and characters' subjectivities.

To further expand on Digital *Mestizaje*, let me first point to how my term borrows from Gloria Anzaldúa's vision of the new mestiza consciousness.

---

[280] The PS4 Pro came with the ability for 4k graphics and contained a one-terabyte memory drive.

Anzaldúa speaks to the mestiza reclaiming the contradictions they inhabit, that "living in the borderlands produces knowledge by being within a system while also retaining the knowledge of an outsider who comes from outside the system" (Anzaldúa [1987] 2012, p. 7). She argues how developing this mestiza consciousness "allows people to navigate these different social contexts and maintain knowledge of what it means to reside in these different social and political interstices" (p. 8). Digital *Mestizaje* acknowledges the borderland between video games and lived reality and posits that video game players experience a mixture between the knowledge they have (from within the hegemonic systems' socialization) and the potential for knowledge / perspective acquired from playing. Players, through Digital *Mestizaje*, may enter into a space of tension with previous knowledge as they begin to learn about their characters' positionality and subjectivity through navigating a game's narrative. This liminal space opens the possibility for players to identify and empathize with characters' struggles as they drive the action of the narrative forward. The tensions between knowledge materializing through the space of Digital *Mestizaje* may create subconscious or conscious change within players. The fusion between cultures is continual, opening spaces of knowledge with the capacity to subvert the socializing effects behind systems of power.

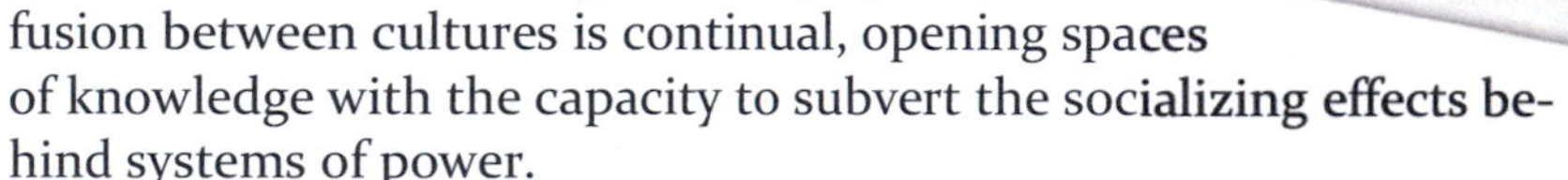

Other key critics and commentators inform notion of the Digital *Mestizaje*: Anastasia Salter's work in her book *What is Your Quest?* adds to the impetus behind Digital *Mestizaje* by referencing the dialogue that occurs with representational elements between players and characters; while Anita Sarkesian, in her web series *FeministFrequency*, proves how video games have the power to produce, reproduce, and maintain culture. In this way, Jennifer Malkowski and TreaAndrea M. Russworm also speak to the importance of analyzing representation in video games and how we can't just focus on a games' hardcore coding dynamics. I also draw inspiration from Adrienne Shaw and how she delves into the relationship between identity, identification, and media representation in her book *Gaming at the Edge*. Scholar Megan Amber Condis explores the process of identification in films and video games, arguing on how players become part of the environments and narratives. I also see my work being in conversation with David Owen's 2017 book, *Player and Avatar: The Affective Potential of Videogames*. In it, he argues video games differ from other mediums because of how "they blur the distinction between audience and performer; ... the player assumes a dual role of both protagonist and audience as opposed to other narrative-driven entertainments"

(Owen 2017, p. 27). Digital *Mestizaje* shares the vision of how subjectivities fuse through the connections via empathy, identification, and the controller(s). All this to say that the scholars above and people like Jan Noël-Thon, Marie-Laure Ryan, Frederick Luis Aldama, Osvaldo Cleger, and Philip Penix-Tadsen have influenced my undertaking of a new way to envision borderlands (*en la* frontera and in video games) and the cultural fusions arising from the playing of these games.

Before moving forward, I must first highlight my goal in studying *Hellblade*, a game illustrating the medium's booming potential as a serious mode of compelling storytelling. My quest involves demonstrating how video game players exist within the liminal third space of Digital *Mestizaje*. The mixture of subjectivities (in this space) occurs as players begin to internalize other perspectives brought on by playing as characters with complex identities such as Senua. This essay seeks to demonstrate how video games generate opportunities for players to identify with and to feel empathic connections with characters and storytelling through Digital *Mestizaje*. Players learn to experience the world as Senua, empathizing with her trials, which, through close reading, may challenge players' preconceptions of how women navigate the world. Here I must state, as a heterosexual, white-passing Latinx cis-man I recognize the privilege of my positionality. I am privy to how people inhabit identities that intersect in various forms, often in ways that disproportionately affect LGBTQ-identifying people, women, and people of color. I have come to this perspective through personal experiences, personal values, and my research which considers how gender, sexuality, and race intersect in media representations. Video games, too, have refracted by perspective in this regard.

When it comes to video games, one must immerse oneself in the text by playing—play is essential to analyzing a video game's content. Indeed, I have fully immersed myself in the study of *Hellblade*, a profound gameplay I have yet to experience anywhere else; and thus, my methodology revolves around my time playing the game. Player 1 refers to my experience of *Hellblade*'s gameplay, and any references to players or a player without this distinction apply to the larger general audience or player experience.[281] My first run-through of the game was spread out over several months with leisurely on-and-off gameplay, while my second attempt at the game spanned two days, amounting to a bit over five hours total. I have also spent extensive time studying recorded gameplay of my second attempt. As I move forward through this chapter, I will refer back to notes taken during both instances of playing *Hellblade* as well as recounting the visceral reactions I often experienced as Player 1.

[281] It is more than possible that other players may experience the same emotions that Player 1 feels. Yet, Player 1 will refer to my particular experience.

Hellblade follows Senua's story, a Celtic warrior woman whose family, lover, and tribe meet their demise at the hands of invading Vikings (referred to as the Northmen). She struggles with "the Darkness" or psychosis, which manifests itself through hallucinations, delusions, memories, and her continual battle with physical manifestations of Northmen and with the voices she hears. Player 1 also hears the voices, especially if they decide to play the game with headphones, which are part of the game's intended gameplay design. Senua's vision quest takes us through a land full of death, disease, and suffering or what the Northmen call Hel—a land painted with the excess of conquest. Before the Northmen's invasion, Senua travels into the wilds to confront the Darkness, only to return to her village to find it and the people in ruins. Here, she finds Dillion, her lover, sacrificed to the Northmen's Gods through what's called a blood eagle—a highly ritualized form of torture/execution. This trauma intensifies her psychosis, leading her to salvage Dillion's skull and to carry it with her as she journeys in search of Hela (the Goddess of Hel). Senua aims to save Dillion's soul by defeating Hela. While on this quest, the player is unable to untangle themselves from Senua's fears, voices, and desperation.

Player 1 begins this journey as Senua does: emerging from the mist in her canoe, navigating the eerily calm waters of her world. In Figure 1, the camera continues to zoom until reaching the third-person perspective[282]—slightly perched over Senua's shoulder.

Figure 1. Senua begins her journey into Hel (*Hellblade*).

---

[282] *Hellblade* provides a third-person perspective, which means Player 1 is constantly seeing the character on their screen—their back specifically. This varies from the first-person perspective where only the character's hands are visible.

This beginning sequence limits the player's control to directional-observations and, combined with the narrator's voiceover, work to insert the player into the game's liminal space. The narrator discloses how Senua's story is already over, but our choice to play retells her story. Consider how Gloria Anzaldúa asserts how "the lifeblood of two worlds [merge] to form a third country—a border culture" (Anzaldúa [1987] 2012, p. 25). By deciding to play, Player 1 enters the border between digital and real. This idea can encompass what happens at the beginning of *Hellblade* and most adventure games where there is an introductory sequence. Player 1 delves into the game via the introduction, inserting themselves into this third space—not our own reality and not merely the reality of the game, but a fusion between worlds, between player and character subjectivity. Thus, as players, we inhabit the border between "non-virtual" reality and virtual realities/subjectivities.[283]

A few moments pass, and the narrator herself speaks to us again, inviting us into Senua's world—one where voices and even the narrator populate Senua's inner subjectivity. She speaks to us softly, "oh how rude of me ... I never told you about the others ... you hear them too right? They've been around ever since the tragedy. Well ... that's not quite true. Some are old, some are new, but they've ... changed" (*Hellblade*). Again, Senua hears voices, and so part of the player's subjectivity fusing with hers involves playing the game with headphones, which become a proxy for entering Senua's mind. For example, scholar Megan Condis writes how, when playing video games, "we forget ourselves and become the other ... we assume for ourselves the identity of the target of our identification ... sharing their perspective and internalizing their view of the world" (Condis 2016, p. 11). From the beginning, Player 1's and Senua's subjectivity fuse as we move from outside the mist—from our outside world/perspective—to inside the game, inside Senua's mind. Headphones (as proxy) facilitate this movement into Senua's mind by plugging players into the game space or as "[Martti Lahti] contends, ... video games epitomize a new cyborgian relationship with entertainment technologies" (Poster 2007, p. 158). Through dialogue between Poster, Lahti, and Donna Harraway, we can see the context for how Digital *Mestizaje* begins with the action of playing, with the action of entering this liminal space.

Indeed, through playing the introduction, there is a shift in subjectivity as Player 1 moves the joystick to look around the river. This fusion exemplifies Frederick Luis Aldama's assertion that "video games rely on the player's doing (purposefully directed movement and action-oriented gestures) to intensify the brain's immersion" (Aldama 2013, p. 251). Player 1 controls Senua's vision here with the tilt of the left joystick, which builds into the immersion Aldama mentions. Player 1's actions of moving the narrative forward triggers the voices to speak: "Look. Can you see them?

[283] Zach Waggoner's distinction that challenges the idea of "lived reality" and nonreal realities of video games. He asserts that video games are indeed real and lived and utilizes "non-virtual" reality to encapsulate his distinction.

Why isn't she looking? Look around you" (*Hellblade*). These requests, because they come in second-person address, simultaneously ask Senua and Player 1 to look, blurring the lines between the game's player and character. The impact of this fusion or blurring can be summed up by Owen, who writes that "the immersive technique of taking the player into the mind of the character can be highly effective (and affecting)" (Owen 2017, p. 252). Indeed, now that Player 1 and Senua are being addressed as the same person through second-person address, the possibility for empathic connections arise. Poster puts it another way, writing, "it is through the generalized experience of technological prosthetic combined with the visual experience of watching and identifying with bodies on a screen that users experience a more interactive form of innervation" (Poster 2007, p. 331). Innervation arises when audiences can pretend to be a character through emotional and physical mimicry (Brookey 2010, p. 326). Thus, players are immediately positioned to identify with Senua because of how they begin to mimic her actions through the action of moving the joystick (the prosthetic attachment) and seeing Senua's body respond by moving her eyes and head in the corresponding direction.

As the introduction ends, the narrator solidifies this fusion by stating "[t]here is nothing to go back to and worse to look forward to. Why don't you join us? Maybe you too have a part to play in this story" (*Hellblade*). This moment of narration brings me to how Penix-Tadsen deploys Alexander R. Galloway's analysis that "if photographs are images, and films are moving images, then *video games are actions*" (Penix-Tadsen 2013, p. 177). Thus, this invitation is another means of identification through Player 1's actions of moving the narrative forward, which allows players "a 'vicarious experience' of the events of the text through the eyes of a character" (Condis 2016, p. 11). Player 1 identifies with Senua not only through seeing as she does, but also through hearing what she hears and feeling what she feels through narrative and controller vibrations. Without a player's involvement, we do not have access to Senua's story, and thus, in order to tell her story, the player or Player 1 must become Senua. Put in yet another way, Ryan claims that "in a narrative game, story is meant to enhance gameplay, while in a playable story, gameplay is meant to produce a story" (Ryan, p. 45). *Hellblade* is ultimately a playable story that through player involvement unfurls the narrative. Anastasia Salter's work on quests in games helps clarify *Hellblade*'s introduction. She writes how players enter "adventure games with an introductory sequence that introduces them to the persona of a specific character as part of enabling a literal out-of-body experience of an entirely different life, that of the avatar" (Salter 2014, p. 44). *Hellblade*'s introduction enables this same "out-of-body" experience, which intensifies as the narrative progresses. As Player 1 learns more about Senua's life, the potential for empathy grows because Player 1 no longer inhabits a fixed, single body.

Video games are unique in their platform; they open themselves up to avenues of identification (and empathy) that may not be available in other modes of representation. However, Adrienne Shaw might disagree, writing how "video games are [not] inherently more interactive than other media. There is a lot to be said for active audiences in all popular culture" (Shaw 2015, p. 24). It is true that popular culture does offer various opportunities for audiences to feel active in what they consume, but Shaw might be too quick to discount how players connect their bodies and minds—via controller apparatuses—to the narratives taking place on-screen. For example, Robert Alan Brookey turns to Jo Bryce and Jason Rutter, who write how players "are more actively engaged than film viewers in both the narrative and the other events [because of] the ability to modify both of these aspects ... [which] shows a level of interaction ... that is not provided by traditional cinema or Hollywood blockbuster movies" (Brookey 2010, p. 33). So far, we've established how the action of looking and how headphones as proxy create this "ability to modify" or to enact *Hellblade*'s narrative. Thus, what film and other media lack are the immersion-inducing prosthetic attachments highlighted by Lahti, Brookey, Bryce, and Rutter.

Salter provides yet another potential counterargument to video games being no more immersive than film or other mediums, citing scholar Jesper Jul. Jul writes how "the relations between reader/story and player/game are completely different—the player inhabits a twilight zone where he/she is both an empirical subject outside the game and undertakes a role inside the game" (Salter 2014, p. 7). Anzaldúa's third space parallels this "twilight zone" sensation—both of which verge on the liminal. Worlds merge because the player is an active agent in developing the narrative. This hyperactivity involves the eyes, mind, and hands, and without an active player, the story (via gameplay experience) cannot be consumed, whereas a film or other media do not necessarily require an active body/participant.

This interactive connection happens quite frequently while playing any game's campaign mode, but there are other moments away from *Hellblade*'s narrative where Player 1 displays full immersion.[284] For example, Condis cites Klimmt et al., who write about how video games display mediated environments where characters perform in ways similar to other media, "but they also enable and invite users to act by themselves in the environment and to become an integral part of the mediated world" (Condis 2016, p. 11). Consider Figure 2 below as one particular moment where Player 1 becomes an integrated part of this mediated world. For example, gaming experiences are now easily shareable across interfaces because of PlayStation 4's share button.[285]

---

[284] The campaign mode refers to navigating the game's single-player narrative content; this content is the quest or quests the character and player must undertake in order to progress in the game or beat the game.

[285] The Sony PlayStation added a new button to their PS4 gaming controllers (not available in PS3 models). This button allows you to upload/share images and video clips from your gaming experience to social media. This option is now built-in to the controller, providing the player complete access to sharing their gaming experience.

Figure 2. Player 1 captures a vista of the setting within the game (*Hellblade*).

Thus, not only can Player 1 play the game, but others can watch that gameplay, increasing *Hellblade*'s immersive potential. In Figure 2, Player 1 stopped to set up a screenshot, capturing the sunlight through the tree line—an illustrative example of Condis's interaction with the mediated world. In addition, this experience can be described as perceptual immersion or "as a shift of attention from the real environment to certain parts of the game" (Thon, p. 270). Game environments, especially ones that are made to be as realistic as possible, create the potential for beautiful vistas that may shift players' attention from the narrative or their immediate environment (the home/play space)—exactly what happens with Player 1 above.

Condis continues to build on identification in video games. She claims, "identification can be encouraged through the suturing of the gaze of the camera (and, therefore, the gaze of the audience) to the gaze of one of the characters" (Condis 2016, p. 11). Although she refers to the gaze in films, the very same process exists within *Hellblade*, especially during the introduction where our gaze becomes Senua's through joystick movements. This fusion between gazes also occurs as Player 1 maneuvers the camera angles (or Senua) to capture the beauty of vistas (see Figure 2). In her book *Playing with Feelings*, Audrey Anable furthers the idea of video games being better suited for interaction with player-audiences. She writes how "pleasure in a game is derived ... through a complex interplay of the body in proximity to the gaming apparatus, the feel of the game controller in the hands, and the images, sounds, and other aesthetic information in the game" (Anable 2018, p. 44). Figure 2 encapsulates the idea of interacting with visual environmental aesthetics. Thus, the game is not only about the quest, but it also involves how players move around in the space and how players interact with environments seen in Figure 2. Consider how Aldama believes emotions arise in gameplay "including the one[s] the player feels when moved by its aesthetic design" (Aldama 2013, p. 253). Setting amplifies a game's immersive potential, which in turn increases the game's ability to move players toward identifying with its content or character. Player 1 can revel in vistas and take pleasure in being part of the game's environment, yet these moments only occur when Senua's fear is in repose. These pleasures come from the fact that when "we play video games, we do so within a real-life safe zone" (p. 251). This safe zone allows Player 1 to take pleasure from these vistas, at the same time revealing how there can be moments of disconnect between how Player 1 shares Senua's experiences.

The setting is not the only factor to contribute to how a player can immerse themselves in video games' liminal third space. Remember how the controller is critical to how players experience said space. As Lahti suggests, this hyperconnectivity between player and controller makes literal Haraway's cyborg. This also allows players to be, as David P. Marshall argues, "moved by electronic games beyond filmic narrative identification into a hybrid state of 'game play' subjectivity" (quoted in Poster 2007, p. 327). Thus, the prosthetic attachment to Senua via controller enables the fusion between "machine and organism, condensed between an image of both imagination and material reality—existing as a border war" (Haraway 1994, p. 292). The border coincides with Anzaldúa's third space; those who play video games are on the cusp, in the liminal cosmos of imagination and reality where the fusion of subjectivities found within the space of Digital *Mestizaje* is not only possible but occurs throughout gameplay experiences.

Figure 3 displays one such instance where the controller becomes an element facilitating Digital *Mestizaje* between Senua's and Player 1's subjectivities. Senua pleads with the darkness and hears her lover Dillion's voice mixing into her thoughts from the past: "Use all your senses. Let the world speak to you. What do you hear?" (*Hellblade*). At this point in her quest Senua faces individual trials; here, she loses the ability to see clearly and must rely heavily on her hearing, which makes Player 1 rely on hearing and on the vibration of the controller. Senua and Player 1 feel the wind escaping from the dark, signaling the correct path. Player 1, if facing the correct direction, hears and feels the wind through the vibrations of the controller. Fusions like Figure 3 reveal "a hybrid of machine and organism" that Senua and Player 1 are "a creature of social reality as well as a creature of fiction" (Haraway 1994, p. 291).

Figure 3. Player 1 relies on sound and controller vibrations (*Hellblade*).

This liminal space allows Player 1 to enter Senua's mind, where Player 1 now battles with the same voices that plague her—fiction and reality fuse. In her exploration of emotional immersion, Ryan argues how players—when feeling empathy for characters on the screen—are "mentally simulating the situation of others, by pretending to be them and imagining [a character's] desires as our own, that we feel joy, pity, or sadness for them" (Ryan, p. 56). Here, Player 1 is made to simulate fear, due to Player 1's able-bodied perspective. Indeed, losing the ability to see makes players reckon with this possibility and the corresponding challenges. Although ephemeral, the loss of vision is palpable because of having to beat this trial without the normativity associated with sight and because of the vibrations (from the controller), which combine to generate space for players to empathize more deeply with Senua's plight.

Now, some game scholars may quickly point to (and rightly so) the mixture of subjectivities being a form of identity tourism. As David Leonard writes, "video games that situate race at their center and that deploy racialized meanings often do so by suggesting that pleasure can be derived from visiting and becoming the racialized other" (Leonard 2003, p. 5). This sort of pleasure derived from becoming the racialized subject is problematic to say the least, and it manifests all throughout video games in often very violent ways (see Leonard's analysis of *Grand Theft Auto 3*). However, *Hellblade* presents a world in ruins (more on this later) that does not center her whiteness, but instead opts to focus on Senua's positionality as colonized subject and her struggle with severe psychosis. When writing about identity tourism, Edward Doran posits "experiencing, even virtually, the marginalization of an alternate identity can provide users with a greater sense of empathy and understanding of what it is like to be someone with less privilege" (Doran, p. 268). Senua's struggles with severe mental illness (a subject that has difficulty escaping the stereotypical representations presented to mainstream audiences) did indeed challenge Player 1's understanding of mental illness and the implications associated with her positionality. Penix-Tadsen believes "[video games] have the capacity not only for bolstering negative and simplistic cultural depictions [but also as Rachel Hutchinson states,] 'creating space for the inversion of stereotype, the subversion of gender roles, and the possible transcendence of the binary system'" (Penix-Tadsen 2013, p. 178). Playing *Hellblade* provides the space to generate empathy, creating possibilities, as Hutchinson suggests, to subvert expectations attached to gender and disability.

Empathy, through the cultural mixtures between a player's subjectivity and Senua's, further cements video games' potential as a highly interactive medium by providing windows into a host of potential subjectivities. Anable points to how video games "are uniquely suited to giving expression to ways of being in the world and ways of feeling in the present" (Anable 2018, p. xii). Player 1 (and potentially players) may challenge their privilege as they experience Senua's trials, reconsidering them as a possible metaphor for how women must often navigate the violence of gender. However, Senua's experience more closely resembles the struggles stemming from psychosis. Iain Dodgeon, creative partnerships manager for the biomedical research charity Wellcome Trust, collaborated with patients who suffer from psychosis and Paul Fletcher, psychiatrist and professor of health and neuroscience at the University of Cambridge. They combined their knowledge to assist *Hellblade* creators in presenting an accurate representation of psychosis. Dodgeon states the goal was "to create a game that provides a fresh perspective on [psychosis] and allows audiences to engage with it in a way that just wouldn't be possible in any other medium" (Ninja Theory). Dodgeon aligns with Doran and Hutchinson, who argue video game spaces allow for identification with nonnormative subjectivities. This dialogue further illustrates how video games contain interactive possibilities to identify and empathize with subjectivities often presented in reductive ways.

There is one particular moment in *Hellblade* that encapsulates the idea of video games being a highly interactive medium that can inspire empathy. In Figure 4, Senua is forced to solve a door puzzle while running away from a fire-possessed spirit; if the spirit sees you, or you get too close, you die and must restart the puzzle in the frenzy of being chased (*Hellblade*). Needless to say, Player 1 died over and over again on the first run-through of the game. Each time, Player 1 ran around scared and confused, standing up while holding on to the vibrating controller with an aimless hope to succeed. Taking a step back for a second, Salter defines an adventure game as involving a user who "is expected to take time to explore the environment and find the elements that make it possible to solve puzzles and advance the story" (Salter 2014, p. 40). *Hellblade* is no exception to this formula. However, the horror in Senua's mind prevents Player 1 from patiently exploring this maze; in fact, it is a daunting experience, one that drives Player 1 to panic. The failure of dying repeatedly leads to frustration, but Jesper Juul, in *The Art of Failure*, argues "every video game is about failure on some level, and as such failure is an essential quality of the medium" (Anable 2018, p. 109). Although dying is natural to video games, on the second run-through of *Hellblade* I attempted a no-death-run.[286]

Figure 4. Senua being chased in a maze by a fire spirit, while trying to solve puzzle (*Hellblade*).

This goal adds depth to the visceral connection (or fusion) of being Senua. I, as Player 1, stood up and got closer to the TV, making Senua run into dead ends, hearing the fire spirit break through wooden panels as it slowly but menacingly made its way towards me. My emotions were doubly felt as I reacted to and internalized Senua's fear of dying, of being caught by this fire spirit. In the midst of a shared anxiety, Player 1 tried to solve the puzzle and run toward the exit all the while holding a vibrating

---

[286] A no-death-run is a successful run-through of a video game's solo campaign or individual dungeons without dying. This task is extremely difficult to attain in most solo campaigns, especially on the harder difficulties. In fact, it is impossible to beat *Hellblade* without dying because death is built into the game's tutorial.

controller signaling the proximity of this fire-spirit. I was afraid of dying, of failing. Expanding on the notion of failure, Anable writes that "[v]ideo games are not distractions from the frustration and failure of our everyday lives; rather, they are intimately linked to how we feel failure—not just how we feel about failure, but how we actually experience the feelings associated with failing" (Anable 2018, p. 119). This moment in the game generates a panoply of emotions which Senua expresses directly through screams and grunting, and which I also felt as Player 1. The anxiety is palpable. Navigating this maze increases our ability to identify with Senua and also our willingness to empathize with how people experience failure, anxiety, and the potential for panic attacks arising from high stress interactions.

Briefly, I touched on how *Hellblade*'s narration speaks to Player 1 in second-person address, creating intriguing moments of empathy and identification. Now let us examine a couple of examples. At one point in the narrative, the woman narrator philosophizes about death: "Have you ever died before? It's a serious question. When the illusion of self is shattered, you simply cease to be. Though it may not seem that way to others, you know it when it is true. You can feel it, a stranger in your own body" (*Hellblade*). The narrator directly addresses Player 1's experience with identity, and how people can continually navigate between constructed illusions of self. According to Doran, identities are "points of temporary attachment' between the subject and the subject position" (Doran, p. 266). Thus, Player 1 as subject inhabits Senua (as subject position) through the temporary act of playing *Hellblade*. The tension arising from the mixture of identities points to how Anzaldúa believes "the woman of color does not feel safe within the inner life of her self" and how she is caught between "*los intersticios*, the spaces between the different worlds she inhabits" (Anzaldúa [1987] 2012, p. 42). Although Senua is not represented as a woman of color, her colonized subjectivity reflects the shattering of selves through how she engages her environment. Thus, players are made to inhabit *los intersticios* between the safety away from Senua's colonized experience and the moments of light she enjoys within and away from her psychosis.

This train of thought brings us closer to a fusion between subjectivities. Consider how "[José Esteban] Muñoz underlines that perhaps in cross-identity identifications individuals do not abandon their own identity as they 'step' into the other person's subjectivity" (Shaw 2015, p. 176). As Player 1 encounters moments of narration like the one above, it is impossible to separate the self from Senua. Player 1 steps into Senua's subjectivity and must navigate *Hellblade*'s narrative on the border between the identities of self and other. Owen provides an example that parallels *Hellblade*, pointing to how screenplay writer Daniel MacIvor (in the film *House*) created "this intentional blurring of fiction and reality … through the use of direct address of the audience in second-person" (Owen 2017, p. 53). *Hellblade,* by utilizing second-person address, achieves similar results through fusing the "you" and the "I," the player and Senua. Thus, Player 1 becomes "a kind of disassembled and reassembled, postmodern collective and personal self" (Haraway 1994, p. 302). This disassembling and reassembling points to the fluidity between subjectivities that occurs both through the controller and through narration.

The narration continues to inform these fusions. For example, the narrator asks,

> what if each one of us is always dreaming even when awake? And we only see what our inner eye creates for us? Maybe that's why people feared seeing the world through her eyes. Because if you believe that Senua's reality is twisted, you must accept that yours might be too (*Hellblade*).

If Senua's reality is twisted, then what does that say about Player 1's journey through her twisted reality and how players relate these travels to their reality? From the onset, Player 1 internalizes Senua's struggles as their own. Therefore, the narrator's challenge to Player 1's perception of reality strengthens the fusion between Player 1 and Senua, progressing to the point where individual realities are no longer easily distinguishable. Indeed, at times the voices in Senua's mind announce, "you're going the wrong way" (*Hellblade*), casting doubt on Player 1's decisions. Haraway claims "[t]he machine is not an it to be animated, worshipped and dominated. The machine is us, our processes, an aspect of our embodiment" (Haraway 1994, p. 315). The embodiment of Senua's pain and experiences occurs only through Player 1's fusion with the machine (controller and console); thus Player 1, as Haraway might argue, is the machine.

Rosemarie Garland-Thomson adds another way of reading the narrator's question above. She writes how disability is a pervasive and "often unarticulated ideology informing our cultural notions of self and other" (Garland-Thomson 2005, p. 16). Thus, the idea that "Senua's reality is twisted" can be read as an already socialized-cultural understanding (of disability) based in heteronormative logic. Garland-Thomson further enlightens us on how the ability/disability binary "produces subjects by differentiating and marking bodies" and "although this comparison of bodies is ideological rather than biological, it nevertheless penetrates into the formation of culture, legitimating an unequal distribution of resources, status, and power within a biased social and architectural environment" (p. 17). The critical idea here is how the formation of culture is not biological but ideological and thus based in normative rhetoric/systems. These ideological and architectural environments Garland-Thomson alludes to manifest throughout *Hellblade* in ephemeral visions of men like Dillion, Druth, and Zymbel, Senua's father and the patriarch of the tribe. Here, players bring cultural inputs which are mixed with the game's cultural outputs, thus a window of empathy materializes through existing in the space of Digital *Mestizaje*.

Here I turn to scholar Evan Watts and his essay "Ruins, Gender, and Digital Games," where he argues how ruins landscapes offer refuge from the organizing logic of binary systems by exploring the power of women who traverse ruins in their quests. He writes that if physical structures were built by "a masculine-dominated society, then the ruins of these buildings symbolize the destruction of that society," becoming "a space that offers freedom from the same gender-oppressive institutions that once permeated them, and thus [becoming] sites of empowerment" (Watts 2011, p. 248). In dialogue with Garland-Thomson, Watts highlights how Senua is already empowered by not being beholden to the ability/disability binary, because, as Watts postulates, that organizing logic no longer exists. Thus, as the sole survivor of the Northmen's conquest Senua remains in a position of power where the existing structures once isolating her are no longer viable. For example, Zymbel is one such patriarchal structure left in ruins. Zymbel exacerbates Senua's mental illness because, as Charles Fernyhough has it, "quite often, the illness comes not from the symptoms but from the stigma, isolation, and mistreatment that comes about from the rest of society" (Ninja Theory). Zymbel isolates Senua from the tribe because of her visions and the fact that she hears voices, and so he is one way the ruins of the patriarchy materialize. Although Zymbel haunts Senua's reality at times, he is still only an apparition of previous structures, which serves to reinforce the idea of Senua being an empowered survivor.

Senua displays empowerment even before the Northmen reduce Zymbel's presence to ruins. She disobeys Zymbel to enter the wilds, so she can confront her struggle with the Darkness. In a vision toward the latter half of the game, Senua revisits this past conversation:

Zymbel: No boy is going to save you. No one can. The Gods can only fix you through my hand. You're going nowhere.

Senua: *unsheathes blade and points it at Zymbel* No! I am leaving.

Zymbel: You cannot escape the darkness. Your curse will make everyone suffer. (*Hellblade*)

Player 1 navigates the ruins of this relationship through ephemeral visions, which allow Player 1 and Senua to fight the voices through the act of persevering in her quest. Senua disobeys the patriarchy to find her own way. Watts points to how gender communicates "through and in connection with ruin imagery [and] must be looked at in terms of both the visual setting of the virtual environment and the potentially gendered nature of the gameplay—in the actions afforded, encouraged, discouraged, and denied to the player" (Watts 2011, p. 248). The setting is limited to traversing Hel, yet Senua navigates Hel through solving puzzles and opening gates; she alters the landscape of the world through vision masks (Figure 5), and she makes connections that are not readily apparent. The fusion between Senua and Player 1 drives the quest forward, demonstrating Senua's abilities and dedication to her mission.

Figure 5. Senua utilizing a vision mask to move between different versions of her world in order to solve puzzles (*Hellblade*).

Characters in video games often display growth while on their quests, and in adventure games this progress usually involves acquiring additional gear or abilities. Anita Sarkeesian points to this idea in her webisode "The Scythian—Positive Female Characters in Video Games." She asserts how "most video game heroes become more powerful as their quest progresses" (Sarkeesian 2015, 4:45). Senua slightly challenges what Sarkeesian alludes to in that she doesn't level up or have more weapons/armor at her disposal, subverting genre and player expectations associated with a hero's tale. In fact, I argue how Senua does not need the luxuries afforded in other adventure games. She is strong enough on her own (of course the coding of the game dictates this, yet the coding is not something Player 1 ever really considers during gameplay). However, Senua's attack combos can also be read as a limitation because you cannot go beyond a specific set of light and heavy attack combinations.[287]

[287] Light and heavy attacks refer to the buttons Player 1 pushes to generate combinations. For example, you can run with [L1] and hit Triangle [△] to make Senua complete a jumping heavy attack combination.

That being said, Watts might see this lack of moves/combos as being consistent with the ruins landscape of *Hellblade*. Once again, Senua demonstrates that she has always had abilities with her essential skill to slow down time with her mirror.[288] This ability further demonstrates her power to subvert adventure game expectations through abilities already possessed. In other words, Senua is already empowered and has no use for new gear/abilities. By including this dynamic in the coding, developers (possibly inadvertently) demonstrate how disability does not equate to a lack of ability.

Salter helps to extend on how Senua subverts expectations attached to gender and disability. She writes how "the focus on character rather than player makes the game 'about' someone with specific gender and class traits—which the player may or may not share—and places the player in dialogue with the storyteller's world and characters" (Salter 2014, p. 6). This dialogue increases the ability for players to feel empathy for the character they play, but also for people who may suffer similar fates, whether that's specific to mental illness or how Senua struggles with pain and loss. The narrator helps garner a sense of empathy within Player 1, but Druth (Senua's spirit guide) also takes part in the dialogue between Player 1, Senua, and the narrator. Narration recounts how "Druth was a troubled man. A scholar turned slave. They tortured him, took him with them on their raids, drove him to madness. Spreading this new form of darkness to new worlds. To my world" (*Hellblade*). Druth's story reminds Player 1 how Senua experiences the Darkness, how it stems from conquest (seen in Figure 6, next page), thus othering Senua's positionality.

Here I argue that Senua possesses what Anzaldúa calls *la facultad*. This subjectivity arises because Senua suffers at the hands of colonization and because she does not "feel psychologically or physically safe in the world" (Anzaldúa [1987] 2012, p. 60). After solving puzzles or advancing within *Hellblade*'s narrative, Northmen appear out of nowhere, manifesting in full attack-mode, which often forces Player 1 to take immediate evasive action. Landscapes can suddenly shift from scenic beauty to horrific darkness and demonic voices. These moments echo the lack of safety mentioned above. Anzaldúa also considers *la facultad* as a way of seeing beyond the surface into deeper realities, to see the deep structures. She states, "It is an instant sensing without conscious reasoning, an acute awareness mediated by the psyche that communicates in images and symbols. The one possessing this sensitivity is excruciatingly alive to the world" (p. 60).

[288] Senua has a mirror ability that charges (through evasive maneuvers and launching attacks), which can then be triggered using the [R2] button, slowing time for Senua to take advantage in battle.

Figure 6. Senua has a vision of the conquest of her people (*Hellblade*).

Figure 7 illustrates how Senua sees beyond the organizing structures of her world, which positions her to solve puzzles through locating symbols in Hel's ruins.

Figure 7. Here Senua finds a symbol hiding among the trees (*Hellblade*).

These symbols appear and disappear depending on Senua's proximity to the symbol's location. Therefore, Senua's *facultad* allows her to outmaneuver the world and the ruins around her by finding clues and utilizing her environment to solve puzzles. I do not intend to conflate psychosis-induced hallucinations with *la facultad*. However, it is clear that Senua does not only see the world through her psychosis, but also through the prism and trauma associated with colonization. Thus, the way Senua solves puzzles indeed stems from the lack of safety associated with psychosis-driven hallucinations, but also with the ever-present effects of colonization.

Hellblade continues to subvert player expectations, especially with game dynamics involving Senua's lack of hypersexualization.[289] In "Lingerie Is Not Armor," Sarkeesian recalls how "[characters] are typically performing activities that call for practical or protective clothing. But when we look at the types of outfits that women characters are made to wear, we can see that they are often both sexualized and completely absurd" (Sarkeesian 2015, 3:29). This assertion proves true more often than not. However, Senua is not hypersexualized; in fact, she dawns reasonable warrior's attire where her butt is not a feature designed to be admired by players.[290] Although Senua fits the standard when it comes to representation of women within video games (a slender, somewhat muscular Anglo woman), her whiteness exists on the margins of conquest. Therefore, her positionality retains our ability to read her as containing subversive power. Additionally, the fact that Dillion is dead forecloses the possibility of a heterosexual union at the end of the game, which furthers Senua's subversive positionality. Sarkeesian also highlights how a character's movement "can resist tired gender stereotypes" (00:42). Senua's movements resist. Her movement is sharp, calculated, and above all, realistic; when she runs, her arms sway rhythmically, highlighting triceps muscles and the broad arch of her shoulders. She walks, sidesteps, tumbles, runs, but does not cater to the male gaze while doing so.

Playing as Senua immerses players in an empowered subjectivity, one that carries with it a host of representational tensions that may often leave Player 1 confused. Anable argues for a productive confusion "between present and past, self and other, and inside and outside that both video games and affect theory might perform" (Anable 2018, p. 12). Confusion arises through the past manifesting in the present through lore stones, which Player 1 unlocks by using Senua's focused vision.[291] These stones display Druth's understanding and interaction with Norse mythology, providing a window into the Northmen's belief system. The self

---

289 Video game creators need to make a more concerted effort to make *Hellblade*'s representation of women the norm. The fact that she is a subversive representation speaks to how much work still needs to be done with representing women in video game design.

290 See Anita Sarkeesian's video "Strategic Butt Coverings" for more.

291 There are forty-four lore stones spread throughout Hel. Several are linked together, recounting a thread of Nordic mythology via Senua's journey and Player 1's willingness to interact with them. Druth narrates these stories, interrupting gameplay, forcing Player 1 to either move away and risk not hearing the story, or stand and listen.

and other fuse in Player 1's interaction from outside the gaming world into the "inside" of Senua's journey. Malkowski and Russworm claim how most video games can be "considered as 'technologies of gender,' which contribute to reproducing, reinforcing, and naturalizing preexisting beliefs about men and women" (Malkowski and Russworm 2017, p. 90). *Hellblade*, through its disorientation and Senua's portrayal, subverts expectations attached to adventure games and women protagonists. Sarkeesian constantly reminds us in her web series *FeministFrequency* how third-person women protagonists are usually "designed to validate [or cater to] the masculinity of presumed straight male players." Therefore, *Hellblade* directly challenges preexisting beliefs associated with portraying women—women are powerful, and Senua exhibits that empowerment, even in the face of psychosis and conquest.

Senua's journey reveals she has *la facultad*, and therefore players can also inhabit that perspective or space. Anzaldúa recalls the impact of this shift in perception and how it deepens the senses, "becom[ing] so acute and piercing that we can see ... [to] the underworld. As we plunge vertically, the break, with its accompanying new seeing, makes us pay attention to the soul, and we are thus carried into awareness—an experiencing of soul (of self)" (Anzaldúa [1987] 2012, p. 61). In the space of Digital *Mestizaje* inhabited by playing *Hellblade*, Player 1 internalizes, or becomes one with, Senua's way of seeing, a subjectivity that opens our awareness up to the soul and the self.

Hellblade provides players the opportunity to temporarily don a woman's perspective, one that may parallel or hint at the constant negotiating of spaces and consequences many women, particularly women of color, must traverse. The fact that Senua's whiteness is othered by conquest deepens *Hellblade*'s potential as a tool for teaching empathy, one where players (especially white heterosexual cis-men), can experience a world full of obstacles—a world that many of them escape. Games like *Hellblade* have the potential to shift the way we think and feel about others, and this is why more studies need to be conducted on *Hellblade*. Thus, here is a standing invitation to take video games—and independent games especially—more seriously. Empathy allows us to occupy the liminal third space Senua inhabits. Furthermore, Senua is like the speaker of Anzaldúa's poem "Una Lucha de Fronteras":[292] "alma entre dos mundos, tres, cuatro / me zumba la cabeza con lo contradictorio. / Estoy norteada por todas las voces que me hablan / simultáneamente" (Anzaldúa [1987] 2012, p. 99). These lines encapsulate Senua's and Player 1's experience. *Hellblade* demands to be studied and written about; it is a game allowing people to identify with a character who suffers from mental illness and colonial trauma, subjects which can and should be elaborated upon further in the studies that will surely follow.

---

[292] Anzaldúa's poem "A Struggle of Borders." Translation: Soul between two worlds, three or four / my head spins with all that's contradictory / I am spun around with all the voices, speaking simultaneously.

References

Aldama, Frederick (2013). *Latinos and Narrative Media: Participation and Portrayal.* In "Getting Your Mind/Body On: Latinos in Video Games," pp. 241–258. New York: Palgrave MacMillan.

Anable, A. (2018). *Playing with Feelings: Video Games and Affect.* Minneapolis: University of Minnesota Press. Project MUSE.

Anzaldúa, Gloria ([1987] 2012). *Borderlands/La Frontera: The New Mestiza.* Aunt Loot Books. 4th edition. Print.

Brookey, Alan (2010). "Playing the Games, Being the Heroes." In *Hollywood Gamers: Digital Convergence in the Film and Video Game Industries.* Bloomington: Indiana University Press.

Bryce, J., and J. Rutter (2002). "Spectacle of the Deathmatch: Character and Narrative in First-Person Shooters." In G. King and T. Krzywinska, eds., *Screenplay: Cinema/Video Game/Interfaces*, pp. 66–80. London: Wallflower.

Chess, S. (2017). *Ready Player Two: Women Gamers and Designed Identity.* Minneapolis: University of Minnesota Press. Project MUSE.

Condis, Megan Amber (2016). "Playing the Game of Literature: Ready Player One, the Ludic Novel, and the Geeky 'Canon' of White Masculinity." In *Journal of Modern Literature* 39 (no. 2), pp. 1–19.

Doran, Edward. "Identity." In Ryan, Emerson, and Robertson, pp. 266–269.

Everett, Anna (2009). *Digital Diaspora: A Race for Cyberspace.* Albany: State University of New York Press.

Garland-Thomson, Rosemarie (2005). "Feminist Disability Studies." In *Signs: Journal of Women in Culture & Society* 30 (no. 2), Winter 2005, pp. 1557–1587.

Galloway, Alexander R. (2006). *Gaming: Essays on Algorithmic Culture.* Minneapolis: University of Minnesota Press.

Haraway, Donna (1994). "A Cyborg Manifesto: Science, Technology, and Socialist-Feminism in the Late Twentieth Century." In Anne C. Herrmann et al., *Theorizing Feminism: Parallel Trends in the Humanities and Social Sciences*, pp. 424–457. Westview Press.

Hellblade: Senua's Sacrifice (2017). Ninja Theory. PS4.

Hutchinson, Rachael (2007). "Performing the Self: Subverting the Binary in Combat Games." *In Games and Culture* 2 (no. 4), pp. 283–299.

Klimmt, Christoph, Dorothée Hefner, and Peter Vorderer (2009). "The Video Game Experience as 'True' Identification: A Theory of Enjoyable Alterations of Players' Self-Perception." In *Communication Theory* 19 (no. 4) , pp. 351–373. Print.

Juul, Jesper. *The Art of Failure: An Essay on the Pain of Playing Video Games.* MIT Press.

Lahti, Martti (2003). "As We Become Machines: Corporealized Pleasures in Video Games." In Mark J.P. Wolf and Bernard Perron, eds., *The Video Game Theory Reader*, pp., 157–170. New York: Routledge.

Leonard, David (2003). "'Live in Your World, Play in Ours': Race, Video Games, and Consuming the Other." In *Studies in Media & Information Literacy Education* 3 (no. 4).

Malkowski, J., and T.M. Russworm (2017).*Gaming Representation: Race, Gender, and Sexuality in Video Games.* Bloomington: Indiana University Press. Project MUSE.

Muñoz, José Esteban (1999). *Disidentifications: Queers of Color and the Performance of Politics (Cultural Studies of the Americas).*

Ninja Theory LTD (2017). *Hellblade Documentary*. England. Digital. PS4.

Owen, David (2017). *Player and Avatar: The Affective Potential of Videogames.* MacFarland & Company, Inc., Publishers. Book.

Penix-Tadsen, Phillip (2013). "LATIN AMERICAN LUDOLOGY: Why We Should Take VideoGames Seriously (and When We Shouldn't)." In *Latin American Research Review* 48 (no. 1), pp. 174–190.

Poster, Jamie (2007). "Looking and Acting in Computer Games: Cinematic 'Play' and New Media Interactivity." In *Quarterly Review of Film and Video* 24 (no. 4), pp. 325–339.

Ryan, Marie-Laure, Lori Emerson, and J. Robertson (2014). *The Johns Hopkins Guide to Digital Media.* Baltimore: Johns Hopkins University Press.

Ryan, Marie-Laure (2009). "From Narrative Games to Playable Stories: Toward a Poetics of Interactive
Narrative." *In Storyworlds: A Journal of Narrative Studies* 1, pp. 43–59.

Salter, A. (2014). *What Is Your Quest? From Adventure Games to Interactive Books.* Iowa City: University of Iowa Press. Project MUSE.

Sarkeesian, Anita (2015). "The Scythian—Positive Female Characters in Video Games." In *Tropes Vs Women in Video Games.* YouTube, *FeministFrequency*, March 31, 2015. www.youtube.com/ watch?v=gXmj2yJNUmQ.

Shaw, A. (2015). *Gaming at the Edge: Sexuality and Gender at the Margins of Gamer Culture.* Minneapolis: University of Minnesota Press. Project MUSE.

Thon, Jan-Noël (2014). "Immersion." In Marie-Laure Ryan, Lori Emerson, and J. Robertson, *The Johns Hopkins Guide to Digital Media*, pp. 269–272. Baltimore: Johns Hopkins University Press.

Thon, J. (2016). *Transmedial Narratology and Contemporary Media Culture.* Lincoln: University of Nebraska Press. Project MUSE.

Waggoner, Zach (2017). *My Avatar, My Self: Identity in Video Role-Playing Games.* McFarland and Company Inc.

Watts, Evan (2011). "Ruin, Gender, and Digital Games." In *Women's Studies Quarterly* 39 (no. 3/4), pp. 247–265. JSTOR, www.jstor.org/stable/41308360.

AFTERWORD
AFTER WORDS

ANTONIO RAFELE

> La seconda delle cose umane, per la quale a' Latini da *humando*, seppellire prima e propiamente vien detta *Humanitas*, sono le *seppolture*; le quali sono rappresentate da UN'URNA CENERARIA RIPOSTA IN DISPARTE DENTRO LE SELVE
>
> Vico, *La Scienza Nuova*, 1744

Following the moving words of my American colleague Frederick Luis Aldama in the opening pages of this critical anthology, I move here, in its final movements, to a discussion of the posture required towards our most current communications, between keywords and method indications—I present to you a collage in which the parts are welded together, bound a bit in the shadows, thickening the plot.

Against skills

> Today in reality there is almost no theory and ideology resonates, so to speak, from the mechanism of a praxis from which one cannot escape, ... [the] non-ideological is instead the thought that does not allow itself to be traced back to operational terms and instead simply tries to help the thing itself to find those words that the dominant language otherwise chokes in its mouth.
>
> Theodor W. Adorno, *Prisms*, 1955

The so-called "operational terms" are empty and misleading because communication belongs to the domain of representation, and its apparent social dynamics—to which it subtly opposes itself in order to destroy them—are instead the dreams that, stratified, connect the text to the reader, the showcases and filters to the modes of being.

It is not by chance that McLuhan's ambition is to achieve a profound renewal of literary criticism by shifting, in the cognitive emergency posed by images, the range of action from sight to touch [Marshall McLuhan, *Understanding Media*, 1964]. Here, communication appears as a very intricate field, between the history of organs and perceptions, and the traumas of technology, on which successive aesthetic and cultural forms are innervated—a codified baggage of gestures leaves us indifferent and leads us astray. The only approach is to start from a fragment to reach, in a work of the mind, temporary but completed states of consciousness, which are, at the same time, the retrospectives of a recent past and the

Teddy told me that in Greek,

nostalgia literally means
the pain from an old wound.

It's a twinge in your heart

affirmation of a second-degree language. A linguistic *koiné*, which carries the refractory trait of singularity; the opposite, therefore, of transparent communication, which presents itself and lets itself be understood without any leap of the spectator.

Which narratives to pay attention to? Where can we see the reflections of vital movements? We turn here, of all people, to Don Draper, played by Jon Hamm:

Create a deeper connection with things. Nostalgia. It is delicate and in Greek it indicates a pain that comes from an old wound. *Carrelata*. It is a heartbreak far more powerful than memory. [Draper *pauses*]

This instrument is not a spaceship; it's a time machine. It takes us somewhere we want to go back. It's not called a wheel, it's called a carousel. *Silence*. It makes us travel the way a child travels. It goes round and round, and then it comes home.

[*Mad Men*, The Carousel, 2009]

When there are no words. It's a good sign because it puts us in the realm of the amateur, always poised between brilliance and chivalry. Like when you listen or read, among romantics or on the radio or on Netflix, the reconstruction, simulated or live, of a life story.

What prevails is the voice that hides an imminent fragility; in the apparent hesitation, it leads the listener in search of words able to grasp the emotional state, revealing an organic relationship, unbuilt or overpowering, with the language.

At one point one approaches the speaker, cancelling the previous distances. At the same time, the words placed in succession—which constitute the vital ganglia of the path, in a co-presence of mnemonic traces that reveal their drive and origin, as laborious and deserved "conquests"—and are seen to be born in the very instant in which they are fixed as images of consciousness. They are therefore authentically alive images.

Dialectical Images

> To write history is, therefore, to quote history. In the concept of quotations, however, it is implicit that the historical object is torn from its context.
>
> Walter Benjamin, *The Arcades Project*, 1922-40

At the centre, then, are the images of the observer, that is, the ways in which history comes to impress the perceptive faculties of the individual. Circumstances come to impress the observer's attention when they provide the often imperceptible experience of the discontinuity of time.

Shock establishes the voluntary (when the observer returns obsessively) or involuntary (when the event evokes a detail that was believed to be dormant) need for memory. The conscience, which aligns itself with the memory, carries out an exhausting analytical work on the scenes and emotional states, which in the beginning dwell disparate or confused; by pushing the matter to the point of breaking down its boundaries, the observer visualizes in an instant the minute links that unite (and interpenetrate) the different parts of the reflection—as Benjamin writes in his *The German Baroque Drama* (1928): "the value of the individual fragments of thought is all the more decisive the less immediate their relationship with the whole, and the splendour of the representation depends on the value of those fragments as the splendour of the mosaic depends on the quality of the molten glass."

Matter and reflection come together in writing, and have no consistency outside this plane. In the same way the observer is an image of the reflective process: the image of an ordinary person sinking into thought and the links between things.

Images are not mere virtual articulations. They become readable at a given moment, and are therefore the reflections of time. Even at the end of the journey, the images remain in power: how could one, on the other hand, block the memory of a lived circumstance or exhaust a priori its meaning and expressive power. It is always in the present that the observer reaches an exposure of the different temporal layers of reflection. At each stage he will present the constellation of problems in which he is immersed under another light, another way of exposing the matter that recomposes the order and value of the parts.

I am writing of the principle of re-reading as an essential moment of knowledge—an "entering" into the "folds" of matter (which is actually a way of "shredding" matter without stopping it) that still moves the angle of observation forward a little bit. *In medias res.*

> No confidence, rather torment, befits those who embark on a journey without reference points, let alone an idea of the conclusion.
>
> Soren Kierkegaard, *The Dialectics of Communication*, 1851

A sort of primitiveness (of the second degree) takes possession of the observer—a willingness to start again from the last occasion. Even at the end of the journey, when the reflection has reached an accomplished exposition, it will bear the distinctive and refractory mark of singularity. The difficulty inherent in indirect communication lies in expecting from the reader an equal willingness—to be so malleable as to be "devoured" by the *koinè* "banished" in the text.

In the same way the reader "grasps" by leap, the link that unites the reflection to the experience: a state of mind due or predisposed by the always renewed topicality that procures the fragmented procession of reflection: each point will bear the sign of what has been said before, so that nothing appears as perfectly finished, until it gives the impression that everything is present at once—in an instant the reader reaches an image of the path taken, and here is, in the gnoseological reflection, the most authentic relationship with history.

Persona, vocations.

> Often when I see clothes with many pleats, galleys and ornaments, which rest beautifully on beautiful bodies, I think they won't stay in that state for long, but they will take pleats... Only sometimes in the evening, when they return late from a party, their face appears worn out, swollen, dusty, seen by everyone by now, and which they can no longer wear.
>
> Franz Kafka, *Clothes*, 1919

The experience of illness perhaps brings to light the imperceptible interval that separates the person from appearance: the blow that it inflicts leaves the person inert and dumb; at the "return" to life, one advances as if suspended in the simultaneity of a double time: the "presence" in ordinary things, but also the sensation of being able to immediately distance oneself from them, as if things concealed from the beginning the emptiness and suffering, the "naked reality."

The "person" exists in a retrospective made through appearances. The clash with pain is something direct and immediate, a suffering suffered in the living flesh and, as the annulment of distance, a frustration of intelligence. Only as an entity immeasurable to the rational cognitive order, does pain come to occupy the scene. But, if on the one hand it breaks the mythical concatenations and the fabulous link between man and nature, on the other hand it fixes itself as an afigural moment. Pain fixes itself as a permanent structure of the imaginary (of consumption), a function of the ego. Isn't it in search of "an outside"—a founding myth ("origin") such as to create a distance from ancient and religious myths—that the first romantic experiments can be traced back to the dark background that preceded the appearance of the poetic word, and later the obsession of media images with "tactility"?